Microbiology of Foods and Food Processing

Microbiology of Foods and Food Processing

John T. Nickerson
Anthony J. Sinskey

Department of Nutrition and Food Science
Massachusetts Institute of Technology
Cambridge, Massachusetts

American Elsevier
Publishing Company
New York Amsterdam London

AMERICAN ELSEVIER PUBLISHING COMPANY, INC.
52 Vanderbilt Avenue, New York, N.Y. 10017

ELSEVIER PUBLISHING COMPANY
335 Jan Van Galenstraat, P.O. Box 211
Amsterdam, The Netherlands

International Standard Book Number 0-444-00124-7

Library of Congress Card Number 77-187687

Manufactured in the United States of America

CONTENTS

PREFACE

There are many books texts and monographs, which deal with the subject of microbiology and especially with bacteriology. There are also some books which cover the subject of microbiology of foods and others which are specifically concerned with food-borne diseases. Few, if any single publication, consider the subjects of the microbiology of foods, the microbiology of food processing, and the microbiology of food-borne diseases. In this volume we have attempted to tabulate the pertinent information concerning these three subjects since we believe that knowledge in these areas is most valuable to food handlers, especially those who are concerned or will be concerned with food handling outside of the home.

It is believed that this book will be of interest and provide valuable information to those who supervise the various aspects of food manufacturing, the catering of foods, or the preparing and serving of foods in restaurants or institutions. Students of Food Science or Technology and students of Hotel Management or Home Economics will also find that the information in this publication is directly concerned with their field of endeavor. No attempt has been made in this volume to tabulate all of the information available in all of the subjects covered. This would not be possible. For instance, separate volumes have been written on thermal processing, the freezing of foods, food-borne diseases, and other subjects covered in this book. However, we believe that sufficient information is given herein to provide for the understanding of derivations and application of methods, for the prevention of food spoilage, and for the use of procedures which would tend to prevent food-borne diseases.

This volume is different from others in the general field in that it covers a wider area of subject matter without including a large amount of material which is interesting but not absolutely essential to the understanding of the various subjects. It is also different in that it covers a wider variety of subjects than is usually attempted in a book of this kind.

For the use of selected material which appears in this book a special note of appreciation is due to the following: The American Public Health Association, Inc. for the MPN tables; Dr. G. G. Knock, *Journal of Scien-*

tific Food Agriculture **5** (1954), for Figures 2–9, 2–10, and 2–11; William Underwood Company for Figures 2–12, 2–13, and 2–14; Oscar Mayer & Company for Figures 7–1a and 7–1b; Kraftco Corporation, Division of the National Dairy Products Corporation for Figures 9–1 and 9–2; Kitchens of Sara Lee for Figure 5–1; C. O. Olson, F. C., *Sterilization in Food Technology,* McGraw-Hill, New York, 1957.

This book is dedicated to better foods and better health.

J. T. R. NICKERSON
A. J. SINSKEY

Cambridge, Massachusetts

CHAPTER 1

Methods Used for the Microbiological Examination of Foods

The methods described in this Chapter are those that would be applicable to the microbiological examination of foods as required for most purposes. It is obvious that all microbiological procedures which might be applied to foods cannot be included. Actually, the methods which have been detailed are mainly bacteriological, since foods are most often spoiled because of bacterial growth, and the common types of food-borne disease are caused by bacteria. However, food spoilage is sometimes caused by yeasts or molds, and molds or viruses may cause some types of food-borne disease. Procedures to determine the presence of yeasts and molds, or their end products, have, therefore, been described.

Sampling of Foods for Microbial Testing

One of the problems associated with the testing of foods for microbial contamination of any kind has been and continues to be the question of how many samples should be examined from a particular lot and also what constitutes a particular lot.

For some types of production past experience has shown that this may be worked out in the plant rather simply by trial and error. For instance, it is the practice in canning plants to set aside 12 cans from each canning line taken 15–20 min after the start of the operation and 12 cans from each line near the end of the day's run (six at each time for No. 10 can sizes). The cans from one-half of each group are incubated at 30°C (86°F), for 10 days and the other half at 50°C (122°F), for 7 days. They are examined daily for swells, and bacteriological examinations of the swelled cans are then normally conducted. At the end of incubation periods all cans are opened and checked for spoilage odors, discoloration, and pH. The spoiled products are examined microscopically (National Canners Association, 1956).[19]

In some plants producing precooked frozen foods in which the product was produced in 400-gal lots by batch processing, it has been found that, if the last carton filled from each batch were examined microscopically and the total number of viable organisms were also determined, a good evaluation of the bacteriological quality could be obtained.

When microbiological sampling of lots of product already produced is to be carried out, as by regulatory agencies, the problem of sampling becomes quite complex. Some suggestions have been made regarding the sampling of such lots. These suggestions are in the following list (Thatcher, 1963):[23]

Canadian Food and Drug Directorate	5 packs per lot or the square root of the number of samples in the lot
England—Imported foods	5 to 10% of the number of packs in the lot
Association of Food and Drug Officials of the U.S. (for precooked frozen foods)	10 packs per lot

It would appear that the AFDOUS system of sampling lots may be more realistic. The examination of only five units per lot would seem insufficient to provide adequate information as to the bacteriological quality. With some lot sizes the square root of the number of units would require a high percentage loss of product and with other lot sizes excessive laboratory work, although the percentage loss decreases as the number of samples in the lot increases. The testing of 5–10% of the units in a lot seems entirely unrealistic both from the standpoint of loss of product and the use of personnel and materials. The "Production Lot" has been defined by AFDOUS as follows: "Refers to a plant's designation as such, usually all of the units of a product of the same size produced under essentially the same condition" (Association of Food and Drug Officials of the United States, 1966).[3] This definition seems to be lacking in specificity and might allow, under some interpretations, days or weeks of production units to be considered as components of the same lot. Under such circumstances the examination of 10 units per lot would not provide for an adequate or representative sampling of production units.

The Plate Count

The equipment and materials required for plate-count techniques include diluent in bottles of 90 or 99 ml, or in tubes of 9 ml, a high-speed blender if solid food is to be examined, pipets 10, 1, and 0.1 ml, petri dishes which may be glass or plastic, a molten culture medium, and an incubator which will maintain the required temperature ± 1°C (1.8°F). In counting colonies on plate cultures a Quebec-type colony counter is ordinarily employed.

All materials contacting the sample in any form, blender tops, bottles, stoppers, tubes, pipets, diluent, and media, must be free from viable microorganisms when used. It is preferable to precool the diluent to 3°C (37.4°F) to prevent excessive heating during blending and to prevent growth of bacteria during plating. Butterfield's phosphate buffer (Butterfield, 1933)[7] or a 0.1% peptone solution are recommended as diluents. Such solutions minimize the dying off of bacteria during plating. The medium should be poured into plates at 44–46°C (111.2–114.8°F); otherwise, it will be too cold and lump during pouring or too hot and destroy some of the heat-labile bacteria. Also, hot media will cause condensation of moisture on the underside of the top of the petri dish which may drip and facilitate the growth of spreaders. Normally, the media is held in a water bath, regulated at the temperatures stated, until used.

The number of bacteria present in foods is usually so large that in order to obtain discrete colonies on petri dish cultures the food must be cultured after dilution. If solid, the food must be diluted for purposes of testing in any case, since it is not possible to make plate counts on solid food and even the plating of low dilutions (1:10) often causes difficulties in distinguishing food particles from colonies.

The first dilution, usually 1:10, may be prepared by placing 10 ml of the food in 90 ml of diluent (11 ml in 99 ml may also be used) if the food is liquid. If solid or semisolid, a 1:10 dilution of the product should be made by weighing a portion into a sterile blender top, adding the required amount of dilution water and blending at high speed for 2–3 min. Portions of 50 g of food and 450 ml of diluent are often used for this purpose. Serial dilutions are made from the 1:10 dilution by adding 10 or 11 ml of this to 90 or 99 ml of diluent (1:100 dilution), 1 ml to 99 ml of diluent (1:1000 dilution) and so on. Dilutions not agitated by blending should be shaken mechanically or by hand (with the hand moving from the shoulder to a horizontal position 25 times) prior to transfer. Either some knowledge of the number of bacteria in the product is required or several dilutions of the product must be plated in order to prepare cultures in such a manner as to have countable numbers of colonies on petri dish cultures after incubation.

The plate culture is usually prepared in one of two ways. In the regular method 1 ml of various dilutions of product is transferred to each of two separate petri dishes after which approximately 15 ml of the molten agar medium is added to each dish. After gentle mixing of the bacterial suspension with the medium, the agar is allowed to solidify. After solidification the plate cultures are inverted to prevent condensation of moisture on the surface of the agar medium during incubation.

In some cases surface counts are made, especially when the predominant flora in the food is strictly aerobic. In such instances the culture medium is first poured into petri dishes and allowed to cool and solidify in such a manner as to prevent condensation of moisture on agar surfaces. These plates are then preincubated to make certain that they have not become contaminated (no colonies on surfaces). In plating, usually 0.1–0.3 ml of the appropriate dilution of the food being analyzed is pipetted onto the surface of the solid culture medium. A sterile bent glass rod (glass hockey stick) which has been previously treated with silicone (to prevent the adherence of bacteria to the rod) is then used to distribute the sample over the entire surface of the medium.

Plate count agar (Difco) or Milk Protein Hydrolysate agar (Difco) have been recommended as culture media for plate counts, but Eugon agar (Difco, BBL) both with or without supplementation with yeast extract, and trypticase agar (a product of BBL) have been found to give higher counts for some foods such as certain kinds of fish. Therefore, it is usual to evaluate the efficiency of various media for a particular food.

Incubation temperatures between 32 and 37°C (89.6 and 98.6°F) have been recommended for plate-count cultures, but 35 ± 1°C (95 ± 1.8°F) is probably most often used. Incubation periods of 48 ± 3 hr are commonly employed (American Public Health Association, Inc., 1960).[1] In making counts on flesh-type foods, the spoilage flora are normally psychrophilic. Higher counts, therefore, can often be obtained by incubating at 20°C (68°F) for approximately 120 hr.

Plate counts should be reported only from plates containing 30–300 colonies. The count is reported as the number of colonies (average of two plates) times the dilution from which the plate cultures were made. Only the first two significant figures are reported, thus, a count of 261 through 264 would be 260, and a count of 265 through 270 would be 270. In cases in which two consecutive dilutions provide plate cultures with between 30 and 300 colonies, they may be averaged according to the count at each dilution provided that the highest count obtained at one dilution is not more than double that of the lower count obtained with the other dilution; otherwise, the lower count is used as such. In those instances in which all dilutions provide plate cultures with more than 300 colonies, the number of colonies should be determined as nearly as is possible and the average number per plate times the dilution reported as an estimate.

Tests for Coliform Bacteria in Foods

In water the quality standards are such that, in effect, the presence of more than one organism of the coliform group per 100 ml is considered as indicating that the water is not potable. In foods, coliform bacteria as a group do not have the same significance as with water. Coliform standards have been set up for some foods, such as raw certified and pasteurized milk (U.S. Department of Health, Education, and Welfare, 1953)[25] and precooked foods used by some military groups (U.S. Army Natick Laboratories, 1965),[24] but these standards are much more lenient than those for water. Many bacteriologists consider that the results of tests for the coliform group in foods have little significance and that the presence of fecal coliforms *(Escherichia coli)* only should be used to evaluate sanitary quality. Even fecal coliform standards for foods are in some instances (U.S. Department of Health, Education, and Welfare, 1953)[25] much more lenient than those for coliforms in water.

Tests for coliforms include presumptive tests, confirmation tests, and completed tests. Recommended methods for the first two tests are commonly applied and agreed upon; however, recommendations for completed tests vary from none to the application of methods which go beyond the regulation completed tests; which is the utilization of the IMViC reactions (Tests to determine production of: indole; acid, as indicated by methyl red; 2,3 butylene glycol, and growth in Koser's citrate broth or utilization of citrate).

Presumptive Tests

Plating Method. In using the plating method as a presumptive test, 1 ml of appropriate decimal dilutions of the water or food is added to petri dishes in duplicate, molten violet red bile agar (Difco, BBL) is added, mixed, and then allowed to harden in the manner described for plate counts. Plate counts are incubated at 35 ± 1°C (95 ± 1.8°F) for 18–24 hr. Positive results are indicated by dark-red colonies with a surrounding zone of precipitated bile at least 0.5 mm in diameter. A dark-field type of Quebec colony counter is recommended for counting colonies. As in plate counting, the number of colonies per plate culture is multiplied by the reciprocal of the corresponding dilution to determine numbers per gram of product.

Plating methods when used as presumptive tests have the disadvantage that low numbers of coliforms cannot always be identified by this procedure since it is not possible to plate out undiluted solid materials in such a manner as to obtain a count, and small numbers of coliform may be overgrown by large numbers of other types of bacteria.

Most Probable Number (MPN) Method. The medium used for this procedure is lauryl sulfate tryptose broth (Difco). Quantities of 10 ml of broth are placed in each tube, and a small inverted vial is also placed in each tube in such a manner that no air bubbles are present in the vial after sterilization of the medium. If 10-ml portions of a particular dilution or of liquid food or water are to be added to a series of tubes, double-strength medium is used for that particular dilution. No fewer than three tubes may be used per dilution, but five tubes per dilution is recommended. The ordinary procedure is to add 10 ml of a dilution of the material being examined or 10 g of product to the first series of five tubes, 1 ml or 1 g to the next series of five tubes, and 0.1 ml or 0.1 g or the equivalent to the next series of five tubes. Tube cultures are incubated at 35 ± °C (95 ± 1.8°F) for 48 hr and examined for gas formation after 24–48 hr of incubation.

The presumptive MPN is determined by selecting the number given in Table 1–1 (Haskins, 1933)[15] corresponding to the number of positive tubes. For instance, if in 10 ml five were positive, in 1 ml four were positive, and in 0.1 ml three were positive, the number in MPN tables corresponding to five, four, three would be selected. The most probable number of coliforms per gram of sample may then be determined as follows:

$$\frac{\text{Number from tables}}{100} \times \text{dilution factor of middle series of tubes} = \text{MPN/gram.}$$

A membrane-filter technique may be used as a presumptive test for coliforms, but this is applicable mainly to water.

The presumptive test for coliforms does not constitute absolute identification of these organisms. All positive presumptive tests for coliforms should be confirmed by further testing if the results are to be considered as valid.

Confirmation of Presumptive Tests for Coliforms

If violet red bile agar has been used as a presumptive medium, select six representative colonies and transfer with a sterile inoculating needle from the center of each colony to a separate tube of brilliant green lactose bile broth. The broth should be in 16 × 150-mm tubes with an inverted vial (10 × 75 mm). A volume of 10 ml of broth should be used per tube, and after sterilization no air should be present in the inverted vials. If the MPN technique using lauryl sulfate tryptose broth has been used as a presumptive test, three loopfuls are transferred from each gas-positive tube of the three

TABLE 1–1

Table of Most Probable Numbers (MPN) of Coliform Bacilli per 100 ml of Sample (Using Five Tubes with 10, 1, and 0.1 Volumes [15]

Positive [a] 10 1 0.1	MPN[b]	*Positive* [a] 10 1 0.1	MPN[b]	*Positive* [a] 10 1 0.1	MPN[b]	*Positive* [a] 10 1 0.1	MPN[b]	*Positive* [a] 10 1 0.1	MPN[b]	*Positive* [a] 10 1 0.1	MPN[b]
0 0 0	0	1 0 0	2	2 0 0	4.5	3 0 0	7.8	4 0 0	13	5 0 0	23
0 0 1	1.8	1 0 1	4	2 0 1	6.8	3 0 1	11	4 0 1	17	5 0 1	31
0 0 2	3.6	1 0 2	6	2 0 2	9.1	3 0 2	13	4 0 2	21	5 0 2	43
0 0 3	5.4	1 0 3	8	2 0 3	12	3 0 3	16	4 0 3	25	5 0 3	58
0 0 4	7.2	1 0 4	10	2 0 4	14	3 0 4	20	4 0 4	30	5 0 4	76
0 0 5	9	1 0 5	12	2 0 5	16	3 0 5	23	4 0 5	36	5 0 5	95
0 1 0	1.8	1 1 0	4	2 1 0	6.8	3 1 0	11	4 1 0	17	5 1 0	33
0 1 1	3.6	1 1 1	6.1	2 1 1	9.2	3 1 1	14	4 1 1	21	5 1 1	46
0 1 2	5.5	1 1 2	8.1	2 1 2	12	3 1 2	17	4 1 2	26	5 1 2	64
0 1 3	7.3	1 1 3	10	2 1 3	14	3 1 3	20	4 1 3	31	5 1 3	84
0 1 4	9.1	1 1 4	12	2 1 4	17	3 1 4	23	4 1 4	36	5 1 4	110
0 1 5	11	1 1 5	14	2 1 5	19	3 1 5	27	4 1 5	42	5 1 5	130
0 2 0	3.7	1 2 0	6.1	2 2 0	9.3	3 2 0	14	4 2 0	22	5 2 0	40
0 2 1	5.5	1 2 1	8.2	2 2 1	12	3 2 1	17	4 2 1	26	5 2 1	70
0 2 2	7.4	1 2 2	10	2 2 2	14	3 2 2	20	4 2 2	32	5 2 2	95
0 2 3	9.2	1 2 3	12	2 2 3	17	3 2 3	24	4 2 3	38	5 2 3	120
0 2 4	11	1 2 4	15	2 2 4	19	3 2 4	27	4 2 4	41	5 2 4	150
0 2 5	13	1 2 5	17	2 2 5	22	3 2 5	31	4 2 5	50	5 2 5	180
0 3 0	5.6	1 3 0	8.3	2 3 0	12	3 3 0	17	4 3 0	27	5 3 0	79
0 3 1	7.4	1 3 1	10	2 3 1	14	3 3 1	21	4 3 1	33	5 3 1	110
0 3 2	9.3	1 3 2	13	2 3 2	17	3 3 2	24	4 3 2	39	5 3 2	140
0 3 3	11	1 3 3	15	2 3 3	20	3 3 3	28	4 3 3	45	5 3 3	180
0 3 4	13	1 3 4	17	2 3 4	22	3 3 4	31	4 3 4	52	5 3 4	210
0 3 5	15	1 3 5	19	2 3 5	25	3 3 5	35	4 3 5	59	5 3 5	250
0 4 0	7.5	1 4 0	11	2 4 0	15	3 4 0	21	4 4 0	31	5 4 0	130
0 4 1	9.4	1 4 1	13	2 4 1	17	3 4 1	24	4 4 1	40	5 4 1	170
0 4 2	11	1 4 2	15	2 4 2	20	3 4 2	28	4 4 2	47	5 4 2	220
0 4 3	13	1 4 3	17	2 4 3	23	3 4 3	32	4 4 3	51	5 4 3	280
0 4 4	15	1 4 4	19	2 4 4	25	3 4 4	36	4 4 4	62	5 4 4	350
0 4 5	17	1 4 5	22	2 4 5	28	3 4 5	40	4 4 5	69	5 4 5	430
0 4 6	9.4	1 5 0	13	2 5 0	17	3 5 0	25	4 5 0	41	5 5 0	240
0 5 1	11	1 5 1	15	2 5 1	20	3 5 1	29	4 5 1	48	5 5 1	350
0 5 2	13	1 5 2	17	2 5 2	23	3 5 2	32	4 5 2	56	5 5 2	540
0 5 3	15	1 5 3	19	2 5 3	26	3 5 3	37	4 5 3	64	5 5 3	920
0 5 4	17	1 5 4	22	2 5 4	29	3 5 4	41	4 5 4	72	5 5 4	1600
0 5 5	19	1 5 5	24	2 5 5	32	3 5 5	45	4 5 5	81	5 5 5	2400+

[a] Number of positive tubes with each of three volumes used.
[b] All figures under MPN may be divided by 100 for reporting MPN per milliliter (or per gram).

highest dilution to a separate tube of brilliant green lactose bile broth with an inverted vial.

All inoculated tubes of brilliant green bile broth are now incubated for 48 hr at 35 ± 1°C (95 ± 1.8°F) and examined for gas in the inverted vial or for effervescence from a cloudy medium. Absence of gas or effervescence is considered to indicate a negative test. The number of coliforms per gram may be determined according to the number of colonies from violet red bile agar plates which gave positive results in brilliant green lactose bile broth. The MPN may be calculated from the number of tubes of lauryl sulfate tryptose broth of various dilutions which gave positive results in brilliant green bile broth.

The confirmation test may also be made on prepoured and solidified eosine–methylene blue agar plates (Difco, BBL). These are streaked using a curved inoculating needle with broth from gas-positive lauryl sulfate tryptose fermentation tubes that were used for the presumptive test. This may also be done by transferring a loopful of the tubes of broth to an eosin–methylene blue agar plate and spreading the broth over the surface of the plate with a glass hockey stick or inoculating needle. The plates are incubated for 18–24 hr at 35 ± 1°C (95 ± 1.8°F).

Completed Test for Coliforms

Colonies which are typical on eosin–methylene blue agar or positive Brilliant Green Lactose Bile broth cultures are streaked and stabbed (butt portion) on a slant of nutrient agar and are also transferred to a tube of lauryl sulfate tryptose broth containing an inverted vial. After incubation for 18–24 hr at 35 ± 1°C (95 ± 1.8°F) the slant is examined for the growth on the surface and in the stabbed portion of the slant. A Gram stain using the Hucker modification is also made on the growth from the slant. The lauryl sulfate tryptose broth is observed for growth and gas formation after 18–24 and 48 hr incubation at 35 ± 1°C (95 ± 1.8°F). A spore stain is also made.

Typical colonies on eosin–methylene blue agar are either deep brown or blue–black with a raised portion at the center and usually, but not always, have a metallic sheen, or are pale pink, lavender, or grayish in color and mucoid, with or without streaks of metallic sheen.

The completed test for coliforms, then, includes culture characteristics which indicate that the organisms are gram-negative, nonspore-forming rods which are mesophilic, will grow either aerobically or anaerobically, and ferment lactose in LST broth within 48 hr.

Test for Fecal Coliforms (U.S. Dept. of Health, Education, and Welfare, 1965)[26]

A portion of three loopfuls of each positive culture of lauryl sulfate tryptose is transferred to a tube of EC medium (Difco, BBL) and incubated immediately. If colonies from violet red bile agar or gas-positive brilliant green lactose bile cultures are to be tested, they should be first transferred to lauryl sulfate tryptose broth and incubated until gas is produced or for 48 hr or less. Incubation in EC broth must be carried out in an accurately regulated water bath at 45.5 ± 0.1°C (113 ± 0.2°F) for 24 hr. The presence of turbidity and gas constitutes a positive test.

Incubation temperatures for this test most commonly used are 44.5 or 45.5°C (112.1 or 113.8°F). It has been found that the lower temperature may produce some false-positive results and the higher temperature some false-negative results.

Detection of Salmonellae in Foods

Pre-enrichment and Sample Size

No standard procedures can be given for pre-enrichment in tests for salmonellae in foods. Some groups (Galton et al., 1964; Lewis and Angelotti, 1964)[12,16] have suggested pre-enrichment in nutrient broth for 24 hr and others have suggested preculturing in lactose broth for 24 hr (North, 1961)[20] prior to subculturing in standard liquid enrichment media. For dried or freeze-dried foods pre-enrichment has been suggested in most cases.

The recommended sample size of food which is being examined for salmonellae by enrichment or pre-enrichment varies. Some investigators suggest 20- or 30-g single samples, others suggest that two 25-g samples be used. The Food and Drug Administration examines a total of 400 g per lot.

Since the AFDOUS (Association of Food and Drug Officials of the United States) procedure (A.F.D.O.U.S., 1966)[2] is that which would probably be used by regulatory agencies, the sequence of methods of testing for salmonellae in foods recommended by this group will be described.

Enrichment of Sample

Two 25-g samples of the food are weighed out aseptically into separate 16-oz screw-capped jars. A 225-ml portion of tetrathionate broth is added to one jar, and the same amount of selenite–cysteine broth is added to the other. To all foods having a high fat content 6 ml of a 10% Tergitol No. 7 (sodium heptadecyl sulfate) solution is also added. The lids of the jars are now tightened, and the jars are shaken vigorously after which they are incubated for 24 hr at 35 ± 1°C (95 ± 1.8°F).

Plating of Enrichment Cultures

After incubation a loopful of each enrichment culture is streaked onto a plate of Salmonella–Shigella agar (Difco, BBL) and onto a plate of brilliant green agar (Difco, BBL). In each case the agar should be streaked in such a manner that some areas of the plate will have isolated colonies. The plate cultures are incubated for 22–24 hr at 35 ± 1°C (95 ± 1.8°F) on Salmonella–Shigella agar. On Salmonella–Shigella agar *Salmonella* colonies are usually colorless and transparent but may have a light-tan, light-pinkish, or yellow appearance. They are usually larger than 5 mm in diameter and may have black centers. On brilliant green agar *Salmonella* colonies are pink to deep fuchsia (red) in color but occasionally are brownish. Also, when coliforms are also present, *Salmonella* may appear to be transparent and green.

Screening

From SS and brilliant green agar five typical colonies are selected, and a portion of the center is transferred in each case to a triple sugar iron agar (Difco, BBL) slant with the slant being streaked and the butt stabbed. The slants are incubated overnight at 35 ± 1°C (95 ± 1.8°F) and then examined. For salmonellae-positive cultures the slant should be alkaline (purplish), the butt should show acid (yellow), and almost always the stab is darkened, indicating the formation of hydrogen sulfide, although a few serotypes may not produce hydrogen sulfide.

Confirmation Tests

Triple sugar iron agar-positive cultures are now examined for their reaction to polyvalent (somatic) antiserum. Some of the growth on the culture is emulsified in a drop of saline on a glass slide, then mixed with a drop of the polyvalent antiserum. Examination of slides for agglutination may be done by eye, but excellent results may be obtained by comparing a control (suspended in physiological saline) and the suspension mixed with antiserum under a wide-field microscope at about 40× magnification.

Those cultures which give positive results with the polyvalent antiserum should now be tested for flagella antigens. In this test the organism is grown overnight in veal infusion broth at 35 ± 1°C (95 ± 1.8°F). The culture is now inactivated by adding an equal volume of formalized physiological saline (6 ml of formalin plus 8.5 g of sodium chloride in 1 liter of distilled water). The polyvalent Bacto *Salmonella* flagella antiserum mixture is now diluted to 1:1000 with physiological saline solution. Portions of 0.5 ml of the formalized culture and 0.5 ml of the diluted antiserum mixture are now added to Kahn-type serological tubes, mixed, and incubated for 1 hr at 50°C (122°F). Agglutination, which is observed by eye, is considered to be indicative of a positive test.

When positive results are obtained, the above series of tests for *Salmonella* bacteria are accepted as proof that these organisms are present in the material being examined. However, if it is considered desirable to send the isolated organism to a *Salmonella*-typing center, for purposes of identifying serotypes, it is suggested that the culture first be tested for certain other biochemical reactions (Association of Food and Drug Officials of the United States, 1966).[2] These biochemical reactions involve tests to determine the production of urease, indole, and lysine decarboxylase and growth in medium containing potassium cyanide. Salmonellae do not decompose urea, do not produce indole, and do not grow in potassium cyanide medium. They decarboxylate lysine, L-ornithine, and arginine.

It has been suggested (Taylor and Silliker, 1958)[22] that a dulcitol–lactose iron agar be used in place of triple sugar iron agar. Dulcitol–lactose iron agar is a phenol red agar with dulcitol as part of the medium. This agar is placed in a tube and allowed to solidify at the base of the slant, the same phenol red agar containing lactose, ferrous ammonium sulfate, and sodium thiosulfate is then layered on top and slanted to form the slant portion of the medium. The slant is streaked and the butt stabbed. This medium is said to differentiate between different types of enteric organisms, the salmonellae producing an alkaline slant (red), hydrogen sulfide in the second medium in the area not slanted, and acid (yellow) and gas in the butt portion.

The standard method of determining the presence of salmonellae in foods involves a number of different tests which are somewhat involved and require much time and materials. It would be distinctly advantageous, therefore, to have some methods of screening out samples to determine those which might be positive and those which might be negative. A fluorescent antibody technique has been developed for this purpose (Georgala and Botthocoyd, 1964).[13]

In preparing the culture for the fluorescent antibody techniques, portions of 25 or 50 g of the food being examined are ground and cultured overnight in 100 or 200 ml of cystine–selinite broth at 35 ± 1°C (95 ± 1.8°F). A 10-ml portion of the culture is withdrawn and centrifuged at 400 rpm for 15 min after which the sediment is resuspended in 0.5 ml of sterile distilled water. The suspension is smeared on a slide, allowed to dry, and then washed consecutively with phosphate-buffered saline, pH 7.4, absolute alcohol twice, a 50:50 alcohol–xylene mixture, and finally with xylene after which the slide is air-dried.

In the direct method the smear is covered with polyvalent salmonellae (somatic) fluorescent conjugate, allowed to stand for 30 min in a moist chamber at room temperature, washed in buffered saline, and air-dried. A control is prepared in the same manner excepting that the regular nonfluorescent polyvalent antiserum is used. The slides are mounted in phosphate-buffered glycerol (pH 7.4) and examined with the microscope using an ultraviolet light source. If salmonellae are present, the cells will appear as fluorescent rods in the fluorescent antibody-treated preparation while the control will show no fluorescence.

With the indirect method the control is treated with normal serum and the test sample with regular *Salmonella* polyvalent antiserum after which both the control and the test smear are treated with a fluorescent antibody to the serum of the animal in which the polyvalent (somatic) antiserum was produced. For instance, the control is first treated with normal rabbit serum, and the test smear is treated with polyvalent (somatic) antiserum produced in a rabbit after which both smears are treated with goat anti-rabbit fluorescent conjugate. A positive control may also be used in this case.

It has been reported that some false-positive results are obtained with the fluorescent antibody technique for salmonellae, but very few false-negative tests result. A technique of this kind may thus be used to screen out negative samples, hence conserve time, materials, and the use of personnel.

Tests for Shigellae

The enrichment procedures for shigellae are the same as are those for salmonellae. After enrichment the cultures should be streaked on Salmonella-Shigella agar and on MacConkey agar and incubated for 24 hr at 35 ± 1°C (95 ± 1.8°F). On these media shigellae colonies are colorless and transparent, except for *Shigella sonnei,* which produces large colonies with yellow centers and irregular edges. Typical colonies are transferred to triple sugar iron agar slants, incubated at 35 ± 1°C (95 ± 1.8°F), and observed after 24 and 48 hr; on this medium the slant is alkaline (purple) and the butt acid (yellow). Positive or suspicious colonies may be transferred to lysine iron agar in which case the slant is streaked and the butt stabbed. These cultures are incubated overnight at 35 ± 1°C (95 ± 1.8°F). With *Shigella* the butt of the lysine iron agar is acid (yellow) and the slant is alkaline (purple). Neither gas nor hydrogen sulfide is produced.

Colonies which give positive tests for *Shigella* may be tested for agglutination with polyvalent antiserum for groups A, B, C, D, and *Alkalescens–Dispar* (A–D). Positive results will tend to confirm the presence of these groups. However, other enteric bacteria may undergo agglutination with these antiserums.

Bacteriological tests for shigellae are far from adequate. There is great need for the development of bacteriological tests which will provide a more positive identification of this group of enteric organisms.

Tests to Determine *Clostridium botulinum* in Foods

Bacteriological Tests

Whereas there are no satisfactory bacteriological tests for the identification of the various types of *Clostridium botulinum* in foods, enrichment cultures may be employed in the tests to identify the presence of the toxin, and the organism may itself be isolated on bacteriological media.

Enrichment Media. Numerous enrichment media have been suggested for the various types of *Clostridium botulinum* (Lewis and Angelotti, 1964).[16] These include glucose–peptone–beef infusion broth with ground meat, with or without 0.1% of starch; pork infusion broth with 0.1% of starch, and 0.05% of sodium thioglycollate; corn steep liquor; casein digest, autolyzed yeast and glucose; trypticase yeast extract–cysteine hydrochloride and glucose; protease peptone–yeast extract–and dextrin; corn steep liquor, glycerol and calcium chloride; liver broth and solid liver; and trypticase–peptone–glucose broth with or without sodium thioglycollate. Probably several types of media should be used. To remove oxygen these media should be preheated in boiling water and then cooled. Since the clostridia are spore-forming bacteria, it is considered desirable to heat-shock the cultures or sample prior to incubation. The heating serves both to remove air from the medium and to eliminate vegetative bacteria which might overgrow the botulinum types. While heat-shocking at 80 or 100°C (176 or 212°F) is suitable for *Clostridium botulinum* types A and B, the spores of type E are somewhat heat-sensitive. This being the case, heat-shocking of enrichment cultures for 15 min at 60°C (140°F) would appear to be preferable. After heat-shocking, liquid media should be stratified with petroleum jelly to prevent the redissolving of oxygen. One set of tubes should be incubated at each of the following temperatures: 35, 30, and 25°C (95, 86, and 77°F). The incubation period is for 5–7 days. Enrichment cultures are tested for toxin for the various types of *Clostridium botulinum* as described below.

Isolation Media (Lewis and Angelotti, 1964).[16] In isolating the botulinum organisms from toxic cultures for types A or B, the sediment of the culture should first be heated for 10 min at 100°C (212°F). For type-E toxin cultures, the sediment should first be treated with an equal volume of absolute alcohol and allowed to stand at room temperature for 1 hr.

The sediments are streaked on blood agar and on reinforced *Clostridium* medium (Oxoid) and incubated at 30°C (86°F) for 48 hr in anaerobic jars. Anaerobic jars should be operated in such a manner that the final gas mixture is about 10% carbon dioxide and no oxygen is present. This may be done by drawing a 28-in. vacuum on the sealed container, allowing carbon dioxide to flow back into the jar to raise the pressure to about 25 in. of vacuum and bringing the pressure back to normal with hydrogen gas. A catalyst in the jar causes the residual oxygen to react with the hydrogen. A much simpler method which works very well is presently available commercially. After the cultures have been placed in the jar, a packet containing chemicals (gas pak, BBL) is opened and moistened with 10 ml of water. It is then immediately placed in the jar, and the jar is tightly closed. The chemicals react producing carbon dioxide and hydrogen which reacts with the oxygen in the jar to eliminate it and provide an adequate concentration of carbon dioxide to stimulate the growth of clostridia.

On blood agar after 2–3 days of incubation, colonies are 1–2 mm in diameter and show a hemolytic zone of 2 mm.

Testing of Materials for Toxin

Food materials are blended with an equal weight of gel–phosphate buffer or with twice the weight of buffer (buffer–200 ml of 0.2 *M* Na_2HPO_4 plus 300 ml of 0.2 *M* KH_2PO_4 adjusted to pH 6.5 and mixed with 500 ml of distilled water in which 2.0 g of gelatin has been dissolved by heating). Cultures may be mixed with the phosphate buffer directly. The extracts prepared in this manner must be centrifuged under refrigeration before testing. It has been found that, with foods, better extraction of toxin may be obtained by holding the blended materials overnight at refrigerator temperatures prior to centrifuging.

The extract is adjusted to pH 6–6.2 after which part of it is trypsinized by adding 0.2 ml of trypsin (1:250 trypsin, Difco) to 2 ml of extract and incubating at 35–37°C (95–98.6°F) for 45 min (Duff et al., 1956).[9] The trypsinized extract is cooled after which 0.5-ml portions of the trypsinized and nontrypsinized extracts are injected intraperitoneally in each case into two protected and two unprotected mice. Since at least mice protected with A, B, and E types of *Clostridium botulinum* antitoxin should be used, a total of 24 mice is required. It is also possible that C, D, and F *Clostridium botulinum* antitoxin-protected mice should be injected. White male mice, about 20 g in weight, are used, and the protection is conferred by injecting intraperitoneally with 0.5 ml of a 1:10 saline dilution of the antitoxin 30 min prior to injection of the unknown samples.

The mice are observed over a period of 96 hr for typical symptoms (crusty eyes, abdominal distention, immobility of limbs, difficulty of breathing, and death). If the unprotected mice die, but the mice protected with a particular antitoxin do not, the test is positive for the particular type of *Clostridium botulinum*. When it occurs, death usually takes place within 24 hr after injection.

If considered desirable, a quantal mouse assay of the toxin or the LD_{50} (mouse unit) dose may be determined. This is defined as the amount of

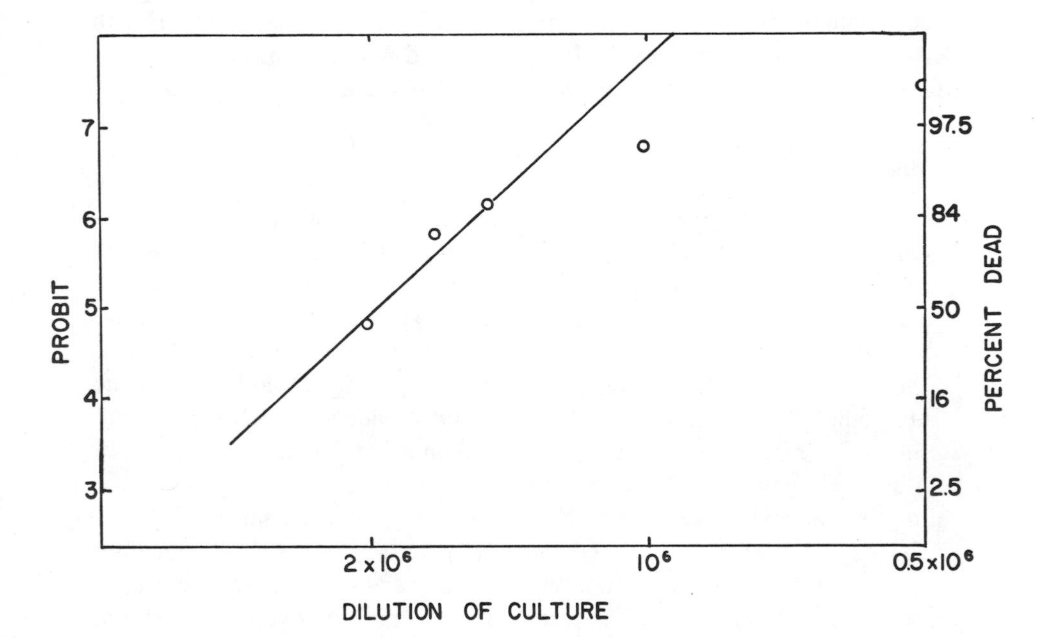

FIG. 1–1. Percentage kill plotted against dose on prohibit-log paper.

toxin that will kill 50% of the injected mice in 96 hr. Serial dilutions of the toxin or of the trypsinized toxin in gel–phosphate buffer are injected intraperitoneally into mice in the usual manner, and the deaths in 96 hr are recorded. Eight or ten mice are usually injected for each dilution. The percentage of kill is then plotted on probit-log dose paper (Fig. 1–1). A rule of thumb method is used to obtain the 0 and 100% kill dose. In this case 0.5 is added to the probit corresponding to N–1 respondents affected (all except one mouse killed) to get the 100% kill dose, and 0.5 is subtracted from the probit corresponding to 1 respondent affected (one mouse killed) to obtain the 0% kill dose.

Tests to Determine *Clostridium perfringens* in Foods

While a standard procedure has not been accepted to determine the presence or numbers of *Clostridium perfringens* in foods, the U.S. Public Health Service, Department of Health, Education, and Welfare, has developed a number of tests to achieve this purpose (Lewis and Angelotti, 1964).[16]

Procedure

Portions of 1 ml of appropriate dilutions of the food are transferred to duplicate culture dishes and mixed with 15–20 ml of sulfite polymyxin–sulfadiazine agar, and the agar is allowed to solidify. The plates are then inverted and incubated anaerobically as described for plate cultures of *Clostridium botulinum.* Incubation is carried out at 35 ± 1°C (95 ± 1.8°F) for 24 hr.

The colonies on this medium are black. The polymyxin and sulfadiazine inhibit other interfering clostridia or facultative anaerobes. Since sodium sulfite and iron salts are present in this medium and *Clostridium perfringens* produce hydrogen, the colonies are black. The hydrogen formed reduces the sulfite to sulfide which reacts with the iron to form the black sulfide.

Typical black colonies are stab-inoculated into nitrate motility medium (beef extract, peptone, potassium nitrate, and agar in water. Agar in a concentration of 3 g liter.) and also transferred to fluid thioglycollate medium. A Gram stain is also made on cells from a colony, on growth in the thioglycollate medium, and on growth in the motility medium after incubation. *Clostridium perfringens* is gram positive. These media are incubated in a water bath at 37°C (98.6°F) overnight. At the end of the 4 hr of incubation, a sporulation medium (trypticase, vitamin-free casamino acids, and sodium thioglycollate in water) is inoculated from the fluid thioglycollate medium and incubated at 35°C (95°F) for 18–24 hr.

After incubation the motility medium is examined for growth. Motile organisms produce growth out into the medium, as well as along the stab, while nonmotile organisms grow only along the stab. Unlike most other clostridia, *Clostridium perfringens* is nonmotile. The nitrate motility medium is also tested for the presence of nitrite with a suitable reagent. *Clostridium perfringens* reduces nitrate to nitrite.

A smear from the sporulation medium is made and from it a spore stain is prepared by the Bartholomew and Mittmer cold method (Bartholomew and Mittmer, 1950).[5] *Clostridium perfringens* shows large rods with blunt ends. The spores are usually centrally located in the cell.

Another test which is often used to partially identify *Clostridium perfringens* is growth under anaerobic conditions on a nutrient medium containing egg yolk. This organism produces lecithinase, and the colonies are circular, slightly raised, and surrounded by a zone of precipitated egg materials.

Tests for Thermophilic Bacteria in Foods

Tests are made on foods to determine four types of thermophiles: total thermophilic aerobic or facultative spores; flat-sour thermophilic spores; thermophilic anaerobes not producing hydrogen sulfide; and thermophilic anaerobes producing hydrogen sulfide (National Canners Association, 1956).[19]

Since thermophiles which cause spoilage of canned foods are spore-forming organisms, the sample preparation involves heating at boiling or slightly higher temperatures.

In examining sugar, 40 g of the sample are placed in a sterile Erlenmeyer flask, and water is added to bring the volume to 200 ml (the flask must be marked to indicate the required volume). This solution is brought to a boil and boiled for 5 min. To each of six petri dishes, 3 1/3 ml of the boiled solution is now added using a sterile 10-ml pipet. About 20 ml of molten glucose–tryptone agar, containing the indicator bromcresol purple, is added to each culture dish and mixed with the sugar solution. The agar is allowed to solidify, and the plates are inverted.

The cultures are incubated at 55°C (131°F) and should be examined after 18, 24, and 48 hr. If frequent examination or counting of colonies is not done, sufficient acid may be produced to cause the entire medium, or large areas of it, to change color (purple to yellow) making the counting of acid-producing colonies difficult or impossible.

In examining the cultures, a total count is made to determine total aerobic or facultative thermophilic spores, and colonies surrounded by a yellow halo or area are also counted to determine flat-sour thermophilic spores. The count per 10 g of sugar is obtained by multiplying the six-plate total count of each type by 2.5.

To determine thermophilic anaerobes not producing hydrogen sulfide, 3 1/3 ml of the heated sugar solution is added to each of six freshly exhausted tubes of liver broth, and the medium is stratified with plain nutrient agar. After the agar has solidified, the tubes are warmed to 50–55°C (122–131°F) and incubated at 55°C (131°F) for 48 hr. The presence of thermophilic anaerobes is indicated by splitting or raising of the agar plug due to gas formation. The tube cultures are recorded as positive or negative.

To determine thermophilic anaerobes producing hydrogen sulfide, 3 1/3 ml of the heated sugar solution is added to each of six tubes of molten freshly exhausted sulfite agar containing an iron nail (this agar contains tryptone as a source of sulfur-containing amino acids and sodium sulfite to keep hydrogen sulfide in the reduced state once formed). The agar is allowed to solidify, warmed to 50–55°C (122–131°F), and incubated at 55°C (131°F) 48 hr. Sulfide spoilage spores are counted as spherical black areas. The total count of sulfide formers must be multiplied by 2.5 to obtain the number per 10 g of product.

Starch is examined in much the same manner as sugar excepting that a slurry is first made with the starch and water; a wide-tip pipet is used for transferring to media, and heating to destroy vegetative bacterial types is carried out after the media have been inoculated with the starch slurry.

Since, because of their excessive heat resistance, thermophiles are most important in relation to canned, heat-processed foods, standards have been set up for canner's grade sugar and starch. Usually five samples of a particular lot of sugar or starch are examined as previously indicated.

Standards for Sugar and Starch

Total Thermophilic Spore Count

For the five samples, no sample shall contain more than 150 spores per 10 g of product, and the average shall not be more than 125 spores per 10 g.

Flat-Sour Spores

The maximum for the five samples shall be not greater than 75 spores, and the average not more than 50 spores per 10 of product.

Thermophilic Anaerobic Spores

These spores shall be present in not more than three of the five samples (60 %) examined, and in one sample not more than four of the six tubes shall contain these spores.

Sulfide Spoilage Spores

These spores shall be present in not more than two of the five samples (40%) examined, and there shall not be more than five spores per 10 g of any one sample.

Lactic Acid Bacteria—Lactobacilli and Streptococci

There are four genera of lactic acid bacteria: *Lactobacillus, Streptococcus, Leuconstoc,* and *Pediococcus.*

Lactic acid bacteria are gram-positive, nonmotile, nonsporing rods or cocci. Usually they can be characterized by the homo- or heterofermentation of carbohydrates. Since these organisms are deficient in cytochromes and catalase, many are microaerophilic.

These organisms are commonly associated with dairy products, pickle manufacture, and the spoilage of carbohydrate and vegetable foodstuffs, as well as cured meats.

Lactobacilli

Lactobacilli are gram-positive, catalase-negative microaerophilic rods. The main product of fermentation of sugars by these organisms is lactic acid. Homofermentative lactobacilli produce primarily lactic acid as the chief fermentation product. Included in this group are *Lactobacillus caucasicus, Lactobacillus bulgaricus, Lactobacillus lactis, Lactobacillus acidophilus,* and *Lactobacillus delbrueckii.* Heterofermentative lactobacilli produce, besides lactic acid, carbon dioxide, and ethanol plus other volatile products. *Lactobacillus fermenti* is heterofermentative and has the additional property of growing well at elevated temperatures (45°C, 113°F).

Enumeration of Lactobacilli

Enumeration of lactobacilli can be useful as a guide to hygiene in the dairy industry, as well as in determining the type of spoilage organisms present in vegetables, certain processed meats, and foods high in carbohydrates.

A suitable sample to analyze is plated on Rogosa's acetate agar[18] at pH 5.0–5.8 and incubated in an atmosphere of 5% carbon dioxide at 32°C (89.6°F) for 5 days. Other media that may also be used are APT media (Difco, BBL) and tomato juice-supplemented agar. Since many of the lactobacilli require growth factors, supplementation is obtained by the addition of such extracts as whey, tomato juice, or even V-8 juice to the medium.

Confirmatory tests for lactobacilli include the Gram stain and the test for catalase. Identification of species is primarily based upon fermentation of sugars and identification of the fermentation products.

Streptococci

Streptococci are gram-positive, nonsporing, nonmotile, aerobic, and facultatively anaerobic cocci that form pairs or chains. Also, they are catalase-negative and all are homofermentative. The streptococci important in foods are the enterococci, lactic, viridans, and pyogenic groups.

One of the most important groups of streptococci are the enterococci. This group consists of *Streptococcus faecalis* and *Streptococcus faecium,* including an acid-proteolytic variety of *Streptococcus faecalis* var. *liquefaciens* and var. *zymogenes* a beta hemolytic variety. Microorganisms of this group are thermoduric and can survive the pasteurization of milk; they are salt-tolerant (6.5%) and can grow at a pH of 9.6 and over the temperature range, 8–50°C (46.4–122°F).

Streptococcus faecalis is common in humans and poultry but rare in the intestines of other animals while *Streptococcus faecium* is present in the intestine of pigs and other animals.

Enumeration of Enterococci

Many investigators have observed that the gram-positive cocci are more resistant to physical agents, i.e, freezing, moderate heat treatments, and thawing, than the gram-negative indicator organisms. Therefore, enterococci may be more useful as indicator organisms for certain processed foods than coliforms. *Streptococcus faecalis* and *Streptococcus faecium* are considered to be of importance in determining the microbial quality of foods and water.

Procedure

For the detection and enumeration of faecal streptococci in foods K-F *Streptococcus* agar (Difco) can be used. This is a selective medium for fecal streptococci employing sodium azide to inhibit gram-negative organisms and containing triphenyl tetrazolium chloride (provides a red color to the colony). Appropriate dilutions of the food sample are prepared as discussed previously, and added to a set of petri dishes. Molten K-F agar at 45°C (113°F) is added, mixed thoroughly with the sample, and after setting the plates are incubated at 35°C(95°F) for 48 hr. Fecal streptococci are observed as colonies that are dark red or have a red or pink central area.

When small numbers of fecal streptococci are suspected, K-F *Streptococcus* broth may be employed as a selective medium. Then the tube dilution method for estimating the most probable number may be used in the manner described for coliforms. Acid production after incubation at 35°C (95°F) for 48 hr. is considered to be positive for enterococci.

Staphylococci

Staphylococci are gram-positive, oxidase-negative, catalase-positive fermentative cocci arranged in clusters. If staphylococci are allowed to grow in foods, certain *Staphylococcus aureus* strains may produce heat-stable enterotoxins which cause food poisoning. Although among the strains of *Staphylococcus aureus* which produce the enterotoxin this property is normally associated with coagulase-positive organisms, recently certain coagulase-negative strains have been shown also to produce enterotoxins. A number of methods involving direct plating and/or enrichment techniques have been proposed for the detection of coagulase-positive staphylococci in foods, but a standard method has not been established.

Procedure

Baird-Parker agar base medium (Difco) (Baird-Parker, 1962)[4] has been shown by some investigators to give satisfactory results. The Baird-Parker medium is a modification of the tellurite glycine agar formulated by Zebovitz et al. (Zebovitz et al., 1955).[27] Baird-Parker modification includes the addition of pyruvate to the medium, the lowering of the tellurite concentration, and the addition of concentrated egg yolk emulsion.

When greater than 100 *Staphylococcus aureus* per gram are expected, direct plating of a diluted sample on prepoured Baird-Parker medium followed by incubation for 30 hr at 37°C (98.6°F) is recommended. Staphylococci grow on this medium as a black colony surrounded by a clear zone.

Confirmatory tests include the utilization of glucose anaerobically, which separates the staphylococci from micrococci, and the ability to produce coagulase which separates *Staphylococcus aureus* from other species of staphylococci. The test for coagulase involves inoculation into blood serum and incubation to determine coagulation of the serum. Also, *Staphylococcus aureus* is separated from other staphylococci based on its property of utilizing mannitol anaerobically. Phage typing also may prove helpful in confirming the identification of the isolated staphylococci.

When counts of less than 100 per gram *Staphylococcus aureus* are expected, the most probable number enrichment technique of Giolitti and Cantoni (Giolitti and Cantoni, 1966)[14] is recommended. Using 0.1- , 1- , and 10-g samples of the food to be analyzed, growth in anaerobic tellurite glycine mannitol pyruvate broth incubated at 37°C (98.6°F) is determined. Positive tubes are plated on Baird-Parker agar base medium and typical colonies are examined for the property of producing coagulase as described above. Other confirmation tests may be used.

Molds and Yeast

Fungi, generally speaking, have an optimum pH much lower than most bacteria. Furthermore, yeast and molds will grow at a much lower water activity than bacteria. The nutritional requirements of many fungi are simple since most can grow on a medium of inorganic salts with the addition of carbohydrate as an energy source. However, some fungi require vitamins—especially the B complex group—and other growth factors which are normally supplied by supplementation of the simple media with 0.1% yeast extract.

Isolation Procedure

Traditionally, molds and yeasts have been isolated from foods on very simple media such as potato dextrose agar, malt extract agar, and Czapek Dox agar. The pH of most of these media is adjusted to 3.5–5.4. Sometimes these media are supplemented with 0.1% yeast extract. Bacterial growth may also be inhibited by the use of such antibiotics as oxytetracycline or other bacterial inhibitors such as dyes (i.e., rose bengal) (Mossel et al., 1962).[18] Under these conditions somewhat higher pH values may be used. The plate cultures are normally incubated at 25°C (77°F).

Molds

The colony characteristics, such as size, surface, appearance, texture, and colony color, are noted.

Wet mounts are normally prepared and examined with the aid of a mordant such as lactophenol which not only stains the molds but also wets the molds and minimizes the entrapment of air bubbles The vegetative mycelium is normally examined for the presence or absence of cross-walls, while shape and characteristics of the asexual and sexual reproductive structures, i.e., sporangia, conidial heads, zygospores, and arthrospores are examined in detail.

Usually such information can lead to the rapid identification of the isolated molds.

Yeasts

The identification of yeasts is more difficult than the identification of the multicellular molds. This is primarily due to the difficulty of inducing some yeasts to produce ascospores on normal media. Special media are required for the production of ascospores. However, microscopic examination can determine the method of division, whether by budding or binary fission, as well as determine the size and shape, all of which provide a basis for the identification of yeasts.

Detection of Aflatoxins

Among the mycotoxins, the aflatoxins produced by various strains of *Aspergillus flavus* have been the most actively studied. Because of the discovery of aflatoxins as contaminants of animal feeds and the fact that the toxicity of the aflatoxins in various experimental animals is quite high, methods have been developed for the determination of aflatoxins in foods. Criticism may be made of all of the reported methods (Broadbent, et al., 1963; Lombard and Vorster, 1965; Campbell and Funkhouser 1966; Eppley, 1966; Eppley, 1968) [6,8,10,11,17] for detecting aflatoxins. For example, it is difficult to obtain representative samples from many food materials such as oil seed and grains.

All of the methods developed for the detection of aflatoxins in foods include the following steps:

1. Extraction with moderately polar solvents such as chloroform, methanol, and acetone.
2. Concentration and removal of interfering substances.
3. Preparation of thin-layer chromatograms.
4. Qualitative identification when the thin-layer chromatograms are exposed to light (365 mμ).

All of the methods require blending, and it appears that a short-term, high-speed blending for extraction with hexane and methanol in water is a suitable procedure.

The cleanup procedures usually involve column chromatography on silica gel or celite using successive solvents for elution. For example, the AOAC method uses hexane as the first eluant followed by chloroform and hexane (1:1). After extraction and cleanup the aflatoxins are spotted on thin-layer chromatograms. Silica gel-R-Hr (usually 250 μ) is the most common support material, while the developer is usually methanol: chloroform (7:93).

Controls are normally run on the thin-layer chromatograms; and, if necessary, confirmatory tests by either a chemical or biological assay may be used.

References

1. Anonymous, *Standard Methods for the Examination of Dairy Products,* Amer. Pub. Health Ass., New York, 1960.
2. Association of Food and Drug Officials of the United States, *Salmonellae,* chapt. 7, *Suppl. issue, Quart. Bull.* (1966), 44.
3. Association of Food and Drug Officials of the United States, The Sampling and interpretation of microbial data on frozen foods. *Suppl. issue, Quart. Bull.,* chapt. 8–11 (1966), 55.
4. A. C. Baird-Parker, An improved diagnostic and selective medium for isolating coagulase-positive staphylococci, *J. Appl. Bacteriol.* **25** (1962), 12.
5. J. W. Bartholomew and T. Mittmer, A simplified bacterial spore strain, *Stain Technol.* **25** (1950), 153.
6. J. M. Broadbent, J. A. Cornelius, and G. Shone, The detection and estimation of aflatoxin in groundnuts and groundnut materials, *Analyst* **88** (1963), 214–216.
7. C. T. Butterfield, The selection of dilution waters for bacteriological examinations, *Pub. Health Rep.* **48** (1933), 681.
8. A. D. Campbell and J. T. Funkhouser, Collaborative study on the analysis of aflatoxins in peanut butter, *J. Ass. Offic. Agr. Chem.* **49** (1966), 730–739.
9. J. T. Duff, G. G. Wright, and A. Yarinsky, Activation of *Clostridium botulinum* type E toxin by trypsin, *J. Bacteriol.* **72** (1956), 455.
10. R. M. Eppley, A versatile procedure for assay and preparatory separation of aflatoxins from peanut products, *J. Ass. Offic. Agr. Chem.* **49** (1966), 1218–1222.
11. R. M. Eppley, Changes in official methods of analysis made at the eighty-first annual meeting, *J. Ass. Offic. Agr. Chem.* **51** (1968), 485–489.
12. M. M. Galton, J. R. Boring, and W. T. Martin, Salmonellae in foods, U.S.D.H. E. W. U.S. Pub. Health Serv. Communicable Disease Center, Atlanta, Georgia, 1964.
13. D. L. Georgala and M. Botthcoyd, A rapid immunofluorescence technique for detecting salmonellae in foods, *J. Hyg.* **62** (1964), 319.
14. G. Giolitti and C. Cantoni, A medium for the isolation of staphylococci from foodstuffs, *J. Appl. Bacteriol.* **29** (1966), 395.
15. J. K. Haskins, The most probable number of *E. coli* in water analysis, *J. Amer. Water Works, Ass.* **25** (1933), 867.

16. K. H. Lewis and R. Angelotti, Examination of foods for enteropathogenic and indicator bacteria, *U.S. Pub. Health Serv. Publ.* 1142, 1964.
17. J. H. Lombard and L. F. Vorster, The estimation of aflatoxin: A critical survey of the available physico-chemical methods. *S. Afr. Med. J.* **30** (1965), 765–766.
18. D. A. A. Mossel, M. Visser, and W. H. J. Mengerink, A comparison of media used for the enumeration of molds and yeasts in foods and beverages, *Lab. Pract.* **11** (1962), 109–112.
19. National Canners Association, "A Laboratory Manual for the Canning Industry," 2nd ed., 1956.
20. W. F. North, Jr., Lactose preenrichment method of isolation of *Salmonella* from dried egg albumin, *Appl. Microbiol.* **9** (1961), 188.
21. M. Rogasa, J. A. Mitchell, and R. F. Weisman, A selective medium for the isolation and enumeration of oral and fectal streptococci, *J. Bacteriol.* **62** (1951), 132.
22. W. I. Taylor and J. H. Silliker, Isolation of Salmonellae from food samples. III. Dulcitol lactose iron agar. A new differential tube medium for confirmation of microorganisms of the genus *Salmonella. Appl. Microbiol.* **6** (1958), 335.
23. F. S. Thatcher, The microbiology of specific frozen foods in relation to public health, Report of an international committee. *J. Appl. Bacteriol.* **26** (1963), 266.
24. U.S. Army Natick Laboratories, "Microbiological Requirements for Space Food Prototypes," 1965.
25. U.S. Dep. HEW, Milk Ordinance and Code, Recommendations of the Public Health Service, 1953.
26. U.S. Dep. HEW, Pub. Health Serv., National Shellfish Sanitation Program Manual of Operations. Part 1. Sanitation of Shellfish Growing Areas, 1965.
27. E. Zebovitz, J. B. Evans, and C. F. Niven, Tellurite–glycine agar: A selective plating medium for the quantitative detection of coagulase-positive staphylococci, *J. Bacteriol.* **29** (1955) 395.

CHAPTER 2

The Destruction of Bacteria by Heat

All living things are killed by heat if the temperature is high enough and is applied for a sufficient length of time. For microorganisms death by heating is probably due to the denaturation of proteins in some molecules required for cell respiration or for cell multiplication.

One explanation for the logarithmic destruction of bacteria by heat is that this is due to the destruction of a single gene essential to reproduction, destruction of such a gene being due to the denaturation of a single protein molecule. Another explanation for the logarithmic killing of bacteria by heat is that in any population of the same organism the heat resistance of all cells is not the same but is normally distributed. However, it has been pointed out that in order for death by heat to be first order, the logarithm of the heat resistances would have to be normally distributed. Still another explanation for the first-order killing of microorganisms by heat is that bacteria, like chemical molecules, must first absorb a certain amount of energy, the energy of activation, before reactions causing inactivation of the cell can be brought about.

Many microbiologists consider that the death of bacteria is logarithmic or a first-order reaction and actual methods of calculating heat-processing times for canned foods are based upon the logarithmic order of death of bacteria by heat. However, as discussed later, there are microorganisms for which a semilog plot of survival at a particular temperature is not linear. Also, for some temperature ranges the thermal death time curve for some microorganisms may be nonlinear. Regardless of these discrepancies, for practical purposes it is necessary to consider the thermal destruction of bacteria as a first-order reaction, otherwise the mathematical calculations necessary for prediction of processing times become very complicated.

The Heating Survival Curve

Survival curves are determined by heating a suspension of cells or spores at a particular temperature, sampling, and determining the number of viable cells at appropriate intervals during heating. The results are plotted as the $\log_{10}$ of the number of survivors versus time of heating at a given temperature.

The rate at which a microbial population will be destroyed by heat depends upon a number of factors. Moist heat is much more destructive than is dry heat; and since food processing to destroy microorganisms usually concerns the application of moist heat, discussion of the subject will be confined mainly to such processes.

In addition to the type of heat applied, the destruction of microorganisms by heat is dependent on other factors. Vegetative types of bacteria are generally much less heat resistant than are bacterial spores; yeasts and molds have a heat resistance which, in general, is greater than vegetative bacterial cells; but bacterial spores are much more heat resistant than are molds, yeasts, or vegetative cells. Bacterial spores may survive higher temperatures than any other known living organism.

In addition to differences in specific microorganisms due to their individual characteristics, other factors affect the heat resistance of a particular organism. The moisture content or water activity of a food affects heat resistance. Generally, microorganisms are much more heat labile in materials having high water activity. Thus, they are more easily destroyed in broth than when suspended in a solid food such as meat.

The fat content of a food may affect the heat resistance of microorganisms. It has been shown (Jensen, 1954)[7] that when bacteria are suspended in fat or oil they are considerably more heat resistant than when suspended in water or broth. This may not affect spoilage of a food after heat processing, since bacteria do not grow in fats or oils if no moisture is present and may remain in the inactive state when suspended in the oil phase of a food. However, such an assumption cannot be relied upon since there is always the possibility that through agitation or some other means viable cells from the oil phase would move into other parts of the food material where conditions are suitable for growth.

The pH of the food or material in which microorganisms are suspended greatly affects heat resistance. Generally, bacteria are most resistant near pH 7.0. It has been shown that in food, bacterial spores exhibit a decreased heat resistance when the pH is lowered beyond 5.5, but at pH 6.0 the heat resistance is not significantly different from that at neutrality (Sognefest et al., 1948).[16] Little attention has been given to foods at pH values above 7.0 since, with the exception of eggs, the pH of strictly fresh foods is on the acid side.

Generally, in neutral foods the 90% destruction time for very heat-resistant spore-forming bacteria will be 4–5 min at 121.1°C (250°F). Less than 0.1 min may be required to bring about an equivalent destruction of less heat-resistant types at this temperature. There may be some destruction of yeasts and molds at temperatures above 65°C (149°F) and most organisms of this type are destroyed at 80°C (176°F). Many vegetative bacterial types die off at temperatures above 60°C (140°F) but some thermoduric types withstand higher temperatures.

Survival Curves for Bacteria during Heating

Since the rate of penetration of heat into foods or other materials depends upon the mass and heat-transfer characteristics of the system being studied, there has always been a problem associated with the time required to bring the product to be sterilized up to heating temperature and the time required to cool it after heating. This is because, although temperatures below that of the final holding temperature do not have the same lethal effect as that of the holding temperature, they do have a definite effect on survival. Also, the problem is complicated by the fact that when survival curves for bacteria are measured, the information is often meant to be applied to specific foods or food products heated by conduction with a limited rate of heat penetration. Because of the limitations imposed by heat penetration into foods, most methods of determining bacterial survival during heating have attempted to minimize heating and cooling lags.

Some description of the various methods is given in the following.

Thermal-Death Time Tube Method

Bigelow and Esty and Meyer (1921, 1922)[2,6] suspended definite numbers of bacterial spores in neutral phosphate buffer (water, culture media, or food may be used) placed 1–4 ml of the mixture in small tubes (7–10 mm i.d.), sealed the tubes, immersed them in a thermostatically controlled oil bath at different temperatures, removed the tubes at definite intervals, and cooled them in water at or below 21.1°C (70°F). Temperatures of heating were approximately 100–121.1°C (212–250°F). After cooling the tubes were opened, and the material was transferred aseptically to a sterile liquid culture medium to determine growth or no growth. In cases in which liquid culture medium was used during heating, the tubes could be incubated directly after heating and cooling. Apparently 10 tubes were used for any heating time, and the times at a particular temperature were adjusted in such a manner that at the longest heating times there would be survival (growth upon transfer or upon incubation of the heated suspension itself) in one tube out of 10. It should be noted that this is really a method of determining a thermal-death time curve rather than that of determining a typical survival curve. Some discussion of the thermal-death time curve will be given later in this Chapter.

The chief disadvantage of the thermal-death time tube method involves the heating and cooling lag. There is really no way of determining what the heating time has been. Certainly the temperature of all parts of the product is not that of the heating bath from the time of immersion in the oil to time of removal. This is the case since it requires time for all parts of the material in the tube to reach heating bath temperature and during cooling, after removal from the oil bath, some parts of the product remain at lethal temperatures for a significant period of time.

Thermal-Death Time Can Method (Sognefest and Benjamin, 1944)[15]

This is also a method of determining thermal-death time curves rather than determining survival or the rate of destruction. In this method a product, usually semisolid, inoculated with spores is placed in small, 208 × 006 (2 8/16 × 6/16 in.), vacuum sealing cans. The cans are then heated in a small retort provided with quick-opening valves to allow steam under pressure to enter for heating or to allow water to enter for cooling. After heating and cooling, the cans are opened and subcultured; or if a gas-forming organism has been used, the cans may be incubated as such since the lid will open or swell if there is bacterial growth and the production of gas.

This method is subject to the same objections as that of the thermal-death time tube method. Since a significant amount of inoculated material must be used in each container, there is a lag in both the heating and cooling of some parts of the material. This may lead to errors in the actual heating time required to inactivate all of the spores present.

The Flask Method (Levine et al., 1927)[9]

A three-necked reaction flask is used in determining bacterial survival by this method. A small mechanical stirrer is introduced through the central neck of the flask, a thermometer through one of the side necks, and substrate and inoculum are introduced and withdrawn through the third neck of the flask. The substrate is presterilized. The flask is immersed to the proper depth in a thermostatically controlled oil bath, and the substrate brought up to the desired temperature while the stirrer is activated by an electric motor. A predetermined amount of inoculum is now introduced; and, while mixing is continued, samples are withdrawn aseptically at definite periods for purposes of making plate counts on a suitable medium.

The disadvantage of this method is that heating temperatures below the boiling point of water must be used since this equipment is not designed to withstand pressures above atmospheric. The method may, therefore, be used only for vegetative bacterial types or for bacteria which have spores of low heat resistance. This method is probably most suitable for calculation of processes involving pasteurization techniques for foods and is generally not applicable to calculations for processes involving "commercial sterility" for canned foods.

Another objection to this method which may lead to errors is that where material gets onto the inside surfaces of the flask at a level above external heating, by splashing of the contents or tripping of the flask, the bacteria which adhere to the flask above this level will not be subjected to high temperatures, hence may survive. This unheated portion may more or less continually feed back into the substrate leading to errors especially in that part of the survival curve where small numbers of viable bacteria should be present.

The Tank Method (Williams et al., 1937)[22]

A cylinder 4 in. in diameter and 4½ in. deep is used in this method to hold the material to be heated. The cylinder is constructed of stainless steel and is enclosed in a jacket of stainless steel in such a manner that steam under pressure can be introduced between the cylinder and outer chamber. The cylinder holds 900 ml of product and is fitted with four sampling tubes at the bottom. The cover of the cylinder, which is clamped on, is fitted with a gasket, an outlet plugged with a rubber stopper through which a thermometer may be fitted, a petcock for exhausting, and a packing gland through which a stirrer is inserted for purposes of agitating the material in the cylinder. The shaft is motor driven and has two propellers which move in opposite directions providing fast and adequate mixing. The jacket has inlet and outlet parts for steam.

In making determinations the inoculated food is placed in the cylinder, the cylinder cover is clamped on, and the food is brought to temperature as quickly as possible. After the heating temperature has been reached samples are withdrawn after definite time intervals for plating on a suitable medium. The number of spores surviving the heat treatment is determined in this manner.

The objections to this method are that 2–3 min are required to bring the substrate to 115.6–121.1°C (240–250°F) during which period the material is at no specific temperature. This leads to errors in calculations. Also, the technique of withdrawing and cooling samples requires great care. This method, of course, cannot be used for temperatures below that of free-flowing steam.

The Thermoresistometer Method (Stumbo, 1948)[18]

The thermoresistometer is a rather complex apparatus consisting of three chambers connected in line and so constructed that they may be sterilized by steam at pressures as high as 50 psig. A carrier plate of stainless steel is used to hold samples to be heated and may be moved back and forth through the chambers or into or out of the chambers. Steam is supplied from an outside source, and temperatures in the chambers may be controlled at ± 0.17°C(±0.3°F). Each chamber may be operated independently. Quick-opening valves provide for quick heating and cooling, and the maximum correction for such lags is said to be only 0.3 sec. The carrier plate has six holes in each of which an aluminum boat which contains the sample may be placed.

In operation all chambers are presterilized after which the center chamber is brought up to the desired heating temperature. The covers are removed from the outside chambers and the carrier plate positioned so that the holes are under the opening in the first chamber, the holes being closed at the bottom by a bronze plate. A boat and a cup containing 0.01–0.02 ml of inoculated sample are placed in each carrier plate hole. The samples are quickly drawn into the center chamber where they are subjected to the temperature under investigation for a period determined by an electric timer graduated in increments of 0.001 min. At the end of the heating time, a lever is thrown which stops the timer, shuts off the steam supply to the center chamber, and opens the exhaust valve for this chamber. When the pressure in the center chamber has reached 1 lb or less, the carrier plate is moved forward in such a manner that the samples drop into tubes of culture medium to be incubated for purposes of determining growth or no growth or into diluent for purposes of plating out on a suitable medium.

The thermoresistometer has little or no disadvantage from the standpoint of heating or cooling lag because of the small amount of sample used. The chief disadvantage of this method is the high initial cost of the equipment. Substances are confined to liquid suspensions or suspensions in homogenates, but this is true for all methods of determining survival curves or thermal-death times. It is said that this method is confined to the study of temperatures above about 101.7°C (215°F).

Capillary Tube Method (Stern and Proctor, 1954; Licciardello and Nickerson, 1963)[10, 17]

This method requires capillary tubes of known diameter and wall thickness, a device for filling the capillary tubes with the spore suspension in buffer or in a food slurry, and special equipment for heating and cooling the samples.

The capillary tubes used are usually 100 mm long and 1.5–2.0 mm in outside diameter. Tubes with an inside diameter of 1.2–1.4 are used, and this is determined by insertion of a metal wire of known diameter. Prior to use the tubes are cleaned in chemical cleaning solution, thoroughly rinsed, and dried.

The tubes are first bent at one end to form a hook which serves to prevent them from falling through the holes of the sample holder. A glass bridge is made at the center part of the tube by fusing the glass with heat. This bridge prevents the sample material from rising above the surface of the heating oil during the test.

The capillary-tube filling device consists of a Warburg manometer calibrator fitted into a plastic adaptor by means of a threaded joint. The plastic adaptor receives a glass adaptor which is clamped on. Connection with a hypodermic needle is again made with a ground-glass joint on the other end of the glass adaptor. The ground-gass joint has a small hollow center to allow

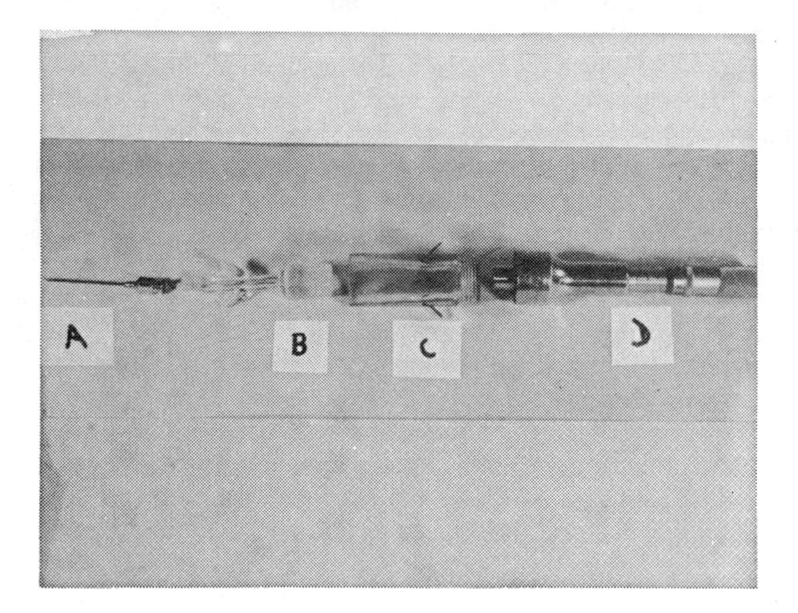

FIG. 2–1. a. Device for filling capillary tubes. (Licciardello, Ph.D. Thesis, M.I.T., 1960) A, Hypodermic needle; B, Glass adapter; C, Plastic adapter; D, Warburg manometer calibrator. b. Capillary tube transfer device: Bath No. 1 is the heating bath and Bath No. 2 is the cooling bath. The internal device is not shown. The transfer arm is shown in the filling position (dashed lines) and in the heating position. (Farkas, Ph.D. Thesis, 1960, M.I.T.)

the flow of sample. Quantities of 0.001 ml may be delivered with the filling device (Fig. 2–1a).

The heating bath consists of a container of a mineral oil heated by electric elements, thermostatically controlled to ± 0.5°C (0.9°F) and agitated by a stirrer. When the sample holder is placed in position so that the sample tubes are immersed in the heating bath, an electric timer which has been preset for a definite time period is started (accuracy ± 0.0025 min). At the end of the heating period a switch is activated which causes the tubes to be carried over and immersed in a bath of ice water. A period of 0.37–0.40 sec

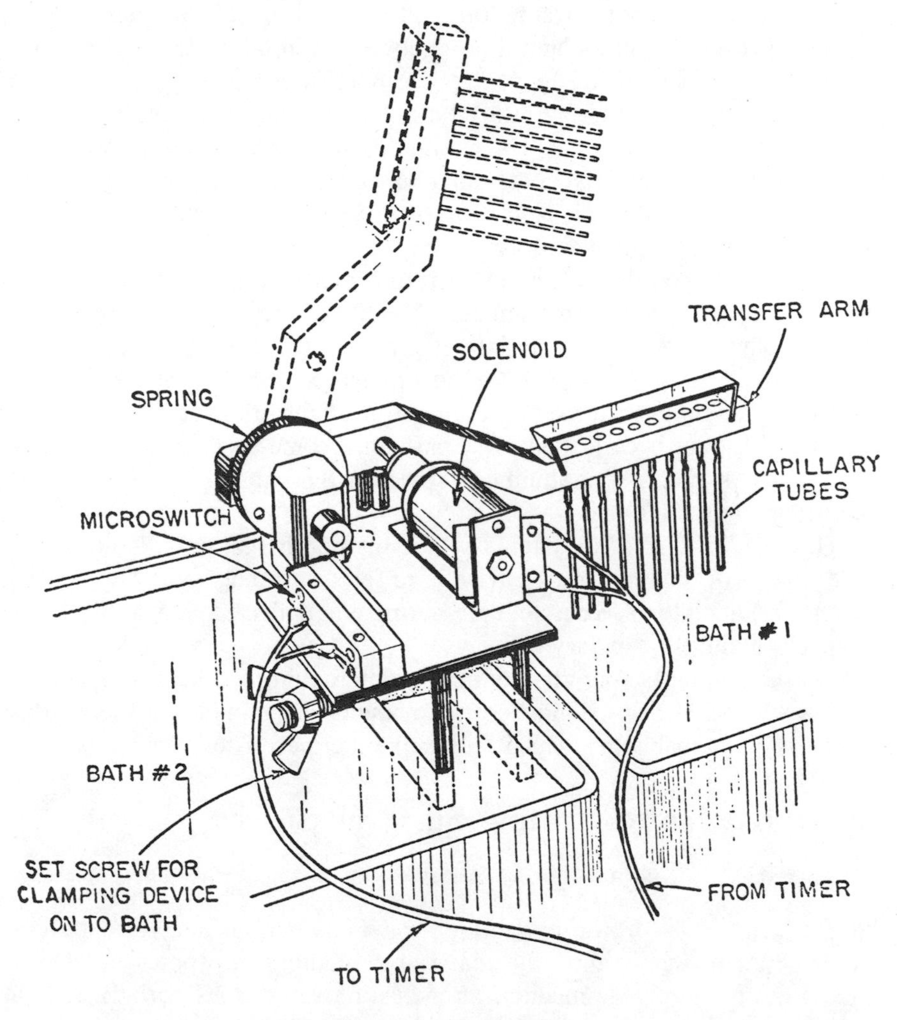

FIG. 2–1. Continued

is required to remove sample tubes from the heating bath and immerse them in the cooling water (Fig. 2–1b).

In operation the tube filler is used to place 0.025 ml of the buffer or slurry containing bacterial spores in each capillary tube adjacent to the glass bridge through the open end of the tube opposite the hooked portion. The open end is now sealed with a gas flame while holding a moistened absorbing paper over the portion containing the spore suspension to prevent heating of the spores. The filled capillary tubes are then heated and cooled as previously described.

After heating and cooling the capillary tubes are wiped dry, rinsed in petroleum ether to remove heating oil, and wiped dry, and then immersed in chemical cleaning solution and thoroughly rinsed in tap water. The capillaries are then broken off at the bridge, the portion containing the spore suspension being allowed to fall into 10 ml of sterile phosphate buffer or other suitable diluent in each case. The capillaries are crushed with a sterile glass rod to liberate the material containing the spores, and, after mixing, the material in the dilution tube is plated out on a suitable agar medium several dilutions being plated when necessary.

Depending upon the heating temperature used and the material used for suspending the spores, approximately 10–20 sec are required to bring the center temperature of the material being heated (as determined by thermocouples) within 0.1°C (0.18°F) of heating temperature. To alleviate the effects of come-up time, therefore, the control or zero time count is made on tubes which have been heated for a particular come-up time and cooled in the regular manner. The equipment used in the capillary-tube method is shown in (Fig. 2–1.)

The disadvantage of the capillary-tube method is that only liquids or dilute slurries of food materials can be used for suspending spores for heating due to the difficulty of accurately measuring and delivering a quantity of the suspension into the capillary tubes.

Considering all factors, the method of determining the heat resistance of spores which will give the most accurate results and provide for the greatest versatility is probably that which utilizes the thermoresistometer.

Interpretation of Heating Survival Curves

Plate-Count Method

Survival curves are drawn for heating for definite periods at a constant temperature. With this method the material containing the spores or cells is plated out on a suitable medium after heating for various periods of time

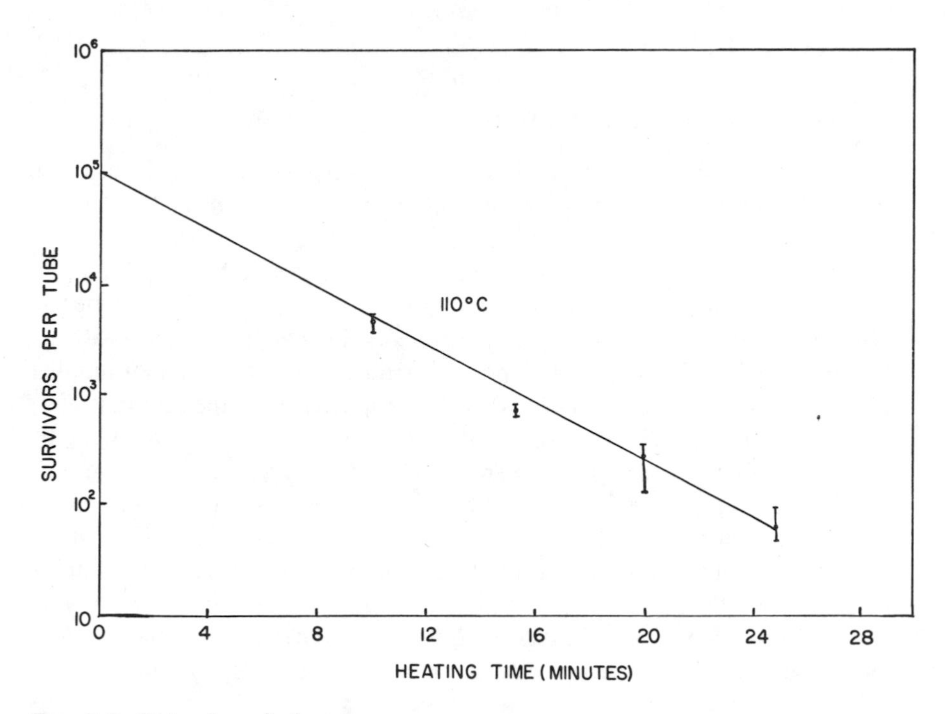

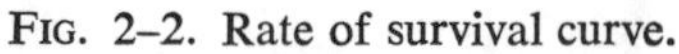
FIG. 2–2. Rate of survival curve.

and cooling. A curve is then drawn by plotting the $\log_{10}$ of the number of survivors versus heating time (see Fig. 2–2). If there is some scattering of replicate data, a regression line is determined by the method of least squares and drawn to represent the relationship between heating time and survival of spores or cells.

The slope *D* of the survival curve has been defined as the time required to traverse one log cycle of decrease in the number of cells. This is actually the inverse slope but is used in this manner because of convenience. If the survival curve is a straight line, and the count at a particular time and the *D* value are known, then the line may be drawn by fixing the known point (cells surviving at a particular time) and locating the second point as one log cycle in numbers above or below this point, at a time the value of *D* before or after that of the known point. The line may then be drawn through the two points.

D value can also be determined by formula if the zero time count is known, and the count after a particular heating time is known. Thus, $U =$

D ($\log_{10} a - \log_{10} b$), where U is the heating time, a is the number of viable cells or spores at the start and b is the number of spores or cells viable after heating time U.

Growth or No Growth Method

With this method the bacterial cells or spores suspended in liquid medium are heated for specific periods at a definite temperature, and the broth is incubated to indicate survival or no survival. Also, the suspension of spores or other material may be heated as indicated then added to a broth culture medium to indicate survival or no survival. An example of the manner in which the results of such tests are interpreted is indicated in Table 2–1.

There are 10 tubes with 1000 spores per tube, hence the original number of spores in a particular series is 10,000. Also, because of the fact that after 25 min only one tube shows growth, it is concluded that this represents only one cell surviving. Using the former equation, $U = D(\log_{10} a - \log_{10} b)$ or $25 = D(\log_{10} 10{,}000 - \log_{10} 1)$ or $D = 4 - 0 = 6.25$ min. Since a is known, the survival after any time of heating can be determined by using a picked time in minutes; and using the calculated value of D, thus assuming a straight line, the survival curve can be drawn. An assumption is made that in a particular culture showing growth, when the other cultures in the series for the same heating time show no growth, only one spore or cell has survived in the entire series of cultures.

It has been observed by a number of investigators (Licciardello and Nickerson, 1963)[10] that survival curves may exhibit a shoulder, or during the first part of heating there is a slow dying off of cells during which survival is not linear, then, after a certain period, there is a logarithmic decrease in the number of cells. This observation would seem to have been made by too many investigators to be ignored entirely. Suggestions have been made as to the cause of nonlinear survival curves of this type, but the actual reason is not known.

TABLE 2–1

Effect of Heating on Outgrowth of Bacterial Spores

Number of tubes	Spores per tube	Heating time (min)	Number of tubes showing growth
10	1000	5	10
10	1000	10	10
10	1000	15	10
10	1000	20	8
10	1000	25	1

It is also a fact that some survival curves exhibit tailing; when heating has been carried out long enough so that the number of cells or spores surviving are comparatively few, the curve tends to become asymptotic. An illustration of this is the heating of the spores of type-E *Clostridium botulinum* in broth. It has been found that when heated at 82.2°C (180°F) the D value for the straight-line portion of the survival curve is only about 17 sec, yet on one occasion a suspension was heated for 45 min at this temperature; the broth was incubated, and it was shown that there was outgrowth of the organisms and production of toxin. This heating time, according to the D value, would be equivalent to more than 150 decimal reductions which would be an impossibility because of the cell concentration required, hence the curve must become nonlinear as lower concentrations of cells are reached. The tailing phenomenon of survival curves has been noted by other investigators (Davis and Williams, 1948; Withell, 1942).[5,23]

Thermal-Death Time Curves

The End Point Method

Thermal-death time curves may be determined by the classical or end point method of Bigelow and Esty and Meyer (1921, 1922).[2,6] In this method a specific number of spores in a suitable suspending medium are heated at various temperatures to the point that, upon culturing, survival is attained in only one of several tubes heated at any temperature. From this data the $\log_{10}$ of heating time is plotted versus temperature. In the original study 6×10^9 spores were used per tube, and after heating, outgrowth took place in the material from only one tube out of the 10 which were heated at a particular temperature. Figure 2–3 shows the classical curve for *Clostridium botulinum* type A with the corrected thermal-death time of 2.45 min 121.1°C (250°F) (the original time found was 2.78 min).

z, the slope of the thermal death time curve, is defined as the degrees Fahrenheit required to traverse one log cycle of time. This is really the inverse slope of the curve, and the slope has been defined as indicated as a matter of convenience. The thermal-death time at 121.1°C (250°F) for this curve is defined as the F value. Given the F value or thermal-death time at some other temperature and the z value, assuming the curve to be linear, it can be drawn as has been explained for survival curves.

The fact that the original thermal-death time curves for type A *Clostridium botulinum* involved a reduction of 6×10^{10} spores to 0.1 spore imposes no particular restriction as to what number of decimal reductions is used for thermal-death time curves set up for other organisms in foods. The choice of the above parameters was an entirely arbitrary one. In preparing

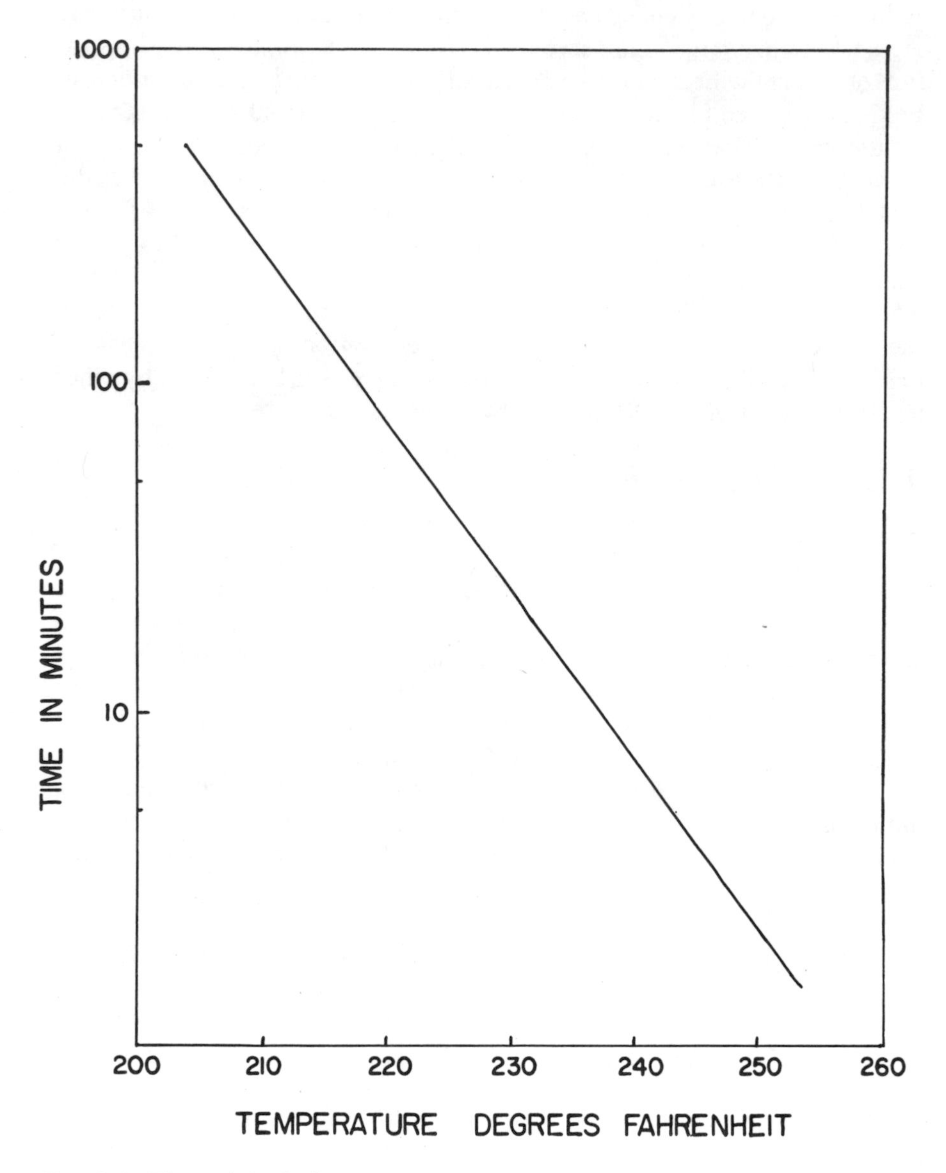

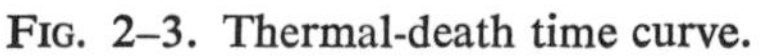
FIG. 2–3. Thermal-death time curve.

thermal-death time curves for other organisms, any number of decimal reductions may be used, according to whatever decrease in number is considered adequate for the particular organism. For instance, with salmonellae something between 4 and 7 decimal reductions is considered sufficient for the heat pasteurization of egg products—to eliminate these organisms to the point where they no longer constitute a public health hazard in these foods.

The Survival-Curve Method

If the D value of the survival curve for a particular organism is known for two different temperatures, then two points on the thermal-death time curve can be drawn if the survival curves are linear. Using the formula $U = D$ $(\log_{10} a - \log_{10} b)$, it is only necessary to solve for U by substituting the D values and whatever values are required for a and b. In the Bigelow and Esty curve, for instance, a would have the value 6×10^{10} and b would have the value of 0.1. Having located the two points, the thermal-death time curve can then be drawn.

For those cases in which the survival curve exhibits a shoulder, the D value may be determined from the regression slope of linear portion of the curve. However, when D values derived in this manner are used to construct thermal-death time curves, a factor which takes into consideration the initial lag in the survival curve must be included. There would appear to be no method of correcting for tailing exhibited by survival curves, when points on the thermal-death time curve are to be located by formula from survival-curve data.

The Calculation of Heat-Processing Times for Canned Foods

In determining heat-processing times for canned food it is necessary to have a thermal-death time curve for whatever organism the process is being set up to destroy. Processes have been so calculated as to at least provide a minimum botulinum kill (adequate to provide close to a 12 log cycle reduction of *Clostridium botulinum* type A). Usually a more severe heating than this is required and is applied because of the presence of bacterial spores of greater heat resistance than *Clostridium botulinum* type A. Other organisms, if not destroyed, might grow out and cause spoilage of the product even though no particular public health hazard might be involved. Since whatever organism is considered in calculation of processing times, the method is the same, the thermal-death time curve for *Clostridium botulinum* (Fig. 2–3) will be used for a description of calculations in this text.

Another requisite for process time calculations is knowledge of the temperature in the can of product, at the slowest heating point in the container, during retorting or during heating by other means. Normally, this is obtained by inserting thermocouples into the product in the cans in such a manner that the terminal is positioned at the slowest heating point, connecting the thermocouples to a temperature recorder by means of wires threaded through a packing gland in the retort, and subjecting the cans to heating at a

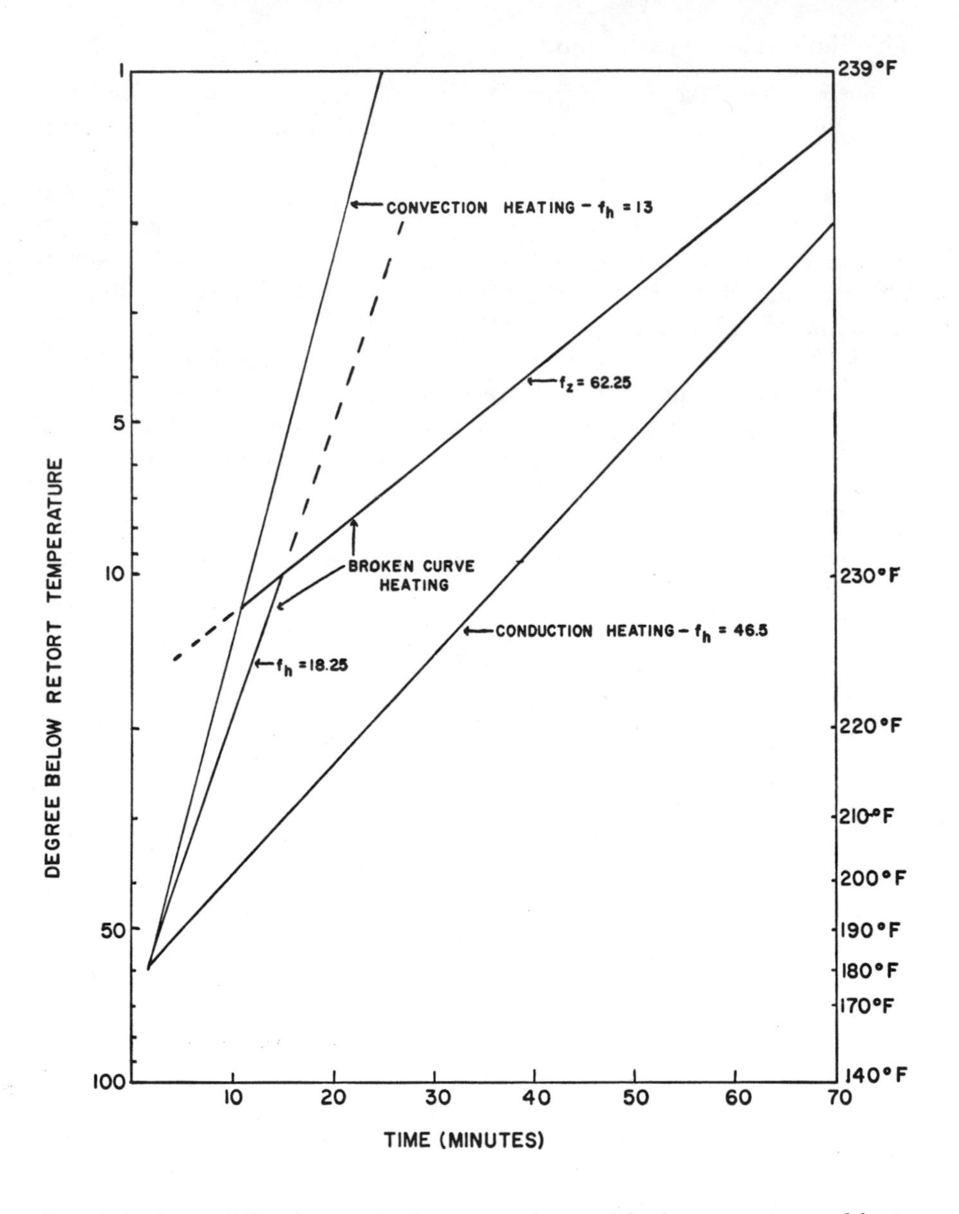

FIG. 2–4. Curves indicating conduction, convection, and broken-curve types of heating in canned foods.

particular temperature (usually 115.5°C (240°F) or above) in the retort under steam pressure. Heating is carried out for periods sufficient to bring all parts of the product to near-processing temperature or even for longer times than this.

Canned foods heat either by conduction or by convection. Thick products, such as concentrated mushroom soup, solid packs of meat or fish, or condensed pea soup, heat by conduction (Fig. 2–4). In this case heat penetrates into the can of product from all sides, and the slowest heating point of the product is on the central axis at the geometric center. The thermocouple in this case, therefore, is inserted into the can in such a manner that the point resides at the geometric center of the container. Less viscous products such as consomme or fluid milk heat by convection. Vegetables in brine are also considered to heat by convection. In convection heating a layer of heated liquid close to the can wall rises along the sides and flows across the top portions as a current of colder material moves slowly down the central axis and across the container near the bottom to join the material rising at the can wall (see Fig. 2–4). In this case the slowest heating point is on the central axis about 3/4–1½ in. above the bottom, depending upon whether the can is small or large. The junction point of the metals of the thermocouple should be placed at the slowest heating point in each case.

Some foods, such as brine-packed spinach or syrup-packed sweet potatoes, exhibit what is called broken-curve heating (see Fig. 2–4). Generally these foods first heat by convection and then by conduction. For such products it is considered that thermocouples should be inserted both at the geometric center and on the central axis near the bottom in order to determine the slowest heating point.

In obtaining the temperature profile of canned foods during retorting, the retort is closed, the steam brought up near operating temperature through a by-pass valve, the by-pass closed, and the retort allowed to operate through an automatically controlled steam valve set to operate at the temperature of processing. The time required to bring the temperature of steam in the retort up to processing temperature may be noted for some methods of calculating processing times. Ordinarily the temperature of the product will be recorded at the start (initial temperature) and for every 5 min during actual heating. When the cans have been processed for long enough to provide data for calculations, the retort is blown down or the steam is turned off, pressure (air) is maintained, and water is added to the retort for cooling. If the retort is blown down, the product is removed to a cooling canal after the pressure has been reduced to atmospheric. In either case, temperatures should be recorded at intervals of 1 min during cooling until the temperature of the slowest heating point in the product reaches approximately 100°C (212°F).

TABLE 2–2
Tabulation of Lethal Rates

Time after steam was turned on (min)	Temperature of product (°F)	(°C)	Thermal death time at temperature indicate	Lethal rate (1/thermal death time)
Heating				
26	210	(99.0)	400	1/400 = 0.0025
31	218.5	(103.6)	138	0.0092
36	224.5	(106.9)	64	0.016
41	229	(109.4)	36	0.028
46	232.5	(111.4)	23.6	0.042
51	234.5	(112.5)	17.7	0.057
56	236.1	(113.4)	14.2	0.07
61	237.2	(114.0)	12.5	0.08
66	238.1	(114.5)	11.2	0.09
71	238.8	(114.9)	10.5	0.096
76	239	(115.0)	10	0.10
Cooling				
78	238	(114.4)	11.4	0.088
80	236.5	(113.6)	13.7	0.074
82	230.5	(110.3)	30	0.033
84	220.3	(104.6)	34.5	0.029
86	208	(97.8)	500	0.002

Having recorded the temperature during processing, a table may now be set up in which the following data are indicated (Bigelow et al., 1920).[3] Time after steam was turned on; temperature of product at that time; thermal-death time at that temperature (this is determined from the thermal-death time curve, see Fig. 2–3); and the reciprocal of the thermal death time which is the lethality rate or the portion of lethality contributed by 1 min at that temperature. Such a table is indicated below from assumed data for processing:

Having obtained the lethal rate as indicated in Table 2–2, the time–temperature effect in respect to lethality may be integrated graphically on rectangular coordinates. What is required is to sum the time–temperature effect, in relation to the destruction of the spores for which the thermal death curve was drawn, to obtain one lethal effect.

If, for the lethality curve (Fig. 2–5) two of the ½ in. squares on the abscissa are made to represent 10 min of heating time, and two of the ½ in. squares on the ordinate are made to represent a lethal rate of 0.01, then four squares under the curve (2 × 2) are equivalent to 10 × 0.01 or 1/10 of a lethal effect. To obtain lethality, therefore, there must be $\frac{1 \times 4}{0.1}$ or forty ½-in. squares under the curve. The area under the curve may be determined with a planimeter or by triangulation.

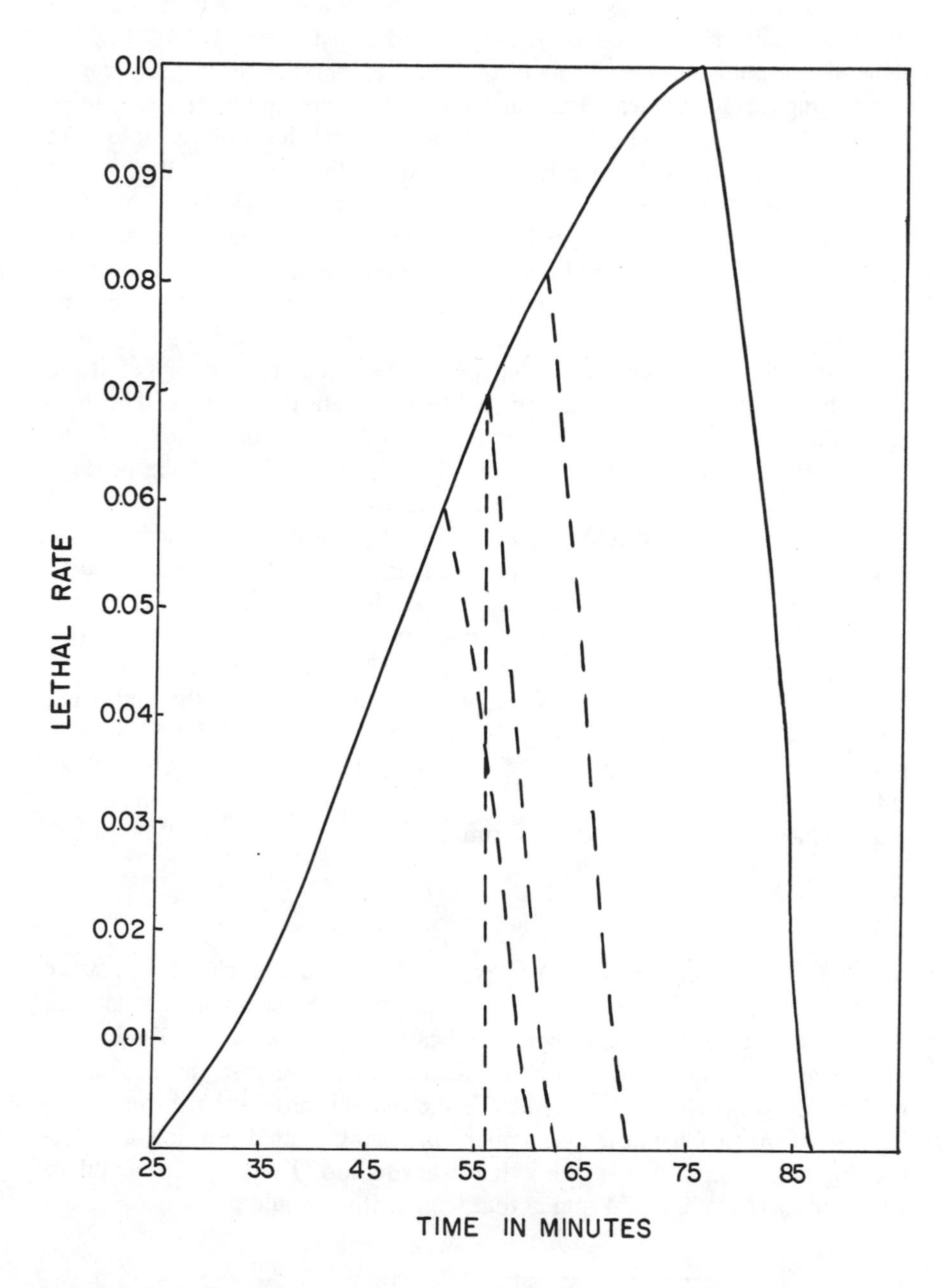

FIG. 2–5. Lethality curve.

Since the curve is drawn to include the lethal effect of cooling after the steam was shut off (in this case at 76 min after the steam was turned on), there are many more squares than 40 included. At intervals less than the time of 76 min, therefore, a curve is drawn to parallel the cooling curve. Continuing this it is indicated that at 56 min after turning on the steam there are slightly more than 43 squares under the curve. If the cooling curve were taken back to 51 min after steam was turned on, there would be fewer than 40 squares under the curve. Therefore, for this particular situation the processing time would be 56 min, steam on to steam off. It should be noted that while by the graphical method the lethal effect of cooling is included in the calculations (area between the straight dotted line and the nearest curved dotted line in Fig. 2–5), cooling time is not included in processing time since processing time represents only that period between turning on the steam and when it would have to be shut off to complete the process. Theoretically, therefore, cooling must be carried out under the same conditions during processing as was used in obtaining data for calculation of the processing time.

Another method of calculating processing by the graphical method is to integrate graphically in terms of the sterilizing value (F_0) of various temperatures and times. The F_0 unit may be considered as the sterilizing effect of a particular time at a particular temperature in terms of the sterilization effect of time, in minutes, at 121.1°C (250°F). If we examine the thermal-death time curve (Fig. 2–3), it will be noted that the thermal-death time at 115.5°C (240°F) is 9 min and that at 121.1°C (250°F) is 2.45 min. The F_0 value of 9 min at 115.5°C (240°F) is therefore, equal to F, or 2.45.

F_0 values may be calculated by formula,

$$F_0 = \frac{t}{\log^{-1}(250-T/z)},$$

where T = the temperature being evaluated, t = time at that temperature z = the slope of the thermal death time curve. Since in integrating the time–temperature effect, one point, indicative of the effect of 1 min at a particular temperature, is plotted at intervals of 5 min, t in the above formula is always equivalent to 1, and T is the temperature in the slowest heating part of the product at the particular time being plotted. For instance, according to the data used in the lethal rate method, T at 76 min would be 239, and the F_0 value for 1 min at that temperature would be

$$\frac{1}{\log^{-1}\frac{(250-239)}{(18)}} = \frac{1}{4.08} = 0.245.$$

As in the lethal-rate method, a table is set up according to the following criteria:

Time after steam was turned on the retort	Temperature at the slowest heating part of the product at that time	F_0 value for 1 min at that temperature

The F_0 values calculated are now plotted on rectangular coordinate paper, F_0 being plotted on the ordinate and time on the abscissa. The remaining procedure is like that of the lethal-rate method excepting that the squares under the curve must sum up to one F value since this is one lethal effect at 121.1°C (250°F) and F_0 is expressed as minutes at 121.1°C (250°F).

Lethal-Rate Paper

It has been shown that the lethal rate at a particular temperature, $L = \log^{-1} \frac{(T-250)}{(z)}$, where T is the temperature being evaluated and z is the slope of the thermal-death time curve. The lethal rate of 121.1°C(250°F) would, therefore, be 1. This being the case, it is possible to prepare a graph paper on which the lethal rate can be plotted directly (Schultz and Olson, 1940).[13] Any line above the abscissa in this case would represent the lethal rate for a particular temperature and would be drawn a distance above the bottom proportional to the lethal rate at this temperature as compared to the lethal rate at 121.1°C (250°F). For instance, assuming a $z = 18$, L at

$$232°\text{F} = \log^{-1} \frac{(232 - 250)}{(18)} = \log^{-1} -1 = 0.1.$$

TABLE 2–3
Lethal Rates for Various Temperatures ($z = 18$)

Time after steam was turned on (min)	Temperature at that time (°C)	(°F)	Lethal rate
31	103.6	(218.5)	0.0178
36	106.9	(224.5)	0.0380
41	109.9	(229)	0.0676
46	111.4	(232.5)	0.1067
51	112.5	(234.5)	0.1377
56	113.4	(236.1)	0.1734
61	114.0	(237.2)	0.1991
66	114.5	(238.1)	0.2183
71	114.9	(238.8)	0.2393
76	115.0	(239)	0.2449
78	114.4	(238)	0.2154
80	113.6	(236.5)	0.1778
82	110.3	(230.5)	0.0867
84	104.6	(220.3)	0.0224

Considering the time–temperature data previously given for the graphical method, a table for the lethal rates at various times when $F = 1$ (a requirement of the lethal-rate paper method) may be set up. In this case $z = 18$ (Table 2–3).

The curve for the data from Table 2-3 on lethal-rate paper is shown in Fig. 2–6.

Use of Lethal-Rate Paper in Process Calculations. The area under the curve on lethal-rate paper is measured preferably with a planimeter. After

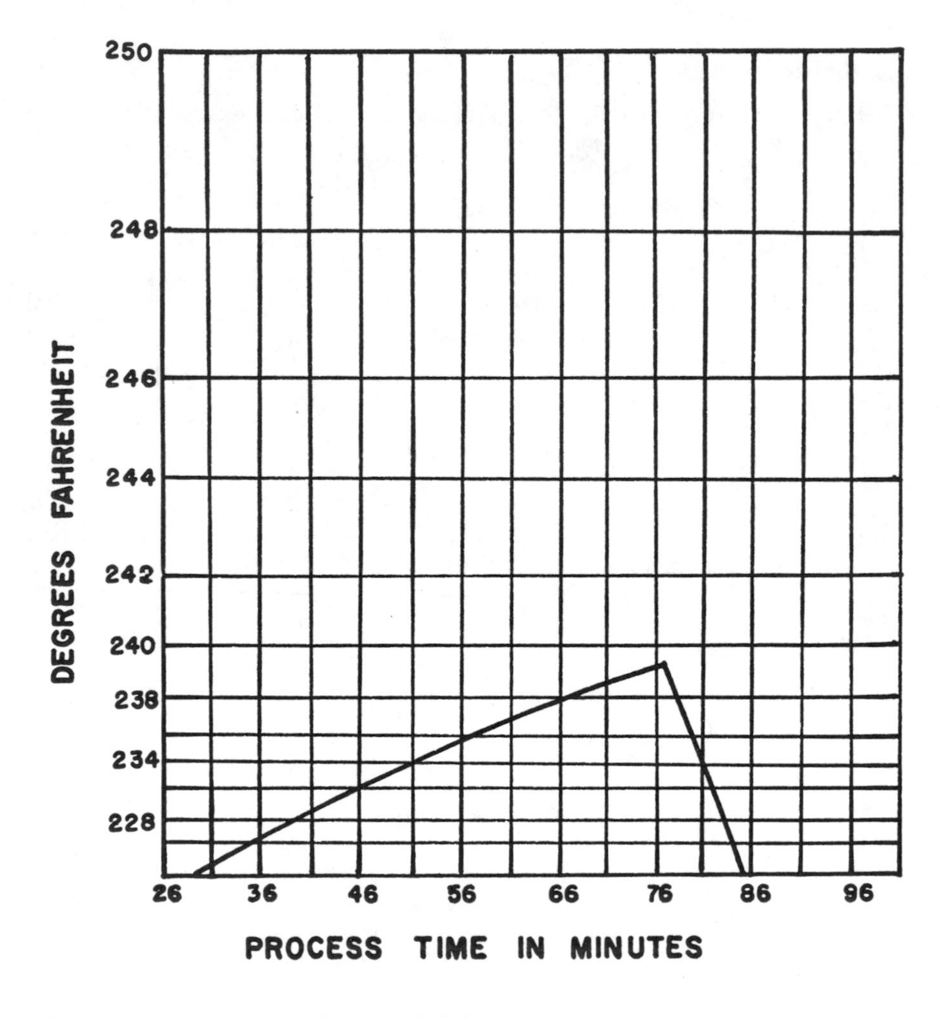

FIG. 2–6. Lethality curve plotted on lethal rate paper.

this has been done, the F_0 or F value for the process is determined by the formula $F_0 = \frac{(MA)}{(10^n d)}$, where M = number of minutes represented by 1 in. on the time scale, A = area under the curve in square inches, n = the number of z changes that must be made to change the highest point on the curve from 250°F to the desired number (this would be the number of 10°C (18°F) changes for the curve in Fig. 2–6). n is positive if the top point is below 250°F, otherwise it is negative; d = the number of inches from the bottom line to the top line. The desired number is 121.1°C (250°F); and since the top line of the paper is 250°F, n equals 250–250 or 0. If the top line were 268°F (one z value above 250°F), n would be -1. If the top line were 232°F (one z value below 250°F) n would be $+1$.

It should be noted that in order to obtain our F value corresponding to that of the thermal-death time for a particular organism at 121.1°C (250°F), for instance, 2.45 for *Clostridium botulinum,* it would be necessary to draw cooling curves, parallel to that originally plotted, for processing times less than that used to draw the original curve and determine the area under the modified curve. This would be done for several processing times to eventually obtain the value of F for which the process was to be set up. In this case for each processing time both A and n would be different from that determined for the original curve.

The Formula Method (Ball, 1923–1924)[1]

The formula method requires the use of a heat-penetration curve. Such a curve is shown in Fig. 2–7. This curve was drawn from the data used to illustrate the lethality curve. It will be noted that this curve has been plotted on semilog paper, but the paper has been turned upside down, and instead of plotting the log of temperature, the log of degrees below retort temperature (actual temperature may be plotted on the right ordinate) has been plotted. Time has been plotted on the abscissa.

If the curve is linear, it will be comparatively flat (numerical value of slope, as defined, will be large) for conduction heating and comparatively steep for convection heating. For broken-curve heating there will be a change of slope in the line at one point. If the heating curve exhibits more than one change in slope, there is no reason to try to draw it since the processing time can be determined only by the graphical method under such circumstances, and temperatures may be determined from the recorded data.

It is necessary to understand certain symbols in order to use the formula method. f_h, the slope of the heat penetration curve, is defined as the time required to traverse one log cycle of temperature (this is actually the inverse slope). If the relationship of time and temperature for all except the very

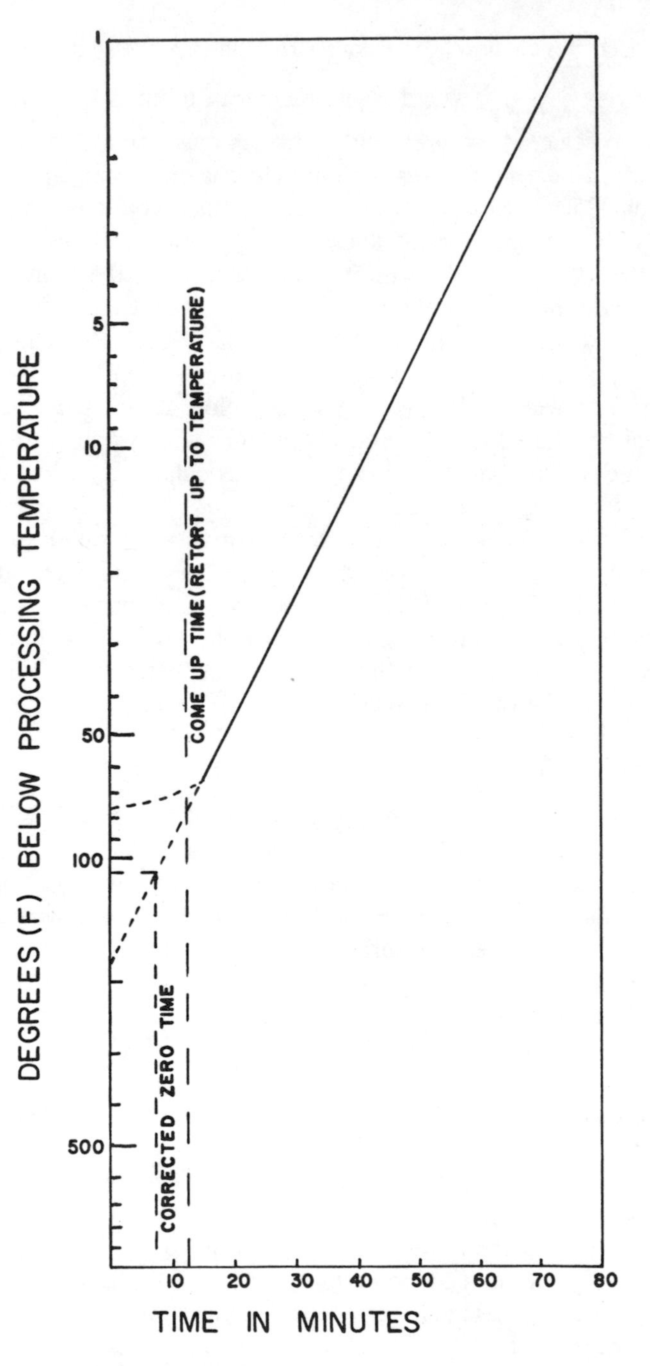

FIG. 2–7. Heat-penetration curve.

first part of the curve is linear, then given the temperature at any part of the straight line portion and f_h, the line can be drawn. The slope of the heat penetration curve in Fig. 2–7 is 49.5 − 16, or 33.5.

It must be explained that when a retort is filled with crates of cans, after the steam is turned on it takes some time for steam in the retort to come up to processing temperature. This is called the come-up time. The come-up time has some sterilizing effect, but it is not equivalent to come-up time at retort temperature. Actually, it has been found that the sterilizing effect in this case is approximately equivalent to 42% of come-up time at retort temperature.

Due to the fact that 42% of come-up time is equivalent to the sterilizing effect of come-up time at retort temperature, the corrected zero time is 58% of come-up time. In the heat-penetration curve shown (Fig. 2–7) come-up time was 12 min. Therefore, the corrected zero time was $0.58 \times 12 = 6.96$, or 7 min.

RT is the retort temperature which, according to the curve, was 115.5°C (240°F).

IT is the initial temperature and in this case was 240 minus 75, or 73.9°C (165°F).

JI is defined as the degrees below retort temperature at which a line drawn vertically from the corrected zero time intersects the extension of the straight-line portion of the heat-penetration curve.

PSIT is the pseudoinitial temperature or is the actual temperature corresponding to *JI*. *J* is the lag factor or time before which there is no increase in temperature at the slowest heating point of the container. *J* may be obtained from the curve as $\frac{JI}{I}$, since $I = RT - IT$, or as $\frac{(RT - PSIT)}{(RT - IT)}$.

U is the thermal-death time at retort temperature obtained from the thermal-death time curve, and in this case it has a value of 8.8.

g is defined as the difference between retort temperature and the temperature of the slowest heating point in the product at the end of the process. Actually *g* is the degrees below retort temperature at which sterility, based on the thermal-death time curve, would just be attained if the sterilizing effect of cooling were included. Since the value of f_h and of *U* are known, *g* may be obtained from $\frac{f_h}{U}$ vs *g* curves (see Fig. 2–8). In setting up these curves a general equation for the sterilizing effect of processing was developed. This contained a constant *C* which included the variables $m + g$ (difference between retort temperature and cooling water temperature) and *z*. By assuming values for these variables it was possible to construct *C* versus *g* curves. Again an equation was developed for the lethal effect at *g*. This

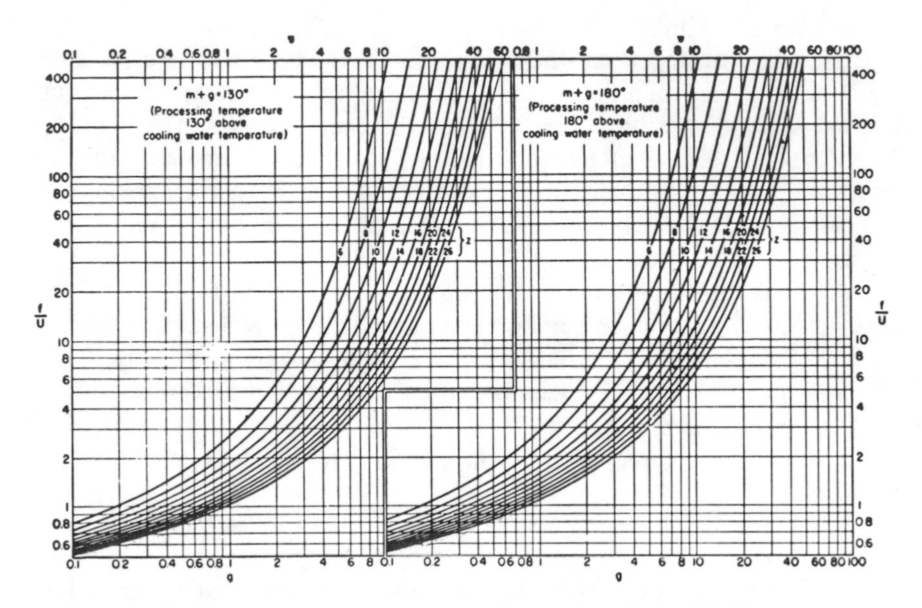

FIG. 2–8. $F/_hU$ vs g curves g values corresponding to the values of F_h/U. (Courtesy of McGraw-Hill Book Co.)

equation contained the variables C, z, and g and could be equated to $\frac{f_h}{U}$; f_h/U vs g curves could then be drawn. Whereas for the data available, considerable mathematical calculations would be involved in using C versus g curves, f_h/U vs g curves are easily applied to solve for g since both f_h and U are known if thermal-death time and heat-penetration curves are available.

It should be noted that since g is dependent on both z and $m + g$, the f_h/U vs g curves have been drawn for specific values of these process-determination parameters.

The formula for determining processing time was developed by applying the general equation for a straight line to the heat-penetration curve. This formula is B_B (processing time) $= f_h$ ($\log JI - \log g$). This formula applies to the time between the corrected zero time and steam off. If, using the data in Figs. 2–3, 2–7, and 2–8, we apply this formula, we have $B_B = 33.5$ $(\log 112 - \log 4.3) = 33.5\ (2.0492 - 0.6335) = 47.4$. To compare with the graphical method the corrected zero time (7 min) must be added making the processing time steam on to steam off 54.4 min. This is a fairly good check with the processing time of 56 min obtained by the graphical method since the time obtained graphically was slightly more than the absolute requirement.

The Stumbo Mathematical Method (Stumbo, 1953; Stumbo, 1965)[19, 20]

It is the opinion of many microbiologists that in order to evaluate the lethal capacity of a heat process in terms of destruction of bacteria in a container of foods, consideration of only the heat treatment received at a single point in the container is not sufficient and that exact evaluation can be obtained only if all points in the container are considered. On this basis a mathematical method of integrating the lethal effects of time–temperature profiles existing throughout a container of conduction-heating foods during heat processing was developed. Improvements in the calculations and analysis have been made with various computer techniques. This procedure assumes the following:

1. That the death of bacteria by heat is logarithmic.
2. That the procedures developed by Ball (Ball, 1923–1924)[1] for evaluating heating effects at the slowest heating point in the container, in terms of the equivalent effects at 121.1°C (250°F), are reliable.
3. That the equations developed by Olson and Jackson (1942)[12] for estimating the value of J, the lag factor for curves representing heating at different points in a container of conduction heating foods, are reliable (Note: Any surface within the container which is iso-F is also iso-J).

In developing the equation for the Stumbo mathematical procedure, the following steps were taken:

1. A number of iso-J regions from the center to the wall of the container were defined according to the Olson and Jackson equation.
2. The volume, V, enclosed by these iso-J regions was determined.
3. The F value, F_λ, representing the heat treatment received by any iso-J region during processing, was characterized, and it was shown that for a particular iso-J region the number of spores surviving (b/a) was proportional to the volume enclosed by that iso-J region.
4. b/a, the fraction of spores surviving, was then equated to

$$\frac{1}{\log^{-1} F/D_r} = 10^{-F_\lambda/Dr,}$$

where F is the equivalent in minutes of heating at 121.1°C (250°F) and D_r is the slope of the survival curve at 121.1°C (250°F). F_λ = F value received at any point in the container except at the geometric center.

5. It was shown that when $F_\lambda - F_c$ was plotted against V a straight line resulted as long as V did not exceed 0.4 (F_c is the F value representing the

heat treatment received at the center of the container in terms of minutes at 121.1°C (250°F)). Thus, according to the equation for a straight line which passes through the origin, $F_\lambda = F_c + mV$.

6. Designating F_s as the F value representing the equivalent of all heat received by the entire container with respect to its capacity to reduce the bacterial population of the container, an equation was set up as follows:

$$10^{-F_s/D_r} \times 1 = \int_0^1 10^{-F_\lambda/Dr} \cdot dv$$

7. By substituting F_λ for certain other terms, a formula was obtained which among others contained the terms V and F_λ. Since the value of F_λ varies with V, the latter term was assigned a constant value of 0.19. This was done for the sake of convenience in using the equation since when this value was assigned $g_\lambda = 0.5\ g_c$ where g_λ is the value of g for a particular iso-J region enclosing a volume of 0.19 of the container contents. g_c is the value of g at the geometric center of the container.

8. With $V = 0.19$,

$$F_s = F_c + D_r\left(1.084 + \left[\frac{\log F_\gamma - F_c)}{D_r}\right]\right)$$

Examples of the manner in which the F_s formula may be applied are given in the following:

Assuming that D_r, the thermal-death time at 121.1°C (250°F) of the organism under investigation when suspended in the food being processed is known. Also realizing that U (the thermal-death time of the organism at retort temperature) $= FF_1$ where F_1 is the thermal-death time at retort temperature when $F = 1$ (This can be seen to be true by drawing a line parallel to the thermal-death time curve with F equal to 1).

For any process in which a heat-penetration curve has been obtained B_B can be determined since f_h and JI can be obtained and g can be obtained from curves (see Fig. 2–8) according to the value of $\frac{f_h}{U}$ which is known.

Having obtained the value of g, it can now be considered to be the value of g at the center of the can, g_c, since F_s applied to conduction-heating products. The value of F_1 can be obtained from the z value of the thermal-death time curve since RT, retort temperature, is known and $F_1 = \log^{-1}\frac{(250 - RT)}{(z)}$.

If g_c is known, then U_c can be determined since f_h is known; and the value of $\frac{f_h}{U_c}$ will correspond to the value of g_c; but $U_c = F_cF_1$; and F_1

is known, hence the value of F_c can be determined. This is one of the unknowns required to solve the F_s equation.

According to the specifications for the F_s formula $V = 0.19$; and when this is the case $g\lambda = 0.5\ g_c$ but g_c is known, hence the value of $g\lambda$ may be determined. It is possible then to determine the value of $F\lambda$ by using $g\lambda$ in the same manner as g_c was used in solving for F_c. Since all the unknowns have now been determined, the value of F_s may be calculated from the formula; but F_s also $= D_r$ (log a − log b) where D_r is known. Also, to use this method, a, the number of spores of the particular organism under investigation present in the unprocessed container, must be known. It is then possible to solve for b, the number of spores surviving per container. Assuming b had a value of 0.15, this is interpreted to mean that in 100 cans there would be 15 which contained a single surviving spore, hence 15% of spoilage of product might be expected under favorable conditions for growth.

In another example, assuming a process at 115.5°C (240°F), with $f_h = 60$, $JI = 70$, $z = 18$, 1000 spores per container, $D_r = 2$. The process is to be taken to 0.01 spore per container.

Then

$$
\begin{aligned}
F_s \text{ (required)} &= D_r (\log a - \log b) \\
&= 2 (\log 1000 - \log 0.01) \\
&= 2 (3 - (-2)) = 10.
\end{aligned}
$$

$$F_1 = \log^{-1} \frac{250 - 240}{18} = 3.59.$$

Assume $F_{c1} = 10$ and $F_{c2} = 15$ since

$$
\begin{aligned}
U_c &= F_c F_1, \\
U_{c1} &= 10 \times 3.59 = 35.9, \\
U_{c2} &= 15 \times 3.59 = 53.9.
\end{aligned}
$$

$\dfrac{f_h}{U_{c1}} = \dfrac{60}{35.9} = 1.67$ and from $\dfrac{f_h}{U}$ vs curves, $g_{c1} = 1.4$ and $g_{\lambda 1} = 0.7$.

$\dfrac{f_h}{U_{c2}} = \dfrac{60}{53.9} = 1.11$ and from curve $g_{c2} = 0.6$ and $g_{\lambda 2} = 0.3$.

If $g_{\lambda 1} = 0.7$, from the curve $\dfrac{f_h}{U_{\lambda 1}} = 1.2$, hence $U_{\lambda 1} = \dfrac{60}{1.2} = 50$ and

$U_{\lambda 2} = \dfrac{60}{0.8} = 75$, correspondingly since $U_{\lambda 1} = F_{\lambda 1} F_1$ or $F_{\lambda 1} =$

$\dfrac{U_{\lambda 1}}{F_1}$, or $F_{\lambda 1} = \dfrac{50}{3.59} = 13.9$ and $F_{\lambda 2} = \dfrac{75}{3.59} = 20.9$ therefore,

$$F_{s1} = 10 + 2\left[(1.084 + \log\left(\frac{13.9 - 10)}{2}\right)\right] = 12.75$$

$$F_{s2} = 15 + 2\left[(1.084 + \log\left(\frac{20.9 - 15)}{2}\right)\right] = 18.10.$$

According to the formula, the

$$F_c \text{ (required)} = F_{c_1} + \frac{(F_{c_2} - F_{c_1})}{F_{s_2} - F_{s_1}} \times (F_s \text{ req.} - F_{s_1})$$

$$= 10 + \frac{(15 - 10)}{(18.10 - 12.75)} (10 - 12.75) = 7.44;$$

then U_c (required) $= F_{creg}F_1 = 7.44 \times 3.59 = 26.7$. Accordingly,

$$\frac{f_h}{U_c} \text{(required)} = \frac{60}{26.7} = 2.25\,.$$

from $\frac{f_h}{U}$ vs g curves g_c required $= 2.25$.

Therefore, processing time

$$\begin{aligned} B_B &= f_h (\log JI - \log g) \\ &= 60 (\log 70 - \log 2.25) \\ &= 89.6 \text{ min.} \end{aligned}$$

Regarding the graphical or formula method of calculating processing times, as compared with other methods of determining the same thing, the two procedures first mentioned are based on the assumption that if the sum of the heat treatment effects received by the slowest heating point in the can is equivalent to the thermal death time of the spores under consideration, then the material in all parts of the container must receive an adequate heat treatment to provide sterility.

If it is considered, as in the F_s method, that the death of bacteria by heat is logarithmic, then the attainment of sterility, or commercial sterility, depends not only upon the heat treatment received at various points in the container but also on the number of bacteria, of a particular heat resistance present at those points. Hence

$$\frac{dn}{dt} = -kn \text{ or } n = n_0\, e^{-kt},$$

where n is the number of spores surviving at time t, n_0 is the number of spores present at time $t = 0$; and k is the death-rate constant for the particular organism in the food product at the temperature of heating.

The F_s and similar methods of calculating processing times are no doubt more accurate than the classical methods. The difficulty is that the level of contamination of the product with spores of a particular heat resistance must be known. This type of information is generally not available to canners. What has been done in most cases, therefore, is to use classical methods to determine the minimum botulinum kill and thereafter double or further increase the process to a point where spoilage due to organisms more heat resistant than *Clostridium botulinum* is at an acceptably or economically low level (see Fig. 2–12). Due to the arbitrary manner in which some processes (times and temperatures) have been chosen, it is almost certain that some canned foods are overprocessed with heat.

A Method of Predicting Flat-Sour Spoilage of Canned Peas

This is a special method (Knock, 1954)[8] applied to thermal processing which will predict flat-sour spoilage in canned peas. By this method it is not practical to estimate spoilage frequencies involving fewer than 1% of the processed cans. It may be considered that a spoilage rate of 1% is so high that from the standpoint of economics no canner could afford it, and this is true. However, it must be remembered that flat-sour spoilage organisms are thermophilic spores of high heat resistance and in those foods in which they may cause spoilage, it is generally not possible to destroy all flat-sour spores in all containers. If all canned foods were heat processed to the extent that all thermophilic bacteria, present therein, were destroyed, the excessive heating would cause great changes in quality. What is done, therefore, is to keep the contamination with spores of thermophilic bacteria as low as possible, heat process to further reduce the number of thermophilic spores to a low level, cool the heat-processed product to a point well below that which is optimum for the growth of thermophiles (a product mass mean temperature of 35–43.3°C (95–110°F) and thereafter hold the product at temperatures well below the thermophilic range whenever possible. In this manner flat-sour and thermophilic spoilage is held to a minimum.

Development of the Method

The method assumes (a) that under suitable conditions one surviving flat-sour spore per container is sufficient to cause spoilage of the food in the container, (b) that *Bacillus stearothermophilus* is representative of the most heat-resistant flat-sour spores found in peas, and (c) that when the number of spores surviving per can is less than one, the percentage of spoiled cans will be equal to the number of spores surviving per 100 cans. Three requirements of the process were: (1) to determine the decimal reduction time of *Bacillus stearothermophilus* at various temperatures, (2) to determine the number of flat-sour spores in the unprocessed can of peas having a heat resistance equivalent or greater than that of *Bacillus stearothermophilus,* and

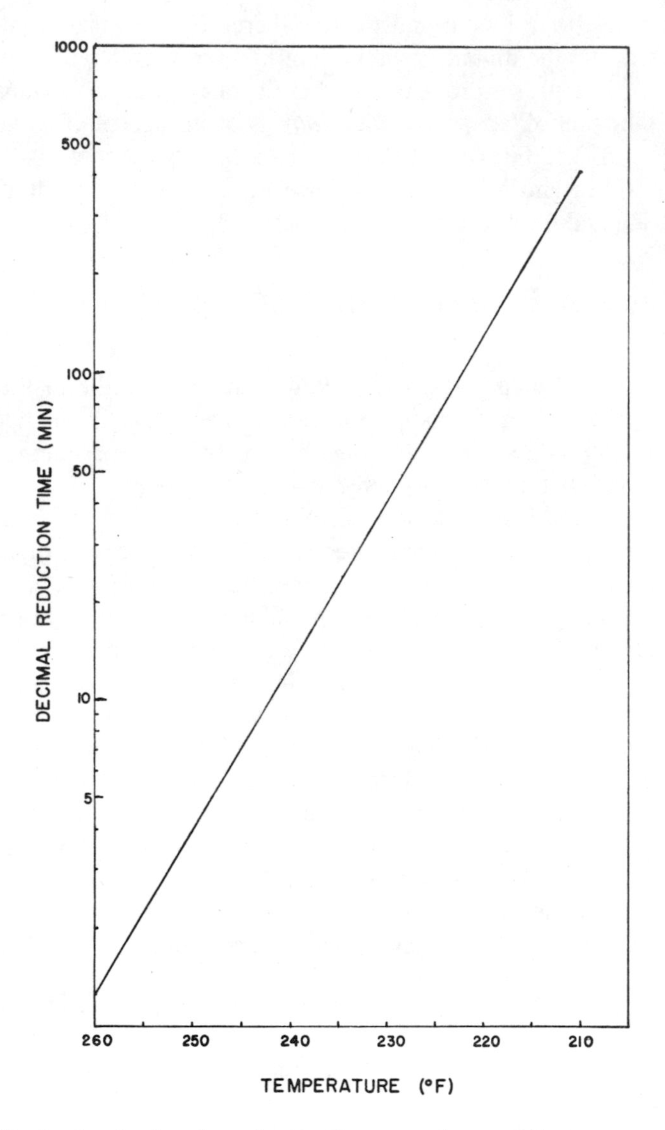

FIG. 2–9. Decimal reduction times for *Bacillus stearothermophilus*. (Courtesy of Dr. G. G. Knock and the Journal of Sci. Food Agric.)

(3) to express the lethal effect of processing times at particular temperatures in terms of the number of decimal reduction times for *Bacillus stearothermophilus*.

The decimal reduction times for *Bacillus stearothermophilus* were determined by suspending the spores in canned pea brine and making counts after heating at various temperatures as is done in determining a survival curve. The log of the 90% or decimal reduction time can be plotted vs temperature (see Fig. 2–9). The decimal reduction times were, for instance, 4 min at 121.1°C (250°F), 12 min at 115.5 (240°F), 36 min at 110°C (230°F), and 124 min at 104.4°C (220°F).

In carrying out the calculations, the number of spores adhering to the peas was related to the number present in 50 ml of unprocessed pea brine. This was determined by experimental tests. For instance, in the No. 1 tall can 301 × 411 (300 g peas to 150 ml of brine) this relationship was 5/2. To obtain the count in 50 ml of pea brine, the unprocessed can was shaken for 1 min after which 50 ml of the brine was pipetted into a bottle of 50 ml of molten, double-strength, dextrose–tryptone agar containing bromcresol purple. This material was mixed and autoclaved for 10 min at 115.5°C

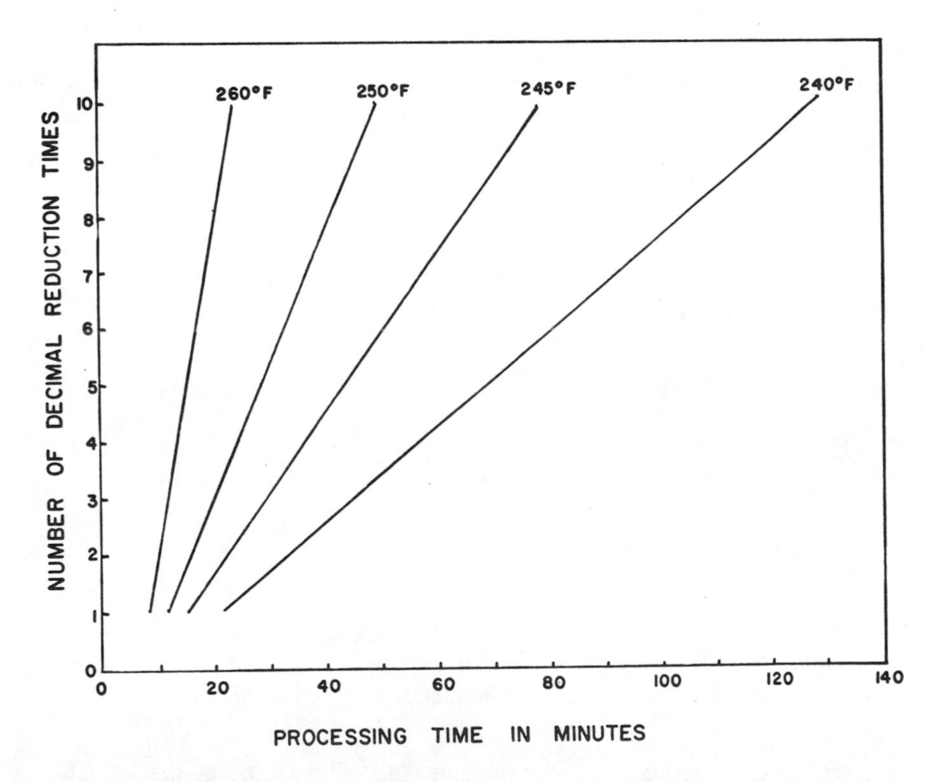

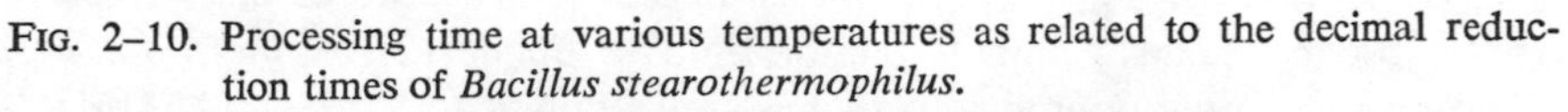
FIG. 2–10. Processing time at various temperatures as related to the decimal reduction times of *Bacillus stearothermophilus*.
(Courtesy of Dr. G. G. Knock and the Journal of Sci. Food Agric.)

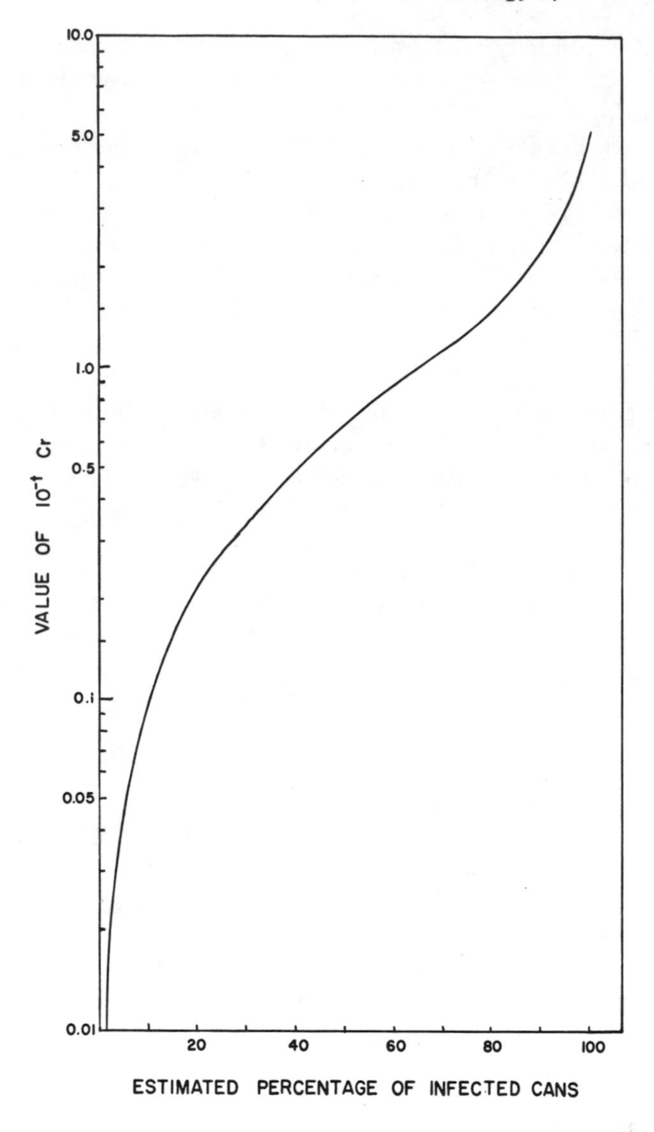

FIG. 2–11. Plot 10^{-t} *Cr* against percentage spoilage.
(Courtesy of Dr. G. G. Knock and the Journal of Sci. Food Agric.)

(240°F) (considering the come-up and come-down time of the autoclave, this is just about one decimal reduction time for *Bacillus stearothermophilus*). The agar was now poured into plates and incubated at 55°C (131°F) to obtain the number of flat-sour spores = *r*. The count in the unprocessed can could now be obtained as $C \times r$ where C is the can factor. The can factor for the number 1 tall would be 3 (150 ml of brine compared to the 50 ml actually counted) $\times$ 5/2 (fraction of spores sticking to the peas) $\times$ 10 (since one decimal reduction time was applied prior to making the count) =75.

Can factors for various sized containers were found to be:

$$\begin{aligned} &\text{No. 1 } (211 \times 400),\ C = 45 \\ &\text{No. 2 } (401 \times 411),\ C = 135 \\ &\text{No. 3 } (307 \times 409),\ C = 90. \end{aligned}$$

The lethal effect of processing at 126.6, 121.1, 118.3, and 115.5°C (260, 250, 245, and 240°F) expressed in terms of decimal reduction times was calculated by a graphical method in which a special graph paper was used. This graph paper was so constructed that the reciprocals of the decimal reduction time for particular temperatures were marked off on the vertical scale, the temperature lines being drawn at distances proportional to the decimal reduction times at the corresponding temperature. Time was plotted on the horizontal scale. Retort come-up time was considered to be equivalent to 42% (of that time) at retort temperature and was so included on the time scale.

A graph of the decimal reduction time values vs time of processing at different temperatures is shown in Fig. 2–10.

The percentage of cans containing viable spores or the number of cans which might spoil was calculated as $N(1 - e^{-x})$ where N represents the number of cans processed and x represents the mean number of spores surviving per can after heat processing.

Also, if r equals the number of spores found in 50 ml of brine, according to the method previously described, and C equals the can factor, then Cr equals the number of spores in the unprocessed can. In addition, if t represents the retort process in terms of the number of decimal reduction times, then $x = 10^{-t}Cr$, and the percentage of cans containing viable spores, hence the percentage of spoilage, would be $100\ (1 - e^{-10^{-tCr}})$. A curve can then be drawn in which the value of $10^{-t}Cr$ is plotted on the vertical scale, and the percentage of spoilage is plotted on the horizontal scale (see Fig. 2–11).

Application of the Method

In using the prediction of flat-sour spoilage method, at various times throughout the day in a pea-canning plant, a can of unprocessed peas is taken from the line after sealing. The can is shaken, opened, and 50 ml of the brine is plated out after mixing with double-strength medium and autoclav-

FIG. 2–12. Filling the retort with cans to be processed. (Courtesy of Wm. Underwood Co.)

FIG. 2–13. Vertical retorts in operation. (Courtesy of Wm. Underwood Co.)

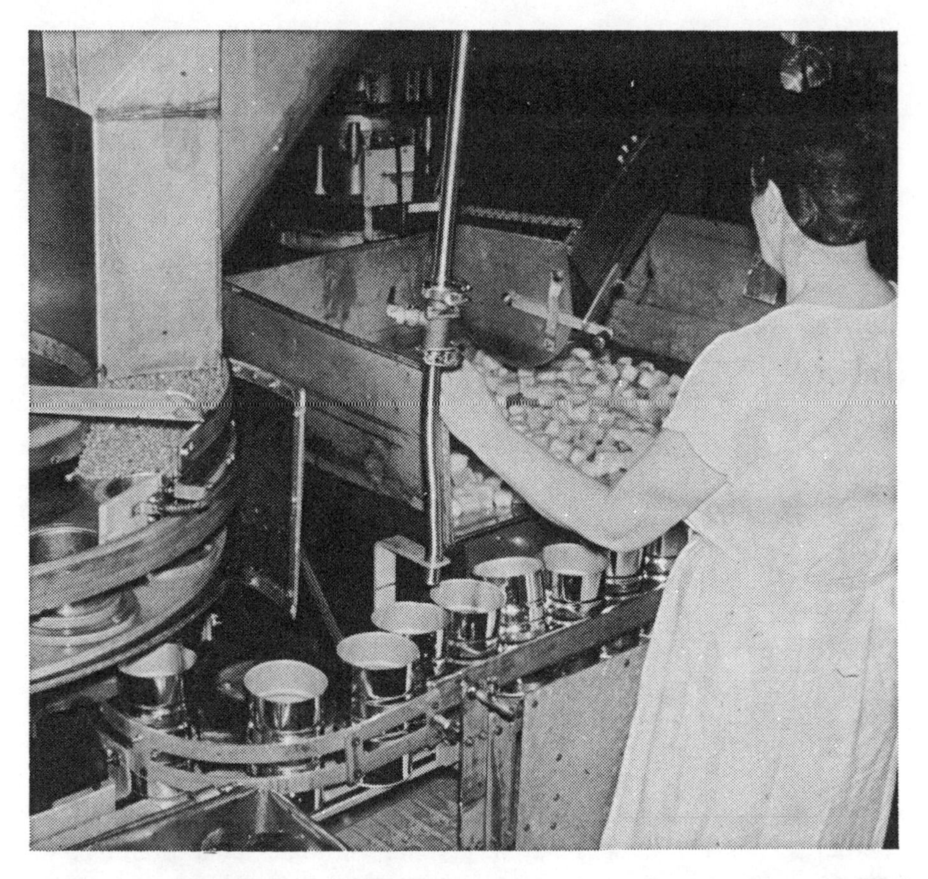

FIG. 2–14. Filling pork and beans into empty containers. Note that beans are filled volumetrically but pieces of pork are filled by hand. (Courtesy of Wm. Underwood Co.)

ing as previously described (10 min at 115.5°C (240°F)). The count r is obtained in this manner; and since C is known, the value of Cr can be calculated. Also, since the retorting time and temperature are known, t can be found from Fig. 2–10. It is now possible to find the value of $10^{-t}Cr$, hence the percentage of infected cans, which would indicate the percentage of flat-sour spoilage at thermophilic incubator temperatures, can be determined from Fig. 2–11.

It is not possible to apply the general principles of the "Prediction of flat-sour spoilage" method to the prediction of other types of spoilage. The reason is that even to predict spoilage at a level as low as 1%, a very small value for r is required. The actual count on 50 ml of brine for most cans in this case must, therefore, be zero, requiring the examination of a large number of cans before an actual count is obtained. Since spoilage for canned foods at ordinary temperatures must be something like 0.001% or lower, the value of r for ordinary spoilage would have to be so low that it would not be possible to examine enough cans of product to obtain a count.

Spoilage of Canned Foods

It is difficult to present a clear picture of the spoilage of canned foods by microorganisms since this depends upon many factors which are interrelated. Among these factors are the numbers of particular organisms surviving the heat process; the temperature at which the surviving organisms will grow, hence the temperature at which the finished product is held, the suitability of the canned food to support growth of the organism; the pH of the food in relation to the growth of surviving organisms; the oxygen tension in the can and the oxygen requirements of the surviving organisms; the presence of added components, such as nitrate, which may provide for the growth of aerobes when no unbound oxygen is present, and other factors.

There is disagreement in the literature as to how often canned heat-processed foods which are "commercially sterile" (do not spoil after heat processing and contain no viable pathogens) actually contain viable microorganisms. Reports of the presence of viable microorganisms in commercially heat-processed canned foods have varied between 100% to none. It can only be concluded that in all probability some percentage of cans of commercially heat-processed foods which do not spoil actually contain viable microorganisms and that this may vary from plant to plant and from time to time.

Whereas the organisms causing canned food spoilage are usually spore-forming bacteria, yeasts, vegetative bacterial types, and especially molds may sometimes grow and cause decomposition, although this situation usually occurs under conditions of leakage or underprocessing.

The bacteria causing spoilage of canned foods may be aerobic, anaerobic, or facultative in oxygen requirements. Generally they are mesophilic growing in the range of temperatures between 20 and 45°C (68 and 113°F) or thermophilic. Obligate thermophiles have been classified as requiring temperatures between 37.8 and 82.2°C (100 and 180°F) for growth; whereas facultative thermophiles have been defined as bacteria which will grow well at 55°C (131°F) but will also grow at temperatures below 37.8°C (100°F) (Cruess, 1958).[4] Actually no good classification of microorganisms can be made on the basis of growth temperature because of the wide range of temperatures over which the various species will grow.

Thermophilic bacteria, especially obligate thermophiles, are generally much more heat resistant than mesophilic types, hence the tendency of the canning industry to control spoilage due to these types by other means than total thermal destruction of these organisms in canned foods.

The special characteristics of the most common types of bacteria causing spoilage of canned foods are listed in Table 2-4.

To indicate the interaction of various factors on the growth of microorganisms in canned foods, some results of tests made with *Clostridium thermosaccharolyticum* are illustrative. This organism was heated at 87.8°C (190°F) in media and in different foods to obtain survival of various numbers of spores after which growth was determined at 36.7°C (98°F) and at 55°C (131°F) the latter being the optimum temperature for this organism. In one artificial medium, pH 4.4, 0.8 spores per milliliter would initiate growth at 55°C (131°F); but when the pH was 4.0, 1000 spores per milliliter were required to initiate growth at this temperature. In pumpkin, pH 5.2, spoilage occurred at 0.8 spores per milliliter when incubation was carried out at 36.7°C (98°F). When the substrate was spaghetti and tomato sauce, pH 4.6, there was no spoilage at 36.7°C (98°F) when the spore concentration was 800 per gram; but there was spoilage when the spore concentration was 8000 per gram. At 55°C (131°F) there was spoilage of this product at spore concentrations of 80 per gram. In tomato juice, pH 4.5, and cottage cheese, pH 4.3, there was no growth of the organism at either 36.7° or 55°C (98 or 131°F) when the concentration of surviving spores was 8000 per milliliter or 8000 per gram.

Canned foods may be spoiled by vegetative types of bacteria, especially when low-heat treatment is given the product or when there is leakage after processing. Among the bacteria causing spoilage in such instances are species of *Lactobacillus* and *Leuconostoc*. This may be due to the fact that these organisms are aciduric and, due to the lower heat resistance of microorganisms at low pH, acid foods are generally processed at much lower temperatures than are the more neutral foods.

Molds, especially those which are heat resistant, sometimes cause spoilage of canned fruits. *Byssochlamys fulva* (Oliver and Rendle, 1934),[11] which sometimes causes swells in canned blueberries, is said to have a heat resistance equivalent to 30 min at 85°C (185°F) in neutral media or 10 min at 87.8°C (190°F) in blueberry juice. The ascospores of this organism have been found to be more heat resistant at pH 5 than at pH 7.0. A heat-resistant *Penicillium* (Williams et al., 1941),[21] which would grow under a vacuum of 25 in. of mercury, has also been isolated from blueberries. The sclerotia of this organism withstood heating in blueberry juice at 85°C (185°F) for 300 min.

TABLE 2–4
Characteristics of Some Spore-Forming Bacteria which Cause Spoilage of Canned Foods

Organism	Approximately lower level of pH for growth	Growth temperature range	Oxygen relationships	Approximate 12 (D_{10}) at 121.1°C(250°F) pH 7.0	Special properties
Bacillus stearothermophilus	6.0	33–70°C(91.4–158°F)	Facultative	48–60	Flat-sour
Bacillus coagulans (thermoacidurans)	4.0	28–60°C(82.4–140°F)	Facultative	0.85	Flat-sour
Bacillus subtilis	5.5	28–55°C(82.4–131°F)	Aerobic	—	—
Bacillus brevis	6.0	28–55°C(82.4–131°F)	Aerobic	—	May Utilize O_2 produce gas
Bacillus circulans	4.8	28–55°C(82.4–131°F)	Facultative	—	—
Bacillus pumilus	5.5	28–50°C(82.4–122°F)	Aerobic	—	—
Bacillus polymyxa	5.5	20–35°C(68.0–95°F)	Facultative	2.4–8.4	Acid and gas
Bacillus macerans	3.8	28–50°C(82.4–122°F)	Facultative	2.4–8.4	Produce acid and gas
Bacillus cereus	5.5	28–45°C(82.4–113°F)	Facultative	—	May utilize bound O_2 may produce gas
Bacillus sphaericus	6.0	28–45°C(82.4–113°F)	Aerobic	—	—
Clostridium botulinum, type A	4.8	12.8–40.6°C(55–105°F)	Anaerobic	2.5	Produces toxin in foods, produces gas

Clostridium botulinum, type E	4.8	3.3–30°C(38–86°F)	Anaerobic	6 min at 82.2°C(180°F)	Produces toxin in foods, produces gas
Clostridium thermosaccharo-lyticum	4.0	30–62.8°C(86–145°F)	Anaerobic	36–48	Acid and gas
Clostridium nigrificans	6.0	30–70°C(86–158°F)	Anaerobic	24–36	Produces H_2S from proteins
Clostridium butyricum	4.0	30–45C°(86–113°F)	Anaerobic	—	Acid and gas
Clostridium bifermentans	5.5	30–50°C(86–122°F)	Anaerobic	—	Produces H_2S acid and gas
Clostridium putrefaciens	6.0	0–30°C(32–86°F)	Anaerobic	—	Some H_2S acid and gas
Clostridium pasteurianum	3.6	20–40°C(68–104°F)	Anaerobic	2–7 40 min at 100–C(212°F)	Produces gas
Clostridium perfringens	5.7	20–50°C(68–122°F)	Anaerobic	—	Acid & gas
Clostridium Sporogenes	6.0	20–50°C(68–122°F)	Anaerobic	0.1 to 1.5	gas

Regarding the classification of foods according to pH, those below pH 4.0 are generally considered to be high in acid. Foods having a pH of 4.0–4.5 are considered as acid, and those with a pH above 4.5 are said to be low-acid foods. In processing it has been shown that the heat resistance of bacterial spores decreases when the pH is lowered below 5.5, and there is a further decrease in heat resistance as the pH is lowered beyond this point. Generally, canned fruits and fruit juices (pH 2.9–4.5) are heat processed in boiling water although tomato products, especially tomato juice, may be given a high temperature–short time treatment [0.7 min at 121.1°C (250°F)], cooled to 93.3–98.9°C (200–210°F), and canned hot. Canned vegetables (pH 4.6–6.5) and meat and fish products (pH 5.8–7.0) are ordinarily heat processed at higher temperatures under steam pressure, an exception being cured meat products which may be processed in boiling water.

General Aspects Regarding Canned Food Spoilage

In the spoilage of canned foods most bacilli attack carbohydrates producing acid but no gas. Exceptions to this are *Bacillus polymyxa, Bacillus macerans, Bacillus cereus,* and *Bacillus brevis. Bacillus macerans* (also called *betanigrificans)* produces acid and gas in the presence of dextrose and iron (the metal container) and has been sometimes associated with the spoilage of commercially canned beets, where it has caused swelling of the container and discoloration (blackening) of the products.

In canned cured meat products *Bacillus brevis* and *Bacillus cereus* have sometimes grown and produced gas under essentially anaerobic conditions. These organisms are able to use bound or chemical oxygen, the nitrate present in the cured meat in this case serving as the source of this element.

Clostridium nigrificans, especially in canned corn, sometimes produces hydrogen sulfide and causes blackening of the product. Owing to the solubility of this gas and its reactivity with the iron of the can, no swelling of the container is ordinarily encountered in this type of spoilage.

In the past a source of thermophiles in canned foods was the starch, used in such products as cream-style canned corn and especially the sugar in canner's brine used for vegetables. Standards for thermophilic bacteria in starch and sugar have been set up and "low-thermophile sugar" for canning is now available. Today it is even possible to purchase thermophile-free starch and sugar as ingredients for canned foods.

Thermophilic bacteria are also present on unprocessed vegetables and may build up on equipment or in materials during processing. The top half of water blanchers has been shown to be a source of contamination with thermophiles. In order to obviate this, steam may be injected into a water blancher, over the water, to raise the temperature to the point where thermophiles will not grow.

Separators for peas or lima beans become a source of contamination with thermophiles if they are not cleaned thoroughly. In order to prevent the growth of thermophilic bacteria therein, hence contamination of the product, canner's brine, in containers or in pipes, should not be held over from day to day in the canning plant. It is preferable to make up new canner's brine daily and to flush the tank and pipes used for holding this material after they have been emptied.

The reservoir on the filler, holding canner's brine should contain heating coils so that the brine when added to the product, will have a temperature of 76.7°C (170°F) or higher, thus preventing the growth of thermophiles.

When heat-processed canned foods are being cooled, at a time when the sealing compound in the top and bottom of the can is still plastic and vacuum is being formed in the container, microscopic amounts of the cooling water may enter some of the cans. This water may contain bacteria which will cause spoilage of the product. That this is the case has been shown by checking the extent of spoilage in air-cooled product as compared with water-cooled products. Spoilage in the latter case has been found to be about 10% greater than in products cooled in air. The water in cooling canals, or used for cooling in retorts, should, therefore, be free of pathogenic bacteria. It is also the case that cooling water should be chlorinated and contain a residual of at least 2–3 ppm of available chlorine. Chlorination to leave a residual of 100 ppm of available chlorine would give better results, but such concentrations will promote outside corrosion of the container; hence, if used, the cans must be further immersed in a solution of ammonia or of sulfurous acid.

It is considered that cooling water having a plate count not greater than 100 viable bacteria per milliliter is suitable for purposes of cooling canned, heat-processed foods.

Pasteurization of Foods

Heat pasteurization is a treatment which is generally supposed to be applied to milk and cream. Heat treatment, which includes destruction of many of the microorganisms present in foods but which does not provide commercial sterilization, is applied to a number of other foods including cooked sausage products, meat loaves, canned and boiled hams, various egg magmas, and cooked crab meat.

Heat pasteurization of milk and cream was introduced to extend the storage life of these products at refrigerator temperature above freezing and not for the purpose of destroying pathogenic bacteria which might be present therein (Shrader, 1939).[14] However, the latter purpose has become one of the main objectives of the process.

It is not possible to state what numbers of pathogens may be destroyed by pasteurization treatment of milk and cream. The times and temperatures for this process seem to have been chosen on the basis of whatever temperatures and times the product could be treated without undergoing extensive organoleptic changes or on the infectivity of inoculated, then pasteurized products, rather than on the basis of how many cells of a particular pathogen might be destroyed during such treatment. That this is the case is indicated by the fact that requirements have in recent years been raised from 30 min at 61.7°C (143°F) and 15 sec at 71.1°C (160°F) to 30 min at 62.8°C (145°F) and 15 sec at 71.7°C (161°F) in order to destroy the rickettsial organism causing Q fever, *Coxeilla burnetti.* There is no doubt, however, that even treatment at the old time–temperature levels have served to destroy most pathogenic bacteria present in milk and cream, and hence such treatment has been responsible for the prevention of much food-borne disease.

In the processing of cooked sausage products, meat loaves, and boiled hams, heating is done to cook these products and convert them to a form in which they can be conveniently handled. Again, for the most part no attempt has been made here to reduce the number of bacteria of any type from one level to another. The regulatory authorities have apparently determined that this type of treatment is such that *Trichinella spiralis,* a round worm which is sometimes present in pork and which will cause trichinosis in man, is destroyed by heating to a temperature of 58.3°C (137°F) or higher. Again the numbers of organisms destroyed and decimal reduction times seem not to have been considered. The heat treatment used for the above products is sufficient to raise the temperature of all parts of the product well above 58.3°C (137°F). Spoilage of cooked sausage-type products is generally due to organisms present because of recontamination during processing after cooking.

Crab meat in cans is sometimes pasteurized by heating to an internal temperature of 77.2°C (171°F). This is done to destroy salmonellae and enterotoxic staphylococci.

The pasteurization of whole egg magma, liquid egg yolk, or egg white prior to further processing (freezing and drying) is carried out to destroy salmonellae which may be present therein. The actual times and temperatures of treatment may vary somewhat, but this treatment is sufficient to provide at least 3–4 decimal reductions of salmonellae of ordinary heat resistance.

Large, refrigerated-type canned hams are processed in boiling water to an internal temperature of 71.1°C (160°F). Such products are not commercially sterile and even vegetative types of bacteria, especially those present in fat, may survive such treatment. Small canned hams and cured luncheon meat-type products are heated at higher temperatures, but again all bacteria may not be destroyed. Generally the latter-type products are commercially sterile and are held without refrigeration. In terms of the reduction of *Clostridium botulinum,* large canned hams receive a heat treatment having an F_0 value of about 0.01 while commercially sterile-type cured products receive a heat treatment equivalent to an F_0 of 0.3–0.6. The salt and nitrite present in these products sensitize microorganisms to heat or prevent outgrowth after treatment, enabling such low-level heat treatments to be used.

References

1. C. O. Ball, The time necessary to process canned foods, *Bull. Nat. Res. Council* **7** (1923–1924), 937 (Parts I, II, III & IV).
2. W. D. Bigelow, the logarithmic nature of thermal death time curves. *J. Infect. Dis.* **29** (1921), 528.
3. W. D. Bigelow, G. S. Bohart, A. C. Richardson, and C. O. Ball, Heat penetration in processing canned foods, *Nat. Canners Ass.* Bull. Res. Lab. No. 16L (1920), 128.
4. W. V. Cruess, "Commercial Fruit and Vegetable Products," McGraw-Hill, New York, N.Y., 1958.
5. F. L. Davis, Jr. and O. B. Williams, Studies of heat resistance—I Increasing resistance to heat of bacterial spores by selection. *J. Bacteriol.* **56** (1948), 555.
6. J. R. Esty and K. F. Meyer, The heat resistance of spores to B Botulinus and allied anaerobes. *J. Infect. Dis.* **31** (1922), 650.
7. L. B. Jensen, "Microbiology of Meats," Gerard Press, Champaign, Ill., 1954.
8. G. G. Knock, A technique for the approximate quantative prediction of flat souring in canned peas. *J. Sci. Food Agric.* **5** (1954), 113.
9. M. Levine, J. H. Buchanan, and G. Lease, Effect of concentration and temperature on germicidal efficiency of sodium hydroxide, *Iowa State Coll. J. Sci.* **1** (1927), 379.
10. J. J. Licciardello and J. T. R. Nickerson, Some observations on bacterial thermal death time curves, *Appl. Microbiol.* **11** (1963), 476.
11. M. Oliver and T. Rendle, A new problem in fruit preservation *Byssochlamys fulva,* J. Chem. Soc. Lon. 53 (1934) 166T.
12. F. C. W. Olson and J. M. Jackson, Heating curves: Theory and practical applications, *Ind. Eng. Chem.,* **34** (1942), 337.
13. O. T. Schultz and F. C. W. Olson, Thermal processing of canned foods in

tin containers III recent improvements in the general method of process calculations, *Food Res.* **5** (1940), 399.
14. J. H. Shrader, "Food Control," Wiley, New York, 1939.
15. P. Sognefest and H. A. Benjamin, Heating lag in thermal deathtime cans and tubes, *Food Res.* **9** (1944), 234.
16. P. Signefest, G. L. Hays, E. Wheaton, and H. A. Benjamin, Effect of pH on thermal process requirements of canned foods, *Food Res.* **13** (1948), 400.
17. J. A. Stern and B. E. Proctor, A micro-method and apparatus for the multiple determination of rates of destruction of bacteria and bacterial spores subjected to heat, *Food Technol.* **8** (1954), 139.
18. C. R. Stumbo, A technique for studying resistance of bacterial spores to temperatures in the higher range. *Food Technol.* **2** (1948), 228.
19. C. R. Stumbo, New procedure for evaluating thermal processes for foods in cylindrical containers. *Food Technol.* **7** (1953), 309.
20. C. R. Stumbo, "Thermobacteriology in Food Processing," Academic Press, New York, 1965.
21. C. C. Williams, E. J. Cameron, and O. B. Williams, A facultative anaerobic mold of unusual heat resistance, *Food Res.* **6** (1941), 69.
22. C. C. Williams, C. M. Merritt, and E. D. Cameron, Apparatus for determination of spore-destruction rates, *Food Res.* **2** (1937), 369.
23. E. R. Withell, The significance of the variations in shape of time-survivor curves. *J. Hyg.* **42** (1942), 124.

CHAPTER 3

Microbiology of Dried Foods

The drying or dehydration of foods is one of the earliest methods of food preservation practiced by man. Many of the traditional methods used for centuries for purposes of drying foods are still practiced throughout the world. Improvements in the procedures used for the dehydration of foods have resulted in shorter drying cycles; also, the new processes have provided for higher quality foods with better rehydrability and storage behavior. Since microorganisms require water for growth and metabolism, removal of water by any method will prevent microbial growth. The water requirements for growth of many microorganisms are defined in terms of the water activity of their environment. Microbial growth does not take place in pure water or in its absence. Therefore, any substrate which supports microbial growth is for all practical purposes an aqueous solution. The lowering of the water vapor pressure by the addition of solutes can be expressed by Raoult's law. Raoult's law is applied to ideal solutions and is expressed by the following relationship:

$$p/p_0 = \frac{n_2}{n_1 + n_2},$$

where p and p_o are the vapor pressures of the solution and the pure solvent, respectively; and n_1 and n_2 refer to the number of moles solute and solvent in the solution. For a 1.0 molal solution of an ideal solute, the vapor pressure is depressed by 0.0177%, therefore, $p/p_0 = 0.9823$. Thus, the a_w of this ideal solution is 0.9823. However, solutions rarely act ideally with nonelectrolytes departing from ideality at low concentrations of solute and electrolytes acting differently than ideal solutions at all concentrations. A 1.0 molal solution of sucrose has a water activity of 0.9806, while that of sodium chloride is 0.9669 at 25°C.

The ratio p/p_0 times 100 gives the relative humidity (relative humidity $= a_w \times 100$); and this term generally refers to atmosphere in equilibrium with the substrate, while p/p_0 or a_w refers to the substrate itself.

Reduction in a_w results from the concentration of the solution which may be achieved either by adding solutes or by removing the solvent water. Thus, addition of salt and sugars leads to an increased concentration of the solution; and this method of food preservation is well-established. Dehydration or evaporation of the solvent is the other common method of food preservation, i.e., that of removing water.

Commonly, the relationship between water content and the a_w of equilibrium for a given food is represented by the water sorption isotherm (Fig. 3-1). At a high a_w the water content is related to food composition, being greatest in foods high in low molecular-weight solutes (Christian, 1963).[4]

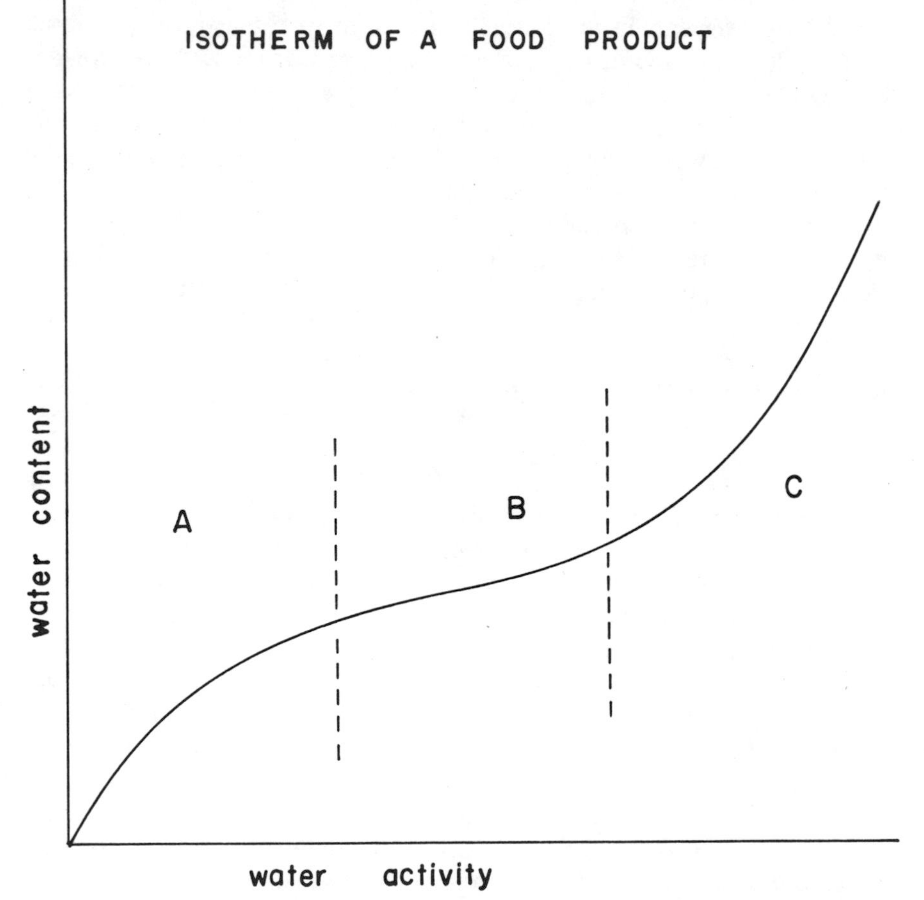

FIG. 3–1. Isotherm of a food product.

For example, it has been shown that at an a_w of 0.70 samples of dehydrated potatoes with 9.7 and 0.7% sugar content contained 15.9 and 12.6% water, respectively. At the lower a_w values (portion A of Fig. 3-1), this part of the curve is generally thought to correspond to the adsorption of a monomolecular film of water. At intermediate a_w values (region B) it is considered that additional layers of water have been adsorbed over the monolayer. Region C is believed to indicate the condensation of water in the pores of the material and the dissolution of soluble material.

Isotherms that are prepared at several temperatures should show a decrease in the amount of water adsorbed with increasing temperatures at a constant water activity. This means that if a food product, for example, is prepared at 20°C (68°F) and then stored at a higher temperature, when stored the water activity of the product will be higher.

Water Activity and Its Effects on Growth of Microorganisms

The effect of water activity on the growth of microorganisms has been reviewed by Scott (1957) and Christian (1963). Generally, there is an optimal a_w for maximal growth and as a_w is reduced, the growth rate decreases. Finally, as the a_w is decreased, a certain level is reached at which microbial growth ceases.

Microorganisms vary in the minimal water activities at which growth will take place. For example, bacteria are, in general, the most sensitive and followed by yeasts and molds. Usually bacteria do not grow at a_w values less than .90, while most yeasts are inhibited at an a_w less than .87, and most molds will not grow at an a_w of .80 (Fig. 3-2). Exceptions and variations do occur.

Staphylococcus aureus, grown anaerobically, is inhibited at an a_w of .90, while grown aerobically, the minimal a_w of *Staphylococcus aureus* is .86. Also, there are extremely halophilic bacteria which will grow at an a_w of .75 and osmophilic yeast and molds that will grow at an a_w of .62. Many factors affect the minimal a_w at which microorganisms will grow. Generally, under conditions of stress, i.e., presence of preservatives, low pH, or suboptimal growth temperatures, the ability of a microorganism to grow at lower a_w values is retarded.

The a_w at which microorganisms are stored will influence their viability when transferred to environmental conditions favorable for growth. Normally, when foods are stored at a_w values which are optimal for the maintenance of good quality, the greatest survival of the microorganisms therein may be expected. The water activity under which these stresses are imposed

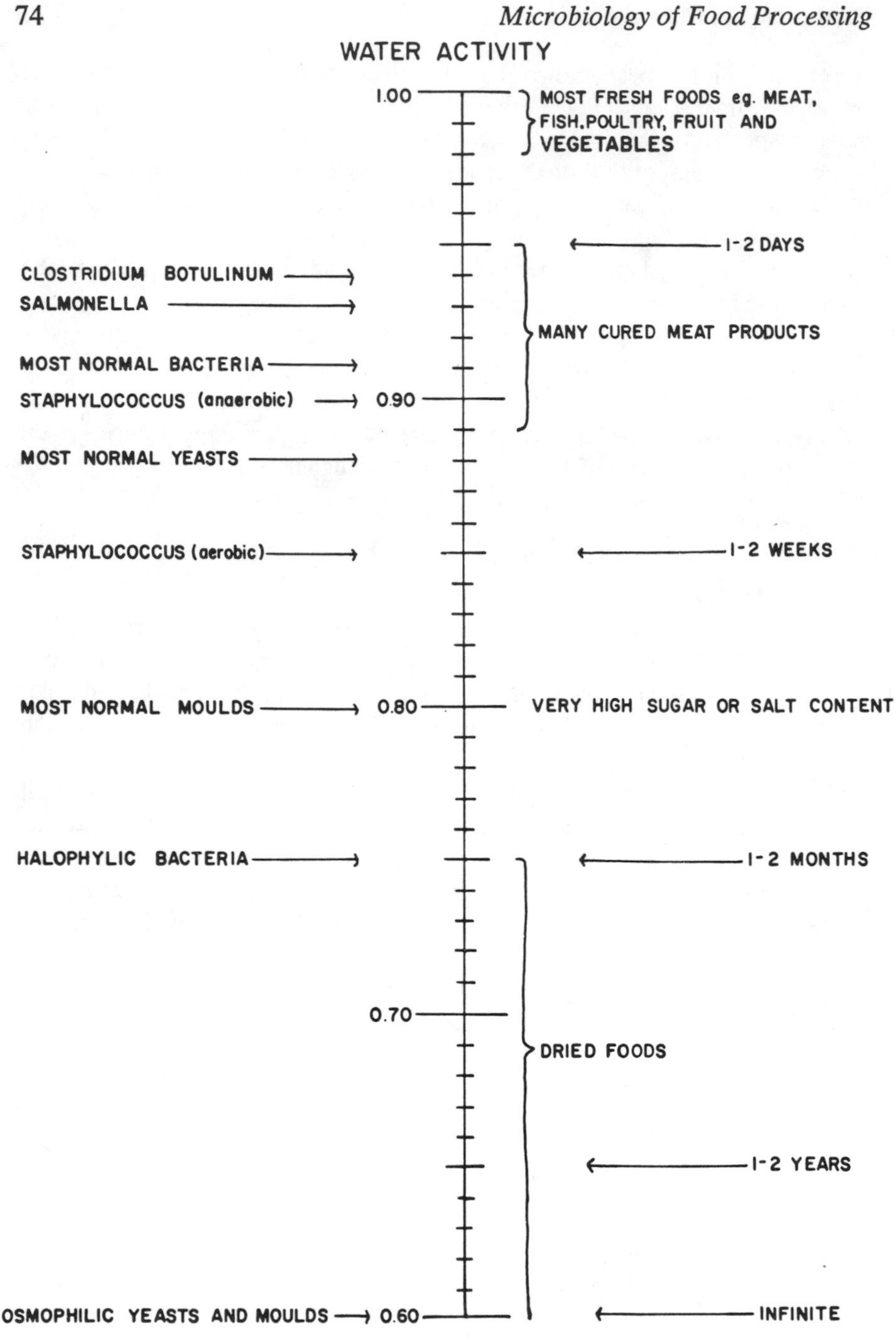

FIG. 3–2. Relationship between water activity and growth of organism.

also affects the heat and chemical resistance of microorganisms. Usually microorganisms are more resistant to heat and chemicals under conditions of low a_w values.

Introduction to the Microbiology of Freeze-Dried Foods

Freeze-drying, also known as lyophilization, has been applied for more than 50 years for the preservation of biological materials. Biological materials preserved by this method include enzymes, bacteria, yeast, fungi, and viruses. Numerous publications (Heckly, 1961; Brady, 1960; Heller, 1941)[2,5,6] are available which discuss the preservation of microbial cultures. For the preservation of microbial cultures high survival is desirable; and, therefore, the freeze-drying operation is conducted in such a manner as to optimize microbial survival. Normally this means that the freeze-drying operation is carried out with the application of a minimal amount of heat for purposes of sublimation; the samples are stored in a vacuum; and low-temperature storage is used. However, when foods are freeze-dried, the opera-

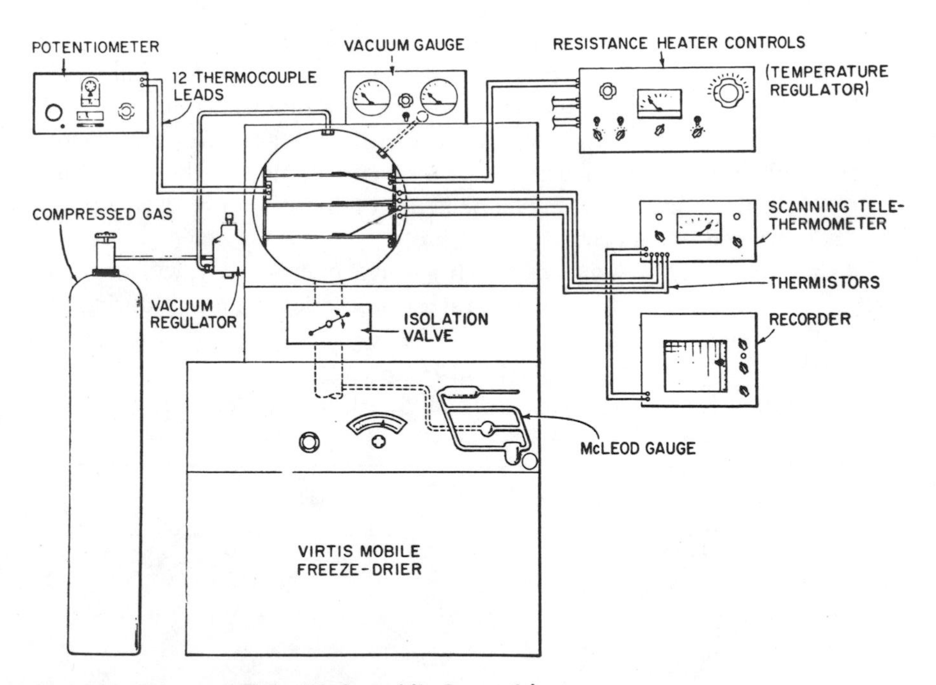

FIG. 3–3. Diagram of the Virtis mobile freeze-drier.

tion is conducted in such a manner as to minimize the drying time while maintaining a high quality final product. Discussion of the engineering principles, including the modes of heat and mass transfer during the freeze-drying operation, have been the topics of numerous papers (Burke and Decareau, 1966; King, 1970; MacKenzie, 1966).[3,7,11]

The Freeze-Drying Process

As a means of food preservation, freeze-drying represents just another link in the food chain from the producer to the consumer. Therefore, a food that is to be freeze-dried will probably be subjected to at least one or more of the more conventional preparation procedures, i.e., selection, trimming, washing, blanching or cooking, and freezing. Certain aspects of the microbiology of these phases of preparation have been discussed elsewhere (Chapter 5). After the freezing step, the food to be freeze-dried is placed in a chamber, where a vacuum is applied and the product heated. During freeze-drying (Fig. 3-3) the heat sublimes the frozen water; and, due to the fact that food material is in a vacuum chamber, water vapor is removed and condensed in the condenser. Consequently, within the food, a dry layer forms due to the receding ice layer. Numerous factors affect the drying time including the temperatures applied, the thickness of the food material, the porosity of the food, and the inherent limitations of the drying equipment used. The freeze-drying operation is necessarily an unsteady-state operation. This means that microorganisms located in different positions within the food will all have different histories with regard to the temperatures and conditions of moisture content encountered. This fact complicates the microbiology of freeze-dried foods and should be kept in mind when one drying operation is compared to another.

Certain of the microorganisms which survive freeze-drying will undergo further changes during subsequent storage in the dried state and during re-

TABLE 3–1

Effect of Freeze-Drying on the Total Plate Count of Commercial Shrimp (Cooked, Peeled and Deveined Prior to Freeze-Drying)

Before freezing (per g)	After freeze-drying and reconstitution (per g)
1,500	340
400	310
4,500	1,200
12,500	570
3,000	1,030
3,200	470

constitution of the freeze-dried food with water. Also, the various species of microorganisms present on foods to be freeze-dried will be in various stages of growth and, consequently, in different phases of physiological development. This fact also leads to variability in the microbiology of freeze-dried foods as is exemplified in Table 3-1.

Influence of Processing Variables on Microbial Survival in Freeze-Dried Foods

Freezing

The microbiology of frozen foods has been discussed in Chapter IV. It should be pointed out that little research work has been conducted on the interactions between the effects of freezing rates and the effects of freeze-drying parameters on microbial survival.

Platen Temperature

The platen temperatures used during the freeze-drying process can be varied depending upon the product being dried. At the start of the freeze-drying cycle, comparatively high temperatures are employed; and, during the drying cycle, the dried surface temperature of a given food product is maintained in the range 40–70°C (104–158°F). Aitken et al.[1] (1962) observed that drying of pork muscle at a platen temperature greater than 60°C (140°F) had an adverse effect on the organoleptic and rehydration characteristics of the product. Lusk et al. (1964),[10] in tests in which shrimp were dried at constant temperatures of 69°C and 52°C (156.2 and 125.6°F), noted that for shrimp dried at 69°C (156.2°F) rehydration was poor when compared to shrimp dried at 52.°C (125.6°F).

From the viewpoint of commercial operation, the primary objective is to produce a dried product that retains as much of the quality of the original product as possible with the minimal operating cost. Therefore, a minimal amount of heating consistent with a reasonable drying cycle is used. The application of such conditions in commercial operations will also lead to a greater percentage survival of the microorganisms present in foods.

The effects of platen temperature on microbial survival in beef, egg melange, and in model system are presented in Table 3-2 Sinskey et al., (1967).[12]

It can be seen that higher platen temperature of 61°C (141.8°F) resulted in a lower survival of bacteria on all products, regardless of whether the organisms were spread on the surface of beef, distributed throughout egg melange, or suspended in the model systems studied (gelatin,

TABLE 3–2

Survival of Four Species of Microorganisms, Freeze-Dried in Food Material and in Model Systems, at Two Different Platen Temperatures

	Platen drying temperature [a]							
	Beef		Egg melange		2% Gelatin		2% Gelatin + 5% dextrose	
	120°F	160°F	120°F	160°F	120°F	160°F	120°F	160°F
	Percentage survival							
Salmonella typhimurium	3	1	33	2	2	0	34	2
Staphylococcus aureus	52	14	51	50	54	17	55	17
Streptococcus faecalis	48	43	93	68	30	11	27	12
Pseudomonas fragi	0	0	1.6	0	0	0	0	0
Final moisture (%)	0.2	0.5	1.9	1.4	3.0	0.2	0.2	0.2
Drying time (hr)	12	8	9	7	17	18	18	14

[a] 120°F = 48.8°C; 160°F = 71.1°C.

gelatin–dextrose, or gelatin–nutrient broth–dextrose). The samples were all dried at a constant platen temperature. Higher survival was observed at a constant platen temperature of 49°C (120.2°F).

In the food and in the model systems studied, survival was higher with gram-positive bacterial *(Staphylococcus aureus* and *Streptococcus faecalis)* than with the gram-negative types *Salmonella typhimurium* and *Pseudomonas fragi).* One conclusion that may be drawn from this investigation is that gram-positive bacteria will survive the freeze-drying process to a greater extent that gram-negative types. Consequently, in determining the sanitary quality of freeze-dried foods it may be preferable to test for gram-positive bacteria (enterococci) than to test for the gram-negative types (coliforms).

Duration of Drying and the Final Moisture Control

The final moisture content of most freeze-dried foods is around 2%. However, since many foods consist of pieces of different sizes, i.e., fish, shrimp, pieces of meat, etc., some portions of the products may be overdried. While excessive drying has been reported to cause a reduction in the viability of lyophilized cultures (Scott, 1958),[14] little work has been carried out on this aspect with freeze-dried foods. One study demonstrated interrelated effects between rehydration temperature, composition of the model system used for drying, and the percentage of final water content in the sample, on the survival of *Salmonella typhimurium* when materials were subjected to freeze-drying at a platen temperatures of 48.9°C (120°F). The results are listed in Fig. 3-4. Samples dried in 2% gelatin exhibited the lowest survival. The addition of glucose to the gelatin menstruum increased survival by a factor of 100–400, depending on the rehydration temperature.

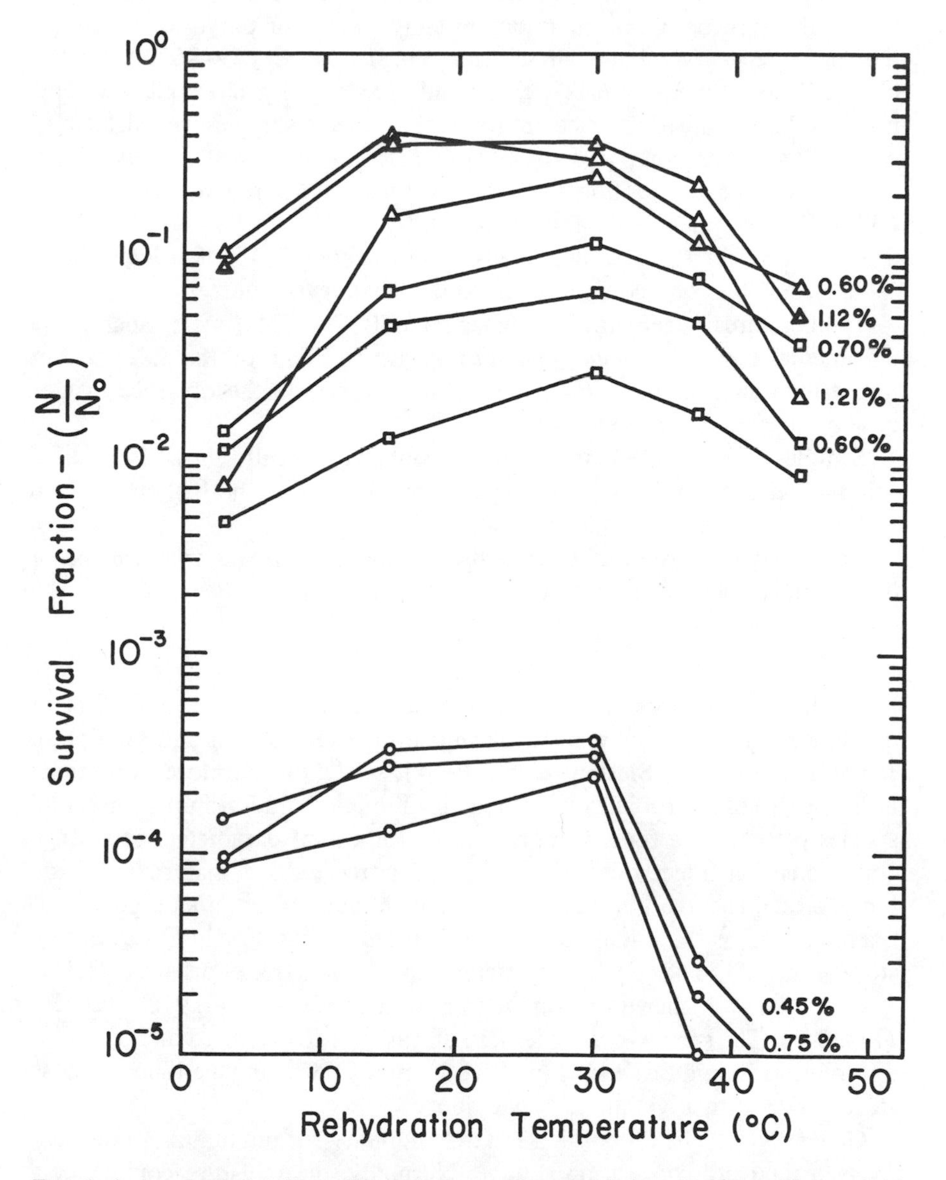

FIG. 3–4. Survival of *Salmonella typhimurium* freeze-dried at 49°C to different final water contents and rehydrated at different temperatures. Freeze-drying media: ○——○, 2% gelatin: □——□, 2% gelatin + 5% dextrose· Δ——Δ, 2% gelatin + 5% dextrose + 0.5% nutrient broth.

Supplementation of the gelatin–glucose menstruum with nutrient broth further improved survival threefold. Independent of the composition of the freeze-drying menstruum, the optimum temperature for rehydration providing maximum survival was in the range of 15–30°C (59–86°F). In the more protective menstrumms, gelatin and glucose, or gelatin, glucose, and nutrient broth, the difference in survival between samples rehydrated at 3°C (37.4°F) as compared to 45°C (113°F) approached approximately one logarithmic cycle. For the 2% gelatin samples rehydration at 3°C (37.4°F) was much less harmful for the freeze-dried cells than was rehydration at higher temperatures, causing only about 3/10 of a logarithmic cycle drop in the total count as compared to that of the control.

Temperatures of rehydration higher than 30°C (86°F) were much more detrimental for *Salmonella typhimurium* freeze-dried in the 2% gelatin menstruum than for the same organism freeze-dried in the presence of glucose or glucose and nutrient broth.

Samples dried to different final water contents generally showed the same optimal range for rehydration temperatures. However, the higher the final water content, the higher the survival.

The osmotic pressure of the rehydration menstruum was found to be of minor importance for the survival of microorganisms.

Storage

Even though microorganisms may survive the freeze-drying operation, a continual "dying off" may occur during storage (Silverman and Goldblith, 1965; Pablo, 1967; Sinskey et al, 1964).[12,15,16] Furthermore, variations in the degree of microbial survival may lead to an alteration in the microbial species present on a food at a given time. Important parameters of storage conditions that affect microbial survival on freeze-dried materials include temperature, relative humidity of the food product, and the storage atmosphere, as well as the initial conditions of freeze-drying. For instance, during storage at 20°C (68°F) in air, bacteria in samples freeze-dried at 71.1°C (160°F) had a higher death rate than in samples dried at 48.9°C (120°F) (Fig. 3-3). The survival of bacteria in all samples freeze-dried and stored at low relative humidities (11% relative humidity) was higher than in those stored at 23% relative humidity and above.

Generally there is an optimal relative humidity for maintaining the quality of a food product during storage. Normally, the moisture content of a food as freeze-dried corresponds very nearly to a monomolecular water content. Furthermore, this water content results in a humidity which provides for a high survival of microorganisms.

Many investigators have demonstrated that survival of lyophilized microbial cultures is higher in the absence of oxygen (Lion and Bergmann, 1961).[9] Storage of freeze-dried samples under nitrogen or other inert gases tends to lead to a higher survival during storage (Sinskey et al., 1964).[16] This is presumably due to the prevention of lethal oxidative reactions.

Rehydration

In the case of dried or freeze-dried foods, rehydration is that part of the food-processing operation that normally takes place in the home or in an institution where foods are prepared for serving. Consequently, this is where microbiological hazards are most likely to become a reality since it is at this point that this handling is most apt to occur. Important questions may be asked as to the possible public health problems which may arise with freeze-dried foods. For instance, it has been suggested that organisms of public health significance could become a potential hazard in dried foods if the normal spoilage flora is reduced, thereby allowing for the growth of pathogens because of lack of interference from the competitive growth of other organisms. It has been shown that the flora of commercially freeze-dried shrimp rehydrated and stored at 4°C (39.2°F) is more psychrotrophic than that of the same product prior to rehydration.

In dried foods organisms which constitute a minor portion of the original flora can also compete after rehydration (Pablo et al., 1967).[12] In rehydrated chicken both *Staphylococcus aureus* and fecal streptococci can compete with a total aerobic flora at temperatures above 20°C (68°F), even though they are present initially as only 0.01% of the total flora. This ability to multiply under such conditions is thought to be due to the fact that the organisms may be growing in discrete colonies; and, thus, no interference (from other colonies), due to nutritional depletion or to toxic metabolic end products, occurs.

A bacterial cell in a completely dried state could be, in a sense, considered as having an infinite concentration of solutes. As solvent water is reintroduced, the solute concentration will decrease rapidly but not uniformly. Thus, extremes of concentration gradients may exist momentarily throughout the specimen. There is no evidence either for or against the adverse effect of such concentration gradients, but their existence appears almost inevitable. Areas of the cell membrane involved in metabolite transfer might be particularly vulnerable to sudden pressure differentials prior to complete restoration of their integrity. The result of this action would be a loss in the control of permeability by the membrane. Permeability alterations have been detected in freeze-dried bacterial cells. Freeze-dried *Escher-*

ichia coli become sensitive to chloramphenicol, streptomycin, and actinomycin D, as well as to 1% sodium desoxycholate (Sinskey and Silverman, 1970).[17] Undried cells of this species are not affected by these compounds upon rehydration. Also, freeze-dried *Salmonella typhimurium* demonstrates an altered permeability of the membrane as measured by increased sensitivity toward antibiotics, and ribonuclease. Changes in the proton permeability coefficient and release of RNA, DNA, and ATP in amounts dependent upon the rehydration temperature also occur. During the recovery process resynthesis of cell-wall and cell-membrane components, as well as reestablishment of transport properties, have been found to be necessary before cellular growth takes place (Lewicki, 1969).[8]

References

1. A. Aitken, J. C. Casey, I. F. Penny, and C. A. Vogel. Effect of drying temperature in the accelerated freeze-drying pork, *J. Sci. Food Agr.* **13** (1962), 439.
2. B. L. Brady, *in* "Recent Research in Freezing and Drying," (A. S. Parks and A. U. Smith, Eds.), pp. 243–247. Blackwell, Oxford, 1960.
3. R. F. Burke and R. V. Decareau, Freeze dehydration of foods, *Advan. Food Res.* **13** (1964), 1–88.
4. J. Christian, Water Activity and Growth of Microorganisms, *Recent Advan. Food Sci.* **3** (1963), 248–255.
5. R. J. Heckly, Preservation of bacteria by lyophilization, *Advan. Appl. Microbiol.* **3** (1961), 1–76.
6. G. Heller, A quantitative study of environmental factors involved in survival and death of bacteria in the desiccated state, *J. Bacteriol.* **41** (1941), 109.
7. C. J. King, Freeze-drying of foodstuffs, *Crit. Rev. Food Technol.* 1 (1970), 379.
8. P. Lewicki, Characterization of injury to *Salmonella typhimurium* subjected to freeze-drying, Sc.D. Thesis, M.I.T., 1969.
9. M. B. Lion and E. D. Bergmann, The effect of oxygen on freeze-dried *Escherichia coli, J. Gen. Microbiol.* **24** (1961), 191.
10. G. Lusk, M. Karel, and S. A. Goldblith, Astacene pigment deterioration in freeze-dried shrimp and salmon during storage, *Food Technol.* **18** (1964), 751.
11. A. P. MacKenzie, Basic principles of freeze-drying for pharmaceuticals, *Bull. Parenteral Drug Ass.* 20 (1966), 101.
12. I. S. Pablo, A. J. Sinskey, and G. J. Silverman, Selection of microorganisms due to freeze-drying, *Food Technol.* **21** (1967), 64.
13. W.J.Scott,Water relations of food spoilage microorganisms, *Advan. Food Res.* **7** (1957) 83.

14. W. J. Scott, The effect of the residual water on the survival of dried bacteria during storage, *J. Gen. Microbiol.* **19** (1958), 624.
15. G. J. Silverman and S. A. Goldblith, The microbiology of freeze-dried foods, *Advan. Appl. Microbiol.* **7** (1965), 305.
16. A. J. Sinskey, A. H. McIntosh, I. S. Pablo, G. J. Silverman, and S. A. Goldblith, Considerations in the recovery of microorganisms from freeze-dried foods, *Health Lab. Sci.* **1** (1964), 297.
17. A. J. Sinskey and G. J. Silverman, Characterization of injury incurred by *Escherichia coli* upon freeze-drying, *J. Bacteriol.* **101** (1970), 429–437.

CHAPTER 4

The Growth of Microorganisms in Foods at Refrigerator Temperatures above Freezing

The holding of foods at refrigerator temperatures above freezing is applied to fruits and vegetables, meats, poultry, fresh milk and dairy products, fish and other marine products, eggs and some egg products, and even to some heat processed canned foods which are given low-level thermal treatment. In addition, many foods processed by other methods may be refrigerated in household refrigerators before or after preparation for serving. Actually the application of refrigeration temperatures to foods is so general that it is often times not thought of as a method of food processing.

It has been estimated that in the United States 85% of all foods are refrigerated at one point or another in the chain of distribution between harvest and ingestion (Anderson, 1953).[2] Considering this, the refrigeration of foods without freezing is an extremely important method of food processing and is possibly the most important method of preserving foods. The principles governing the microbial spoilage of foods during storage at refrigerator temperatures above freezing are, therefore, of great consequence.

While large volumes of foods are cooled and held at low temperatures above freezing, in general, food processors have not applied refrigeration of this type in a manner which would provide for the least amount of microbial and enzymatic deterioration in foods. This Chapter has been written with the objective of delineating some of the factors governing the growth of microorganisms at low temperatures and to indicate procedures by which the growth of psychrotrophic and psychrophilic bacteria in foods may be controlled.

In order to explain the effects of low temperatures on the growth of microorganisms, it is desirable to first consider the bacterial growth curve.

If a pure culture of bacteria is grown in liquid media and viable plate counts are made periodically, it will be found that, when the $\log_{10}$ of the number of cells is plotted on the ordinate versus time on the abscissa, a curve such as that shown in Fig. 4-1 results. This curve may be divided into at least seven phases; however, only the four general phases will be considered here. Except under special conditions, bacteria inoculated into fresh media do not start to multiply at once; and the period in which there is no increase in numbers and until the time that there is a regular and periodic increase in numbers is called the "lag phase" or the "phase of adjustment." Actually, during this phase some of the bacteria may die. The lag phase does not represent a situation in which the cells are biochemically inactive. During this period there is an increase in metabolic activity, enzyme systems are formed or repaired, the cells become larger, and there is an increase in deoxyribonucleic acid. Gradually the cells start to multiply, and there is a rounding off of the curve as more cells multiply until the log or exponential growth phase is reached.

During the logarithmic or exponential phase of growth the organisms grow at a constant rate, and there is a periodic doubling of the number of

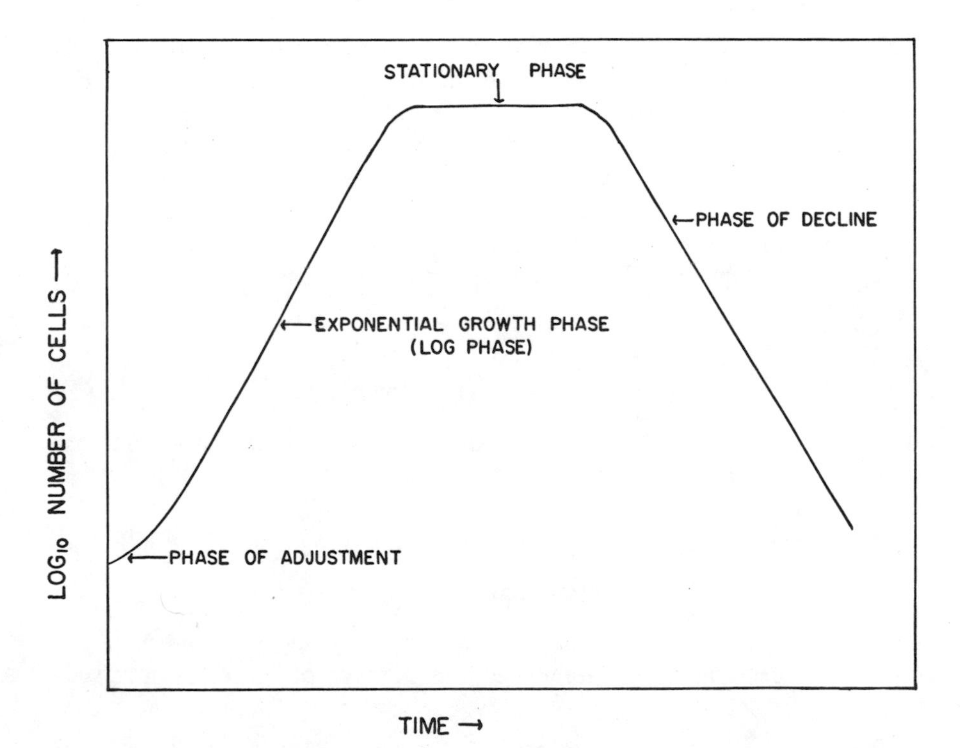

FIG. 4–1. Bacterial growth curve.

viable cells. During this phase the cells usually are most susceptible to undesirable environmental conditions such as the presence of chemical inhibitors, unfavorable temperatures, and so on. As the log phase is continued, environmental conditions become more unfavorable for growth; and the curve rounds off to become essentially horizontal, the stationary phase having been reached.

The stationary phase is not a situation of no growth. In this phase as many cells are dying off as are being formed, so that the total number of cells (dead and living) are increasing, but the number of viable cells remains essentially constant. A particular species of bacteria grown strictly under the same conditions will always reach approximately the same maximum number of viable cells, and this has been called the M or maximum concentration. The M concentration can vary, since different population densities are reached for different conditions of growth. It has been shown, for instance, that for aerobic growth the factor which probably governs the maximum concentration may be the availability of oxygen (Smith and Johnson, 1954; Tynell et al., 1958).[21,23]

As cells are held in the stationary phase, they eventually start to slowly die off in excess of multiplication, and the phase of decline is reached. For most organisms the phase of decline does not represent a logarithmic decrease in the number of cells, and there may even be spasmotic periods of increase in the number of viable cells during this period. If, however, the cells produce a considerable amount of acid, hydrogen peroxide and no catalase, or some other toxic material, the death rate during the phase of decline will be logarithmic. This situation may be represented as

$$-k = 1/t \log a/b$$

or

$$t = -1/k\,(\log a - \log b), \tag{1}$$

where k is the velocity constant (slope of the line), t is time, a is the number of viable organisms at time zero, and b is the number of viable cells at the end of a particular period of time.

Considering the exponential phase of growth:

$$\log b = \log a + N \log 2, \tag{2}$$

where a and b have the same meaning as previously indicated and N = the number of generations (doublings of the number of bacteria) involved. Hence,

$$N = \frac{\log b - \log a}{\log 2} = \frac{\log b - \log a}{0.301}. \tag{3}$$

If *b* was determined at *T* minutes after *a* was determined, then the generation time *G* may be found from $G = \frac{T}{N}$ Generation times vary with the species of organism and, of course, with environmental conditions, favorable or otherwise.

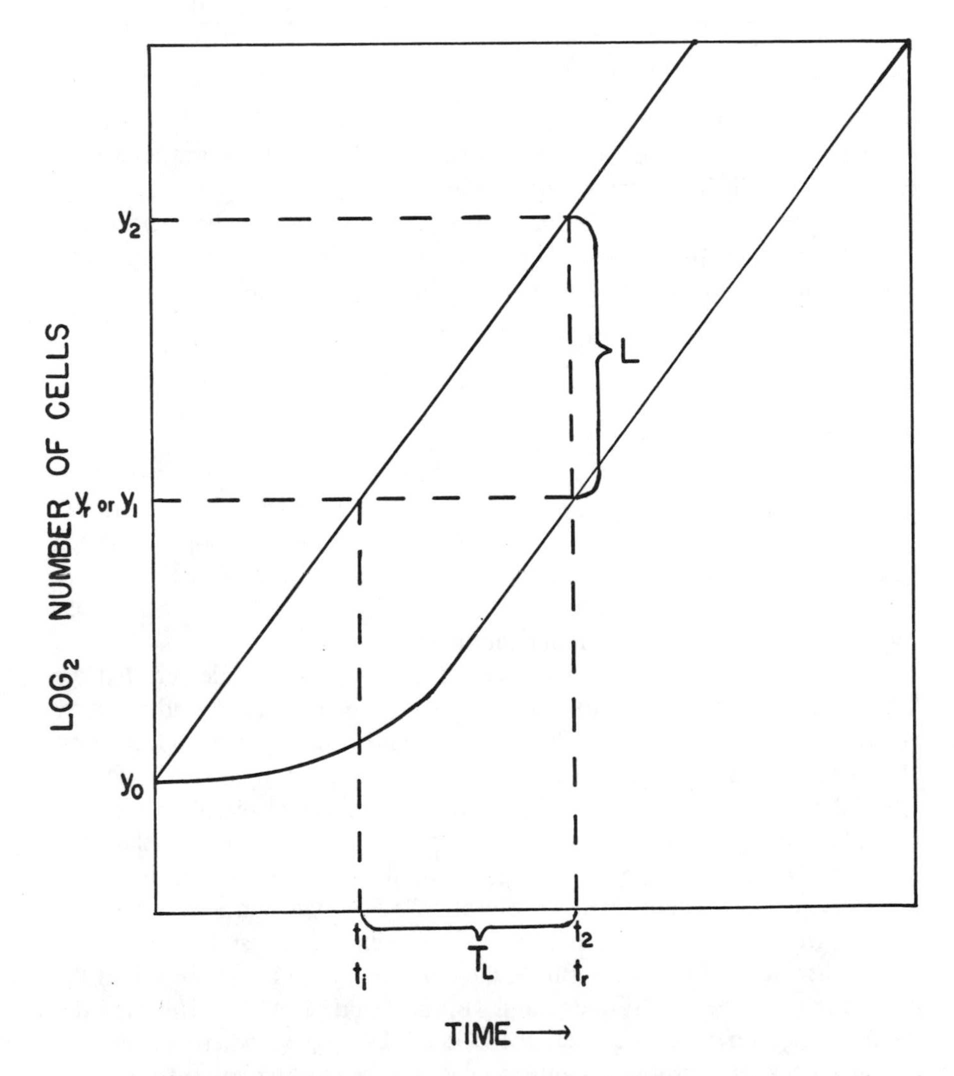

FIG. 4–2. Lag phase and exponential growth phase of bacterial growth curve plotted as $\log_2$ vs time.

The Eq. (2) in which $N \log 2$ was included supposes that all of the bacteria present in the exponential phase of growth are multiplying periodically and that none are dying off. Some bacteriologists have suggested that this is not the case. In such a situation the number 2, in the above equations, would have to be replaced by a factor P.

$$P = \frac{\text{The number of viable cells at the end of a particular time} - \text{the number of viable cells at the start of that time}}{\text{The total cells at the end of a particular time} - \text{the total cells at the start of that time}} + 1. \tag{4}$$

Instead of plotting base 10 logarithms as indicated in Fig. 4-1, it is theoretically advantageous to plot the number of viable cells as the logarithm base 2 ($\log_2$). Such a curve is shown in Fig. 4-2.

Considering the exponential growth phase in this case, the exponential growth rate, R, is the tangent of the angle which the extended straight line makes with the abscissa, L/T_L, or it may be determined as

$$\frac{\log_2 Y_2 - \log_2 Y_1}{t_2 - t_1}. \tag{5}$$

Also, the generation time is now $\frac{1}{R}$ or $\left(\frac{1}{L/T_L}\right)$ or $\frac{T_L}{L}$.

In addition to measuring exponential growth rates and generation times, it is often advantageous to have some measure of the duration of the lag phase. Actually there is no good way to do this, but several methods have been proposed which, at least, may be applied to a single organism when comparing growth under different conditions.

One suggestion has been that an expanded $\log_2$ plot be made and that the time from the start to some arbitrary number of generations, as indicated by viable count, be used to determine the duration of the lag phase (Lockhart, 1960).[13] This involves frequent counts during the early growth period. Also, if some of the organisms die off during the lag phase, then a viable count indicating one or two generations actually represent more than the number of generations that it is supposed to signify. It is evident, however, that this method of measuring the duration of the lag phase may be used without great error as long as it is applied to a single organism.

Another suggestion for measuring the duration of the lag phase is that the lag constant T_L be used (Lodge and Hinshelwood, 1943).[14] This was defined as the difference between the observed time, t_r, when the culture reached a certain density of numbers (this number to be chosen from some density of population within the exponential phase of growth) and the ideal time, t_i, at which the same density of number would have existed had the exponential growth rate prevailed from the start. Therefore,

$$T_L = t_r - t_i$$

or

$$T_L = t_r - \frac{\log_2 Y_r - \log_2 Y_0}{R}. \tag{6}$$

The above method has been criticized on the basis that only cultures having the same exponential growth rate may be compared when the lag constant T_L is used. On this basis it has been proposed (Monad, 1949)[17] that the number of generations between the ideal population density and the observed population density be used to measure the lag period. This is the number of generations which would have been present had the exponential growth rate existed from the start, minus the number of generations present at a time corresponding to some point on the exponential growth curve. This would be L and can be determined as

$$T_L \times R. \tag{7}$$

It would appear that both of the latter two methods [eqs. (6) and (7)] are subject to the same objection and that neither actually measure time of growth.

Growth of *Pseudomonas fragi* at Low Temperatures

In investigations with *Pseudomonas fragi,* (Duncan and Nickerson, 1961; Duncan and Nickerson 1962)[5,6] a psychrophilic bacterium which grows well at 0°C (32°F), the effect of the size of the inoculum, growth temperature of the inoculum, and the physiological age (Cells of different physiological age included those taken from the lag phase, taken directly from the exponential growth phase, taken from the stationary phase or resting cells, i.e., washed and stored cells from the stationary phase) was investigated. In this case the lag phase time was determined from an expanded $\log_2$ plot of growth as a doubling of the initial inoculum.

Regarding the duration of the lag phase, it was found that lag phase time increased as the growth temperature of the culture was decreased from 10 to 0°C (50 to 32°F). Also, when the growth temperature of the inoculum was substantially different from that of the culture, the duration of the lag phase was increased. This was especially the case when, compared with the growth temperature of the culture, the growth temperature of the inoculum was high. The physiological age of the inoculum greatly affected the duration of the lag phase of the culture. Thus, inoculum cells taken from the exponential phase of growth showed little, if any, lag when the growth temperature of the inoculum and that of the culture were the same. The lag phase was greatly extended when cells from the stationary phase were used as inocula and further extended when resting cells (washed and stored stationary phase cells) were used as inocula. The size of the inoculum (increased by 10 or 100 times) had no affect on the duration of the lag phase at any temperature.

Regarding the exponential growth rate, neither the size of the inoculum nor the physiological age of the culture had any effect. Growth temperature of the culture had highly significant effects on the exponential growth rate of the culture, the rate of growth being faster in each instance as the temperature was raised from 0 to 10°C (32 to 50°F).

The temperature coefficient for growth Q_{10} may be determined as $Q_{10} = \frac{K_2}{K_1}$ where K_2 and K_1 are the generation times (doubling times) at temperatures differing by 10°C (50°F). The Q_{10} for a range of temperatures may be calculated as

$$Q_{10} = \sqrt[\Delta T]{\left(\frac{K_2}{K_1}\right)^{10}} = \left(\frac{K_2}{K_1}\right)^{10/\Delta T}.$$

The minimum generation times at different temperatures and the Q_{10} for various temperature ranges determined in the experiments with *Pseudomonas fragi* (previously cited) are listed in Tables 4–1 and 4–2.

According to the results of experiments with *Pseudomonas fragi*, certain inferences may be made and applied to the refrigeration of foods. It is evident from lag phase times and exponential growth rates at different temperatures that a very significant extension of storage life of flesh-type foods may be obtained by holding such products as near the freezing point as is possible without actually freezing. It is also indicated that, as far as storage life extension is concerned, greater advantages will be obtained by lowering the

TABLE 4–1
Growth of *P. fragi*

Incubation temperature (°C)	Average exponential growth rate (generations/hr)	Minimum generation time (doubling times) (hr)
0	0.0885	11.30
2.5	0.1292	7.74
5.0	0.2016	4.96
7.5	0.2860	3.50
10.0	0.3807	2.63
20.0	0.9195	1.09

TABLE 4–2
Temperature Coefficients (Q_{10}) for the Growth of *P. Fragi*

Temperature range (°C)	Q_{10}
0 – 2.5	4.57
2.5 – 5.0	5.92
5.0 – 7.5	4.06
7.5 – 10.0	3.14
10.0 – 20.0	2.42
0 – 5.0	5.20
5.0 – 10.0	3.56
0 – 10.0	4.31

temperature from 4.4°C (40°F) to near 0°C (32°F) than would be the case were the temperature lowered from 10°C (50°F) to near 4.4°C (40°F).

In the experiments with *Pseudomonas fragi,* an organism which grows exceptionally well at 0°C (32°F) was used. It is entirely probable that some of the bacteria which contribute to the decomposition of flesh-type foods at low temperatures (psychrotrophic bacteria) will not grow as well at temperatures below 4.4°C (40°F) as does *Pseudomonas fragi.* Such organisms would be affected to a greater extent by the application of temperatures near the freezing point for purposes of holding foods than would the psychrophilic types.

Certain other implications may be derived from the research with *Pseudomonas fragi.* The results of factors affecting the lag phase of growth indicate that if a refrigerated food product is being processed at low temperatures (such as is often done in cutting meat or fish, skinning frankfurters, etc.), then the equipment used (knives, saws, skinning machines, etc.) should be kept scrupulously clean, otherwise the bacteria thereon will be in the exponential phase of growth, and when transferred to the product there will be no lag phase, and the organisms will continue to grow in the food at an exponential rate. Also, considering the fact that bacteria transferred from growth at a higher temperature to growth at a lower temperature have an extended lag phase, it is implied that for refrigerated products it may be better from the standpoint of spoilage to process at a high temperature. The organisms on equipment in this case will have been transferred from ambient temperatures (the equipment) to a low temperature (the product), hence the lag phase should be extended. In addition, at the higher temperatures organisms present on equipment may not even be the type (psychrophiles) which will grow at low temperatures. The use of ambient temperatures for the processing of refrigerated products can be applied with good results provided that the equipment is kept reasonably clean; there is a continuous flow of product to prevent holding at the high processing temperatures for long periods of time, and relative humidities are held at a low enough level to prevent condensation of moisture on the product or sweating.

Whereas it was shown with *Pseudomonas fragi* that the size of the initial inoculum did not affect either the duration of the lag phase or the exponential growth rate, it is evident that the number of psychrophilic bacteria present in a food, or with which it is contaminated at the start, will have a very great effect on the storage life time. This may be simply illustrated as follows: Assuming that the spoilage cell concentration is 10×10^6 per gram and that at the start sample (a) contained 1000 cells per gram and sample (b) 100,000 cells per gram. Under such a condition between 13 and 14 doubling times (doubling of population) would be required for spoilage of sample (a) and only 6 to 7 doubling times would be required for sample (b) to reach the same condition.

Microorganisms and the Spoilage of Refrigerated Foods

It is considered that psychrophilic bacteria are the chief cause of spoilage of flesh-type foods (meat, fish, and poultry) and of eggs and that molds and yeasts usually cause spoilage of vegetables and fruits. Psychrotrophic bacteria may play some part in the spoilage of foods, especially when comparatively high refrigeration temperatures are used (Panes and Thomas, 1968).[18]

Psychrophilic microorganisms have been defined in various ways as organisms which grow below 20°C (68°F), organisms which grow below 15°C (59°F), organisms which grow well at temperatures below 5°C (41°F), and as organisms which grow well at 0°C (32°F). There is no exact definition for these types of microorganisms but the latter two definitions are most often accepted today. Actually some microorganisms will grow at temperatures well below 0°C (32°F) in unfrozen materials. While psychrophilic bacteria may grow well at 0°C (32°F), the fastest growth of such organisms may take place at 20–25°C (68–77°F).

Psychrotrophic bacteria have been defined (Panes and Thomas, 1968) as organisms which will grow below 10°C (50°F) and may grow at temperatures as low as 3°C (37.4°F) but which rarely become the dominant spoilage flora at temperatures below 10–8°C (50–46.4°F). There are some instances in which the pretreatment of refrigerated foods may have destroyed psychrophilic bacteria and psychrotrophic types become the predominant flora and cause spoilage. Pasteurized milk, for instance, may be spoiled by psychrotrophic *coli-aerogenes* bacteria or these organisms may contribute to spoilage (Panes and Thomas, 1968).[18] Cooked sausage products are sometimes spoiled by species of *Leuconostoc* or *Lactobacillus,* the minimum growth temperature of which is said to be 3.5°C (38.3°F).

The reasons why certain organisms will grow at comparatively low temperatures and others will not are not entirely clear. There is some evidence that the decrease in growth rate, caused by the incubation of organisms at temperatures below the optimum, may be due to the reversible or irreversible denaturation of certain rate-limiting or other enzymes or that low temperatures cause a general suppression of enzyme synthesis (Ingraham, 1962).[10] With certain mesophilic bacteria it is known that at low temperatures the carrier mechanism which transports solutes across cytoplasmic membranes is inactivated while this does not happen in the case of psychrophiles (Rose and Evison, 1965).[19] It is also known that this is not the mechanism which prevents the growth of some mesophiles at low temperatures.

Minimum growth temperatures for microorganisms in foods provide a certain amount of useful information, but such data can rarely be associated with the prediction of what types of microorganisms may be expected to cause spoilage of foods. This is the case since the suitability of the food to support growth of a particular organism, the growth rate of the organism in foods at low temperatures, the cell concentration at which organoleptic changes occur in the food product, and the occurrence of surface growth which may cause visual changes in the food such as discoloration or sliming, are all factors concerned with spoilage.

Some minimum temperatures for the growth of microorganisms in food have been reported and are listed in Table 4-3 (Michener and Elliot, 1964).[16]

It has also been reported that the following bacterial species or groups may grow in some foods at the following minimum temperatures: *Escherichia coli* 3.1–5°C (37.6–41°F), *Aerobacter aerogenes* 0°C (32°F), coliforms 1.8–4.4°C (35.2–40°F), enterococci 0°C (32°F). While this may be the case, it is quite certain that these organisms ordinarily do not cause spoilage of foods held at temperatures below 4.4°C (40°F). An exception to this may be the enterococci which have sometimes been implicated in the spoilage of certain refrigerated, cured meat products.

The spoilage of meat, cured or cured and cooked meat, fish, shellfish products, cooked shellfish meats, milk, and butter during holding at low temperatures above freezing is generally due to the growth of bacteria. Fruits and vegetables may be decomposed by the growth of yeasts or especially molds, although there are some instances of bacterial spoilage as in the case of lettuce where rots of bacterial types may occur *(Pseudomonas,*

TABLE 4–3
Growth of Microorganisms in Foods at Low Temperatures

Type of food	Type of organism	Genus	Approximate minimum temperature of growth
Meat	Molds	*Mucor*	−1°C(30.2°F)
	Molds	*Cladisporium*	−6.1°C(21.0°F)
	Molds	*Sporotrichum*	−7.8°C(18.0)
	Molds	*Thamnidium*	−7.8°C(18.0°F)
	Yeasts	—	−5.5°C(22.0°F)
	Bacteria	*Pseudomonas*	−3.0°C(26.6°F)
	Bacteria	*Achromobacter*	−3.0°C(26.6°F)
Cured meats	Bacteria	*Leuconostoc*	3.5°C(38.3°F)
	Bacteria	*Lactobacillus*	3.5C(38.3°F)
Fish	Bacteria	*Pseudomonas*	−6.7°C(20.0°F)
Oysters	Yeasts	—	−17.8°C(0.0°F)
Milk	Bacteria	*Pseudomonas*	−1.0°C(30.2°F)
Fruit	Molds	*Penicillium*	−3.9°C(25.0°F)
	Molds	*Monilia*	0.0°C(32.0°F)
	Molds	*Cladisporium*	−7.8°C(18.0°F)
	Molds	*Mycoderma*	−7.8°C(18.0°F)
	Yeasts	*Torula*	−3.9°C(25.0°F)
Vegetables	Bacteria	*Pseudomonas*	−3.9°C(25.0°F)
	Bacteria	*Lactobacillus*	−3.9°C(25.0°F)
	Molds	*Sporotrichum*	−7.8°C(18.0°F)
	Molds	*Cladisporium*	−7.8°C(18.0°F)

Xanthomonas or *Erwinia* species). Infection, in this case, occurs in the field and must be controlled by treatment of seed, selection of resistant varieties or other methods common to horticultural practices. However, the spoilage or deterioration of fruits and vegetables during storage is more often due to physiological changes since many of the cells in the tissue of such commodities are still living and respiring. Enzymatic action causes different types of changes such as loss of sweetness, wilting, discoloration, and softening. Shell eggs are sometimes decomposed by the growth of molds. During long storage eggs may also undergo chemical and physical changes which alter the consistency of white and yolk portions and also cause off-flavors (storage flavors) due to reactions not associated with the growth of microorganisms.

Some Characteristics of Bacteria Causing Spoilage of Refrigerated Foods

The Pseudomonads. As previously indicated, the nonpigmented *Pseudomonas* species are the bacteria most often causing spoilage of refrigerated, flesh-type foods and the strains of *Pseudomonas fluorescens* and *Pseudo-*

monas ovalis, which produce fluorescent pigments, are most often the cause of spoilage of shell eggs.

The pseudomonads are gram-negative, vegetative rods 0.3–0.5 × 1.0–4.0 μ in size. It is probable that all members of this genus in nature are motile with polar flagella but motility may be lost as the organism is held in culture. These organisms are strict aerobes, catalase and oxidase positive, and deaminate arginine. Some species attack carbohydrates oxidatively, others do not. All species can be adapted to reduce nitrates and many species reduce trimethylamine oxide to trimethylamine. Some species produce a green (fluorescent), reddish-brown, yellow, or blue pigment. Many *Pseudomonas* species produce no pigment. These organisms are considered to be soil and water types but some are plant pathogens. In general, the pseudomonads are not heat resistant, very large numbers being destroyed by heating at 55°C (131°F) for 1 hr. They are penicillin resistant and are not affected by 2.5 IU of this antibiotic (To determine penicillin sensitivity a nutrient agar is seeded with the organism after the medium has been melted and cooled to 45–46°C (113–114.8°F). The still liquid medium is poured into petri dishes and allowed to harden. Circular discs are now cut out of the agar to form wells. The wells are now filled with 2.5 IU of penicillin in solution and the cultures are incubated. Under these conditions the pseudomonads will grow up to the edge of the well with no zone of inhibition.

Those species of *Pseudomonas* which cause spoilage of refrigerated foods have an optimum growth temperature between 20 and 25°C (68 and 77°F), and many of them grow well at temperatures near 0°C (32°F), and some of them will grow at temperatures below 0°C (32°F). *Pseudomonas* species grow poorly or not at all at 37°C (98.6°F). Species of *Pseudomonas* which have been associated with the spoilage of foods are: *ambigua, coharens, convexa, fluorescens, fragi, incognata, ovalis, perolens,* and *putrefaciens.* These organisms are found in soil or in fresh or salt water as a natural habitat.

The Achromabacter. The *Achromabacter* are a poorly defined group of bacteria belonging to the family *Achromabacteriaceae.* These organisms are short, stout, gram negative, nonmotile, vegetative rods or cocco-bacilli approximately 0.8–1.0 x 1.5 μ in size. They do not produce pigments and are catalase positive and oxidase negative. Some species attack sugars oxidatively. Some species reduce nitrates, trimethylamine oxide or choline, but most species do not. These organisms are mostly aerobic. All species are sensitive to penicillin. The *Herellea* and *Moraxella* species have been classified as belonging among the *Achromabacter.* These species are found on the human body and at times become pathogenic causing a mild conjunctivitis. They may also grow at comparatively low temperatures and become the predominant flora in foods subjected to certain processes and held under refrigeration (Idziak and Incze, 1968; Davies, 1966).[4,9] The optimum growth temperature of the *Achromobacter* is between 20 and 35.5°C (68 and 96°F), but some species will grow at 0°C (32°F). The natural habitats of most *Achromobacter* species are soil and water. Whereas the *Achromobacter* species generally make up a part of the spoilage flora of fresh refrigerated flesh-type foods, they usually contribute less to off-odors and flavors than the pseudomonads in such products. This is because of the fact that the *Achromobacter* do not produce sufficient end products of the right composition to cause off-odors and off-flavors in foods until they have grown to cell concentrations of 100×10^6 or higher, while when *Pseudomonas* species cause spoilage of foods, organoleptic changes are noticeable at cell concentrations of approximately 10×10^6. It has been suggested (Thornley, 1967)[22] that the *Achromobacter* be placed in a genus called *Acinetobacter*, this genus being made up of five phenons or groups. Included in these groups would be those organisms presently classified as *Agrobacterium, Achromobacter, Herellea* and *Moraxella* and some species of *Alcaligenes, Diplococcus* and *Neisseria.*

The Lactobacteriaceae. As a group the *Lactobacteriaceae* are rod or coccal forms which are nonmotile, gram positive, and catalase negative. They obtain energy only by anaerobic metabolism but are indifferent to oxygen and will grow in its presence. When grown aerobically in the presence of certain carbohydrates the *Lactobacteriaceae* convert these compounds to trioses, obtaining no energy in the process. They must metabolize the trioses anaerobically to obtain energy in such instances. However, when carbohydrates are metabolized aerobically by these organisms, hydrogen peroxide is formed; and since they do not produce catalase, hydrogen peroxide accumulates in the growth substrate and eventually inhibits cell growth. Hydrogen peroxide, produced in this manner in some foods (cooked meat products containing nitrite) may cause a green discoloration.

The *Lactobacteriaceae* are either homofermentative, with lactic acid as the sole end product of glucose metabolism, or they are heterofermentative, with lactic acid, acetic acid, ethyl alcohol and carbon dioxide being formed from carbohydrates.

The particular species of the *Lactobacteriaceae* which cause spoilage of refrigerated foods are usually salt tolerant, growing under conditions in which 3% or more of sodium chloride may be present. In some cases these organisms may be acid tolerant. In many instances the *Lactobacteriaceae* which cause spoilage of refrigerated foods are somewhat heat resistant surviving heating at 60°C (140°F) for more than 30 min. This is not always the case, however, since certain non-heat-resistant streptococci *(cremoris* and *lactis* types) may cause souring of raw milk.

The special properties of the *Lactobacteriaceae* which enable them to become the predominant spoilage flora of some types of refrigerated foods under certain conditions are: comparatively high heat resistance for vegetative-type bacteria, minimum growth temperatures in the psychrotrophic or psychrophilic range, growth in the presence or absence of oxygen, tolerance to sodium chloride, and growth at high or low pH levels.

Four genera of the *Lactobacteriacea* may be the cause of food spoilage during holding at refrigerator temperatures above freezing.

The Streptococci. These bacteria are spherical or ovoid in shape and are found in short or long chains or in pairs. They are homofermentative. On the basis of their antigenic properties, the streptococci have been separated into a number of groups. Among the streptococci, the enterococci are the food-spoilage types. These organisms usually occur as diplococci. Their natural habitat is the intestine of the human or warm-blooded animals but they have been also isolated from vegetable matter free from animal excretions.

The enterococci are usually nonhemolytic but may exhibit either α or β types of hemolysis (partial or complete hemolysis of blood cells). They are relatively heat resistant and will grow at temperatures above 45°C (113°F) and as low as 7.2°C (45°F). The enterococci will also grow in media, the pH values of which may be as low as 4.5 or the sodium chloride content of which is as high as 6.5%. Some strains of this group produce dextran from sucrose, hence they may be the cause of ropy curing brines. The chief species of the enterococci are *Streptococcus faecalis* (varieties *liquefaciens* and *zymogenes*) and *Streptococcus faecium* (variety *durans*). *Streptococcus bovis* and *Streptococcus equinus* belong in the same serological group (Lancefield's D group) as do the enterococci.

Streptococcus lactis and *Streptococcus cremoris* are related to the enterococci and have sometimes been the cause of spoilage of fresh milk. These organisms are less heat resistant and their growth characteristics are such that they are unable to compete with other organisms under unfavorable environmental conditions (high salt content, temperatures below 10°C (50°F)).

The Leuconostoc. The *Leuconostoc* are heterofermentative members of the *Lactobacteriaceae* which are ovoid cells, in pairs, short or long chains. The organisms are 0.9–1.2 μ in diameter. The optimum temperature of growth for the *Leuconostoc* is 20–25°C(68–77°F) but some strains are able to grow at temperatures below 4.4°C (40°F). The *Leuconostoc* have a high salt resistance since they are able to grow in some types of curing brine used for fish or meat. Since they can produce dextran from sucrose, the *Leuconostoc* may be the cause of ropy brines. The natural habitat of this group of organisms is dairy products or plant materials.

The Pediococci. Members of the *Lactobacteriaceae* known as *pediococ-*

cus are spherical cells occurring singly, in pairs, as tetrads, or as short chains. The optimum growth temperature for these organisms is 25–32.2°C (77–90°F) but they also grow at temperatures below 7.2°C (45°F). The pediococci are homofermentative. The natural habitats of these organisms are plant materials and dairy products.

The Lactobacilli. The lactobacilli are large rods occurring singly or in chains having as many as four cells. Some members of this group are homofermentative, others are heterofermentative. Only occasionally do these organisms produce pigmented colonies under conditions which have not been delineated. Some of these organisms have optimum temperatures as high as 60°C (140°F) and others as low as 30°C (86°F), but a number of species of this group are able to grow at temperatures below 4.4°C (40°F). Many species of the lactobacilli are salt tolerant and can grow in materials containing 5% or more of sodium chloride. Some species will also grow in foods the pH of which is 4.0.

The lactobacilli have been divided into three groups: a thermobacterium group which contains homofermentative species which will grow at comparatively high but not at low temperatures (below 10°C (50°F)); a streptobacterium group which is homofermentative and which will grow at temperatures below 4.4–10°C (40–50°F), but will not grow at temperatures above the mesophilic range; and a betabacterium group which contains only heterofermentative species some of which grow at high but not at low temperatures; others of which grow at low but not at high temperatures.

Changes in Flora During Refrigerated Storage of Foods

The spoilage of flesh-type foods is in general a surface phenomenon. Bacteria grow aerobically on surfaces and produce end products which cause off-flavors and odors. Since moisture is required for the growth of microorganisms, surfaces covered with fat or unbroken skin (certain areas on a side of meat, for instance) are more resistant to bacterial growth than cut surfaces which cannot be dried.

At the time of slaughter the flora on surfaces of the carcass of warm-blooded animals consist mostly of mesophilic organisms (Elliot & Michener, 1965).[7] The spoilage organisms are, usually, psychrophilic. In red meat at the time of spoilage the vast majority of the cells which make up the spoilage flora are the nonpigmented pseudomonads.

In poultry, some mesophilic bacterial types, some *Pseudomonas* species, and some *Achromobacter* species will be present after slaughter, dressing, and evisceration. Also, the pseudomonds present at the start are mainly of the pigmented types. At the time of spoilage, however, the flora consists mainly of the nonpigmented *Pseudomonas* species, some *Achromobacter* or *Alcaligenes,* and few *Aeromonas* being present. It should also be pointed out that organisms identified as *Achromobacter* may sometimes be *Moraxella* or *Herellea* species (Idziak and Incze, 1968; Davies, 1966).[4,9]

With fish taken from cold marine waters the chief flora, when taken from the ocean, consists of the pseudomonads. Since fish aboard boats are frequently held in contact with ice and since this ice contacts the walls of the pen in which the fish are held, it has been found (Shewan and Hobbs, 1967)[20] that as microorganisms multiply on fish surfaces the flora changes from that of pseudomonads of marine types to one of nonpigmented fresh water strains of pseudomonads. *Pseudomonas fragi* has been cited as a typical fish-spoilage organism but to what extent this has been verified is not quite certain. *Pseudomonas putrefaciens* doubtlessly plays a significant part in fish spoilage as do the pigmented members of this genus.

Fish processing-plant equipment also becomes contaminated with fresh water pseudomonads; and when the fish is cut into fillets or steaks, these organisms are transferred to the product and will eventually cause spoilage.

The spoilage of eggs in the shell occurs differently from the spoilage of meats or fish. Most eggs are sterile when laid, while a small percentage contain viable microorganisms, not necessarily of the type which will grow therein and cause spoilage (Lorenz et al., 1952).[15] However, after the egg is laid the shell sometimes becomes contaminated with bacteria and molds, and, especially if the shell surface is moist as might be caused by droppings from the hen, these organisms may penetrate the pores of the shell and shell membranes, contaminate the material within the shell, grow, and cause spoilage. As in the case of meats and fish, eggs are usually spoiled because of the growth of pseudomonads. However, the situation is somewhat different than with flesh-type foods. It has been found that about 82% of spoiled eggs will show fluorescence under uv light (Ayres, 1960).[3] Whereas other types of pseudomonads, besides *Pseudomonas fluorescens* and *Pseudomonas ovalis,* produce pigments, these other pigments do not fluoresce under uv light. It is probable, therefore, that *Pseudomonas fluorescens* types are the chief cause of spoilage of shell eggs. Other types of pigmented pseudonomonads, *Proteus,* and some molds sometimes cause spoilage of shell eggs.

Molds may grow on the surfaces of meat, especially on sides of beef which are held for long periods at 0°C or slightly below. This holding at low temperatures is done for purposes or ripening or for tenderizing. The molds which grow in this case have been found to be *Thamnidium, Rhizopus,* and *Mucor.* Whereas this mold growth produces no particular off-flavor, the meat must be trimmed to remove mold before it can be eaten, hence there may be excessive trimming losses in some instances.

It is worthwhile to speculate as to why, at the time of the usual type of spoilage in refrigerated flesh-type foods, the flora consists largely of the pseudomonads. It has been shown that the *Achromobacter* may grow as fast at low temperatures as do the pseudomonads (Lee, 1965).[12] The factors that govern spoilage flora, therefore, may be: (1) the larger proportion of pseudomonads present at the start and (2) the production of off-flavor and off-odor components in foods by pseudomonads at a lower cell concentration than is the case with the *Achromobacter*. As an illustration of the latter point, in fish fillets which are untreated definite off-flavors and odors will be present when the total viable count is approximately 10×10^6 per gram, and the microbial flora is mainly pseudomonads. On the other hand, in fish treated with 100–200 krad of ionizing radiations, which largely destroys the pseudomonads, *Achromobacter* species eventually cause spoilage, but at the time spoilage flavors and odors occur, the total viable count will be somewhere between 100×10^6 to 600×10^6 per gram.

Effect of Some Types of Processing on the Spoilage of Refrigerated Foods

Heat. A number of foods which are afterwards held at refrigerator temperatures above freezing are given a heat process sufficient to destroy most vegetative types of bacteria or at least to reduce them to very small numbers, but not sufficient to destroy all bacteria which might be present. Generally the only psychrophilic bacteria which will survive a pasteurization-type treatment are some of the *Lactobacteriaceae,* and these only in instances in which the heat treatment may not have been too severe or the organisms may have been protected to some extent by fat or oil [It is known that bacteria immersed in fat or oil are much more difficult to destroy by heat than when in contact with water) (Hersom and Hulland, 1964)].[8] Examples of this type of survival may be found in such products as large cooked sausage (bologna) or large canned hams.

It is usually the case that cooked-type foods, which are to be held under refrigeration after the heat treatment is applied, contain few viable bacteria and few, if any, psychrophilic bacteria immediately after heating. Such products are often recontaminated with psychrophiles after heat processing. Recontamination of heated foods occurs during postheating processing. Such operations as the skinning of cooked sausage, the slicing of cooked sausage, meat loaves, or ham, the removal of cooked crustacean meats from the shell, the deboning of cooked chicken, the bottling of milk, and similar procedures are not aseptic and usually result in recontamination of the product. The bacteria with which such products are recontaminated must be psychrophilic or psychrotrophic types, or bacteria which will grow at relatively low temperatures if they are to cause spoilage during refrigerated storage of the particular food. For this reason the spoilage flora of cooked refrigerated foods usually consists of the *Pseudomonas, Achromobacter* species, various members of the *Lactobacteriaceae,* or psychrophilic yeasts.

Salt. Most foods do not have a high sodium chloride content but cooked, cured meat products may contain 1.8 to more than 3% of this salt and dried cured meat or fish products may contain much higher concentrations (10–12%).

The chief psychrophilic bacterial spoilage group, the pseudomonads, will not grow well in concentrations of salt ordinarily found in cooked cured meat products. It is true that marine pseudomonads will grow well and may even grow best in media containing about 3% of salt, but such organisms are not normally present in meat-processing plants. The fresh water pseudomonads must sometimes come in contact with cured, cooked meat products but appear not to be the cause of spoilage of such foods.

Since sausage products are sometimes spoiled because of the growth of *Achromobacter* (Alm et al., 1961),[1] it must be presumed that these organisms have some salt tolerance. No data have been found which indicate the salt tolerance of pseudomonads of the soil or fresh water types.

Many species of the *Lactobacteriaceae,* especially species of *Leuconostoc, Pediococcus,* or *Lactobacillus,* are able to grow well in the presence of concentrations of sodium chloride of 3% and higher. It may be for this reason that these organisms are so often associated with the spoilage of cured, cooked meats.

In high salt concentrations, such as may be used for some types of salted fish *Achromobacter* and *Leuconostoc (Lactobacteriaceae)* species have usually been associated with spoilage causing ropy brine. However, this type of spoilage is encountered under conditions of holding at temperatures above 15.5°C (60°F) and certainly above 10°C (50°F); and there is no indication that these organisms will grow in foods at sodium chloride concentrations above 5% when temperatures of 7.2°C (45°F) or below are maintained.

Oxygen. The pseudomonads and the *Achromobacter* species are aerobes. If there are some facultative species of these groups, they grow much faster under aerobic conditions than in the absence of oxygen and apparently cannot compete successfully with organisms which grow well under anaerobic conditions or under atmospheres in which a low oxygen tension prevails. This may be illustrated by the floral types causing spoilage of ground beef under different methods of packaging. Held at relatively high refrigerator temperatures, 7.2°C (45°F) or even at 0°C (32°F), ground beef packaged in such a manner as to allow access to oxygen will invariably spoil due to the growth of the pseudomonads. If beef is ground, packaged under vacuum, and held at these temperatures, spoilage will be caused by the *Lactobacteriaceae* (Jaye et al., 1962).[11] The latter organisms are essentially anaerobic but are indifferent to oxygen and grow in its presence or absence. Ordinarily at the start, after grinding, there should be many more of the pseudomonads than of the *Lactobacteriaceae* present in beef; but held under anaerobic conditions, the latter types outgrow the pseudomonads.

Acid. Whereas some of the psuedomonads and some of the *Achromobacter* species will produce small amounts of acid from certain sugars, these organisms do not grow well at pH values below 6.0. However, some of the *Lactobacteriaceae,* especially some of the Streptococci and Lactobacilli, will grow at pH values as low as 3.5. Actually this characteristic is employed in the preservation of some vegetables or meats (dried-type or summer sausage) by fermentation. However, these organisms are rarely the cause of spoilage in such products as pickled fish to which acid (vinegar) and salt have been added provided that the food is held at temperatures of 7.2°C (45°F) or below.

References

1. F. Alm, I. Ericksen, and N. Malin, The effect of vacuum packaging on some sliced processed meat as judged by organoleptic and bacteriologic analysis, *J. Food Technol.* **15** (1961), 199.
2. O. A. Anderson, "Refrigeration in America," Princeton Univ. Press, Princeton, N.J. 1953.
3. J. C. Ayres, The relationship of organisms of the genus *Pseudomonas* to the spoilage of meat, poultry and eggs, *J. Appl. Bacteriol.* **23** (1960), 471.
4. R. Davies, A taxonomic analysis of the microflora of refrigerated haddock preserved by irradiation or carbon dioxide, M.S. Thesis, Massachusetts Institute of Technology, 1968.
5. D. W. Duncan and J. T. R. Nickerson, Effect of environmental and physiological conditions on the exponential growth phase of *Pseudomonas fragi* (ATCC 4973), Low Temperature Microbiology Symposium Campbell Soup Co., Camden, N.J., 1961, p. 253.
6. D. W. Duncan and J. T. R. Nickerson, Effect of environmental and physiological conditions on the phase of adjustment of *Pseudomonas fragi, J. Appl. Microbiol.* **11** (1962), 179.
7. R. P. Elliot and H. D. Michener, Factors affecting the growth of psychrophilic microorganisms in foods, *U.S. Dep. Agr. Tech. Bull.* No. 1320 Agr. Res. Serv., 1965.
8. A. C. Hersom and E. D. Hulland, "Canned Foods—An Introduction to their Microbiology," 5th ed. Chemical Publ., New York, 1964.

9. E. S. Idziak and K. Incze, Radiation treatment of foods. 1. Radurization of fresh eviscerated poultry, *Appl. Microbiol.* **16** (1968), 1061.
10. J. I. Ingraham, Newer concepts of psychrophilic bacteria, Low Temperature Microbiology Symposium Campbell Soup Co., Camden, N.J., 1961.
11. M. Jaye, R. S. Kittaka, and Z. J. Ordal, The effect of temperature and packaging material on the storage life and bacterial flora of ground beef, *J. Food Technol.* **16** (1962), 95.
12. J. S. Lee, Effect of irradiation on the microbial flora surviving irradiation pasteurization of seafoods, Progress Report, May 1 - May 31, 1965, A.E.C. Contract AT(04-3)-502.
13. W. R. Lockhart, Measurement of generation lag, *Can. J. Microbiol.* **6** (1960), 381.
14. R. M. Lodge and C. W. Hinshelwood, Physiological aspects of bacterial growth. IX. The lag phase of *Bacterium Lactis aerogenes, J. Chem. Soc.* 231, 1943.
15. F. W. Lorenze, P. B. Starr, M. P. Starr, and F. X. Ogasawars, The development of pseudomonas spoilage in shell eggs. 1. Penetration through the shell, *J. Food Res.* **17** (1952), 351.
16. H. D. Michener and R. P. Elliot, Minimum growth temperatures for food poisoning, fecal-indicator and psychrophilic microorganisms, *Advan. Food Res.* **13** (1964), 349.
17. J. Monad, The growth of bacterial cultures, *Amer. Rev. Microbiol.* **3** (1949), 371.
18. J. J. Panes and S. B. Thomas, Psychotrophic coli-aerogenes in refrigerated milk: A review, *J. Appl. Bacteriol.* **31** (1968), 420.
19. A. H. Rose and L. M. Evison, Studies on the biochemical bases of the minimum temperature for growth of certain psychrophilic and mesophilic microorganisms, *J. Gen. Microbiol.* **38** (1965), 131.
20. J. M. Shewan and G. Hobbs, The bacteriology of fish spoilage and preservation, *Progr. Ind. Microbiol.* **6** (1967), 171.
21. C. G. Smith and M. J. Johnson, Aeration requirements for the growth of aerobic microorganisms, *J. Bacteriol.* **68** (1954), 346.
22. M. J. Thornley, A taxonomic study of acinetobacter and related genera, *J. Gen. Microbiol.* **49** (1967), 211.
23. E. A. Tynell, R. E. MacDonald, and P. Gerhardt, Biphasic systems for growing bacteria in concentrated culture, *J. Bacteriol.* **75** (1958), 1.

CHAPTER 5

The Preservation of Foods by Freezing

The freezing preservation of foods has become an important method of food processing. This is primarily due to the fact that, when properly prepared and stored, many frozen foods retain most of the "fresh" properties of the original food for extended lengths of time (see Fig. 5.1). The development of adequate procedures for freezing and bulk handling and storage in retail outlets in a manner which allows for minimal temperature changes has made this system a successful method of food processing.

Although the public health record of frozen foods is good, the frozen food industry has stimulated some applied and basic research concerning the effects of freezing on microorganisms.

In the chain of operations, including manufacture to consumption, a frozen food goes through at least five operational steps. These are prefreezing cooling, actual freezing, frozen storage, defrosting, postthawing treatment. The effects of all of these procedures on microorganisms cannot be entirely separated but an effort will be made to discuss them individually.

Prefreezing Cooling

The manner in which microorganisms respond to temperatures associated with the prefreezing cooling stages can be considered to be similar to that for refrigerated foods. This has been discussed elsewhere.

Actual Freezing

When the freezing process is studied, two important variables are the freezing rate and the freezing temperature. In commerce the temperatures used for freezing food products may vary from approximately −15 to −40°C (5 to −40°F) (Tressler and Evers, 1957).[34] Freezing rates may be defined as follows: Slow: 1°C (1.8°F)/50 min; rapid: 1°C (1.8°F)/min to 100°C (180°F)/min; ultrarapid: 5°C (9°F)/sec to 100°C (180°F)/sec.

FIG. 5–1. The giant holding freezer is the focal point of the revolutionary materials-handling system being employed at the kitchens of Sara Lee for the first time. Nearly 8 million cakes will be automatically handled at a temperature of −10°F. Computer-operated cranes stack pallet loads of cakes in this room automatically and retrieve them on computer command to fill customer's orders. Courtesy of Sara Lee Kitchens.

Currently, two theories are commonly used to explain the nature and kinetics of the death and injury of bacteria at low temperatures. One theory proposes that cellular damage is due to the increase in the concentration of intracellular solutes as ice separates out during the progressive freezing of an aqueous solution (Smith, 1961).[33] The other theory suggests that injury is associated with the formation of ice crystals and specifically with intracellular ice crystals (Mazur, 1960).[20]

Damage by intracellular ice formation depends partly on the rate at which internal water is able to pass through the cell membranes and partly on the degree to which internal water undergoes crystallization within the cell.

In foods and in bacteria, under most conditions of commercial freezing, about 80% of the water is drawn out of the cell and freezes on the outside. Ultrarapid freezing results in intracellular ice formation, and the cells retain their normal size (Mazur, 1961; Nei 1960).[21,29] Slowly cooled cells, on the other hand, shrink and their volume is significantly less than that of cells which are cooled rapidly (Mazur, 1961).[21] Electrical resistance measurements have shown that solutes remain in the cell during freezing and that the cell membranes remain intact during warming until the temperature rises to about −4°C (24.8°F). At that temperature permeability barriers break down, and the intracellular solutes move rapidly out into the surrounding, partially melted, medium (Mazur, 1963).[22] Calculated freezing rates of about 50°C (90°F)/min and less are said to result in no intracellular ice formation. Very rapid cooling or freezing of the concentrated solutions may lead to an inadequate nucleation rate which results in the formation of amorphous and vitreous ice. Furthermore, rapid freezing may cause hindered crystal growth which in turn will produce numerous tiny ice crystals. A subsequent temperature increase or conditions of fluctuating temperatures will produce vapor-pressure differentials and a growth of crystals in the regions of low temperature at the expense of crystals in the regions of higher temperature. Intracellular freezing is not the sole cause of the death of microorganisms at low temperature. The presence of large, external, ice crystals almost invariably appears to be lethal to microbial cells.

Freezing conditions commonly applied to foods result in extracellular ice formation. The amount of water removed from a solution upon freezing is dependent on the final temperature, and the lower the temperature the greater the amount of water removed (Wood and Rosenberg, 1957).[42] Most of the freezable water is removed at temperatures somewhere between 0 and −10°C (32°F and 14°F) (Meryman, 1960).[27] The ice crystals growing in the extracellular space can exert shearing and pressure forces on a living cell, and some cell injury may be directly related to the formation of solid ice. On the other hand, a progressive disappearance of water results in an increased solute concentration and shrinkage of the cell causing a loss of vitality.

Concentration of solutes during freezing results in physical and chemical changes within the cell in such properties as pH, vapor pressure, freezing point, surface and interfacial tension, and oxidation-reduction potential. Except for viscosity, these properties are undoubtedly affected more by electrolytes than by nonelectrolytes. At high salt concentrations a reversible or irreversible denaturation of polymers such as RNA, DNA, proteins, and polypeptides can be expected.

High ionic strength can alter the p*K* values of ionizable groups (Fasman et al., 1964),[4] increase the activity coefficients of peptide bonds (Nagg and Jencks, 1964),[28] and may modify the solvent structure and protein–solvent interactions (Von Hippl and Wong, 1964).[38] Specific ion-binding effects can also take place (Fridovish, 1963).[5]

Studies have been conducted on pH and composition changes in buffer solutions during freezing (Van den Berg, 1959).[37] Buffers, consisting of various combinations of sodium and potassium phosphates, generally exhibited large pH changes upon freezing, sometimes in excess of 1 pH unit. In some cases the change in pH was caused solely by the removal of water in the form of ice, whereas in other cases the precipitation of one or more of the salts was involved. The addition of other salts to the buffer system studied brought about a significant alteration in the pH changes. Similar changes could take place within bacterial cells during freezing.

In living cells the increased solute concentration and changes in pH brought about by freezing will influence the conformation and the association state of macromolecules. For example, lactic acid dehydrogenase undergoes denaturation and renaturation upon freezing and thawing. Low salt concentrations produce hybridization and loss of activity of enzymes, the extent of which depends upon the rates of freezing and thawing. Lactate dehydrogenase loses significantly more enzymatic activity after slow freezing than during rapid freezing (Chilson et al., 1965).[1]

Concentration of salts not only inactivates some enzymes, but can also stimulate the action of certain enzymes. Ribonuclease and deoxyribonuclease have been reported to be stimulated by an increased concentration of many divalent metal salts (Eichhorn, 1969).[2] In freezing this increased activity of enzymes will be balanced to some extent by decreased temperatures. Nevertheless, such conditions can bring about an imbalance of cellular metabolism which may be detrimental to the living cell.

The above considerations indicate that freezing, per se, can produce complex and multifunctional changes in the bacterial cell leading to its injury or death. The relative proportions between injury and death will be a function of the freezing rate and the composition of the freezing menstruum.

Generally, higher percentages of cells survive when frozen slowly than during fast freezing. However, under these conditions there may be higher levels of cell injury (Mazur et al., 1957; Heckly et al., 1958; Meryman, 1956).[8,23,26] When cells of *Escherichia coli* were frozen at −79°C (−110.2°F), there was less survival than when they were frozen at −26°C (−14.8°F). However, the suspension frozen at −79°C (−110.2°F) was stable upon storage while at −26°C (−14.8°F) a continuous decrease in viability was observed (Lindeberg and Lode, 1963).[18] Similar effects of freezing temperature have been observed with cells of *Pseudomonas fluorescens.*

Some Findings Which Have Been Reported Regarding the Effect of Freezing, Frozen Storage, and Defrosting on Microorganisms

It was shown (Keith, 1913)[15] that the survival of bacteria when subjected to freezing temperatures depended on the species investigated and the menstruum in which they were held at a particular temperature. When *Escherichia coli* was frozen and held at −20°C (−4°F) in tap water, less than 1% of the cells survived a storage period of 5 days. A large proportion of bacilli in milk were viable after being subjected to the same treatment. Keith was apparently the first to discover that when glycerol was present in the suspending menstruum in quantities of 5–42%, there was a much greater survival of microorganisms *(Escherichia coli),* during extended periods of storage. It was also found that sucrose solution (10%), added to the surface of agar slant cultures, enhanced the survival of *Bacillus subtilus, Bacillus megaterium, Proteus,* and *Sarcina* during holding at −10°C (14°F). It was also determined in this investigation that the rate of cooling, the rate of defrosting, and the storage temperature all affected the degree of survival of a particular bacterial species when subjected to freezing.

In other experiments on the freezing destruction of bacteria (Hilliard and Davis, 1918)[10] cells of *E. coli* were suspended in tap water and in sugar solutions (of sufficient concentration to prevent freezing) and in milk and cream and held at temperatures between −0.5°C and −10°C (31.1°F and 14°F). It was found that survival was greatly diminished when conditions were such that the suspending menstruum was solidly frozen and lethality was decreased when the organisms were suspended in milk or cream as compared to ice water. Repeated freezing was found to be more lethal than holding in the frozen state. The death of *Escherichia coli* during freezing was considered to be due to a mechanical crushing caused by extracellular ice.

The Avian tubercle organisms, which had been grown on nutrient agar, was frozen in liquid air or in air at −40°C (−40°F) and defrosted at 40°C (104°F) (Kyes and Potter, 1939).[16] The number of cells were not counted but the infectivity of the frozen material was determined. These organisms survived holding at −192°C (−313.6°F) for from 2 to 52 days and freezing and defrosting for as many as 200 times, although repeated freezing reduced infectivity. The best recovery was obtained with methods employing fast defrosting.

Young cells of *Escherichia coli* (log phase) were subjected to changes in temperature from 45°C to 10°C (113 to 50°F) (Sherman and Cameron, 1934).[32] It was found that approximately 95% of the cells were killed by this treatment. When young cells were transferred from 1% peptone to 1% peptone containing 5% sodium chloride 62–91% of the cells were killed. Droplets containing the cells of *Pseudomonas pyocyanea, Staphylococcus aureus, Escherichia coli, Achromobacter, Bacillus cereus,* and *Bacillus mesentericus* were frozen in broth at −5, −20, and −70°C (23, −4, and −94°F) and survival was determined. Rapid freezing (rapid defrosting is assumed) destroyed from 80% of the cells in the case of sensitive organisms such as *Pseudomonas pyocyanea* to none in the case of spores *(Bacillus megaterium).* The rate of freezing was found to have little, if any, effect on the degree of survival. The various organisms were also held from 10 to 30 days at −1, −2, −3, −5, −10, and −20°C (30.2, 28.4, 26.6, 23, 14, and −4°F) and it was found that, with *Escherichia coli* and *Pseudomonas pyocyanea,* the higher the storage temperature, the greater was the destruction. It was also shown that when the proteins from the dried and ground cells of *Pseudomonas pyocyanea,* dissolved in ice water, were held at −2°C (28.4°F), there was considerable precipitation during periods of 1–12 days storage, while little precipitation of this type of material occurred during holding at −20°C (−4°F).

The size of unfrozen, frozen, and frozen and defrosted yeast cells *(Saccharomyces cerevisiae)* was measured (Haines, 1938).[6] No difference in average size was found, indicating that there was no intracellular ice formation. During freezing or frozen storage denaturation of the protein was believed to occur in the cells of some bacteria, and this was considered to be the cause of death.

Tissue containing *Treponema pallidum* and *Treponema pertenue* were frozen at −78°C (−108.4°F) and held for 1 year at this temperature and found to be infective (Turner, 1938).[35] These organisms apparently did not survive freezing at −78°C (−108.4°F) and storage at −10 or −20°C (14 or −4°F) for 2 months. Cell counts were not made.

The spirochetes of relapsing fever were fast or slow frozen in blood plasma and also held in storage (Turner and Brayton, 1939).[36] Storage at −78°C (−108.4°F) (4 days to 6 months) had no effect. Slow cooling, 0°C to −78°C (32°F to −108.4°F) in 6 hr was only slightly less lethal than fast freezing −78°C (108.4°F). Warming from −78 to 0°C in 6 hr killed most of the cells while fast warming did not. There was a gradual decrease in infectivity during storage at −12 or −20°C (10.4 or −4°F). Rapid cooling and thawing was considered to give the greatest survival. Repeated freezing and thawing for four cycles destroyed all of the organisms. As in the previously cited paper, cell counts were not made.

It was found that with *Escherichia coli* there was slight to marked initial sensitivity to cold 0°C (32°F) when suspensions of the cells were held without freezing, after removing from the optimum temperature (Hegarty and Weeks, 1940).[9] Destruction, or sensitivity to cold, depended on which part of the logarithmic phase of growth the cells were in when placed at 0°C (32°F). Sensitivity to cold was believed to be related to cell division.

Yeasts have been found to survive better in tap water or 10% sucrose solutions than in 2% sodium chloride or in cider (pH 3.2) during freezing and to survive better during storage at −20°C (−4°F) than at −10°C (14°F) (McFarlane, 1940a; McFarlane, 1940b).[24,25]

Washed cells of *Escherichia coli* were frozen in 1% peptone (pH 7.0) slowly at 0 to −20°C (32 to −4°F) and fast at −30 to −78°C (−22 to −108.4°F) or at −190 to −195°C (−310 to −319°F) (Weiser and Osterud, 1945).[41] It was found that there was a marked death of cells at temperatures just below freezing, at which temperature intracellular ice would not form, and a greater decrease in viable cells with slow freezing than with fast freezing. The greatest destruction of cells took place during the final stages of the formation of extracellular ice and it was concluded that cell destruction was due to the mechanical action of extracellular ice. There was a decrease in the rate at which cells died during storage at temperatures of −30°C (−22°F) or below and a protective effect of fluctuating temperatures between −30 and −78°C (−22 and −108.4°F).

Suspensions of *Escherichia coli* in 10% sucrose solution were frozen by immersing tubes in liquid nitrogen (vitrofreezing without crystal formation), defrosted fast in buffer solution (vitromelting without crystal formation), frozen slowly with crystal formation, and defrosted slowly with crystal formation (devitrification) (Weiser and Hargiss, 1946).[40] Vitrofreezing and vitromelting were more lethal than slow freezing (crystal formation during freezing) and devitrification (crystal formation during melting) was found to be more lethal than either of the other procedures.

During the storage of frozen foods it has been found that there is a greater decrease in viable microorganisms during storage at −10°C (14°F) than during storage at −20°C (−4°F) (Weiser, 1951).[33]

The number of viable bacteria in commercially packed beans, corn, and peas was determined immediately after freezing and after 6 and 18 months' storage at −12.2°C (10°F), −17.8°C (0°F), and −23.3°C (−10°F) (Hucker et al., 1952).[13] The greatest percentage reduction of bacteria occurred in high count products and generally the higher the storage temperature, the greater was the destruction of bacteria.

Treponema pallidum, Escherichia coli, Diplococcus pneumoniae, and *Rhodospirillum rubrum* in serum saline, with and without 15% of glycerol, was frozen as small quantities by immersing tubes in dry ice and alcohol. Storage tests were made by holding at −70, −40, and −15°C (−94, −40, and 5°F) (Hollander and Nell, 1954).[11] Defrosting was carried out at 30°C (86°F). The results indicated that glycerol protected these species against freezing destruction and it was concluded that death during freezing is due to mechanical compression caused by the expansion of ice crystals. Glycerol was toxic to *Rhodospirillum rubrum.* In saline with glycerol *Treponema pallidum* survived better during storage at −70°C (−94°F) than at −40 or −15°C (−40 or 5°F). During repeated freezing and defrosting the number of viable cells of *Escherichia coli* and *Diplococcus pneumoniae* decreased exponentially, the latter being the most susceptible.

Suspensions of *Escherichia coli, Microbacterium flavus, Lactobacillus fermenti,* and *Bacillus pumilus* in water were subjected to freezing at −22°C (−7.6°F) and defrosting at 35°C (95°F). *Escherichia coli* and *Microbacterium flavus* were also subjected to mechanical abrasion by shaking with glass beads (Harrison and Cerroni, 1956).[7] It was found that there was greater destruction of *Escherichia coli* than of *Microbacterium flavus* during mechanical shaking with glass beads. *Escherichia coli* was less susceptible to freezing damage when aerated. No correlation between physical structure and susceptibility to freezing destruction was found. According to these findings, mechanical damage of bacteria during freezing due to extracellular ice crystals probably does not occur.

The washed cells of *Escherichia coli, Serratia marcescens,* and *Lactobacillus fermenti* were suspended in 5 ml of distilled water or of various media in tubes and subjected to freezing (Harrison and Cerroni, 1956).[7] Freezing was carried out at −40°C (−40°F) then at −22°C (−7.6°F) or at −22°C directly. Storage was carried out at −22°C (−7.6°F). These organisms survived better in distilled water than in media; hence, it was considered that crushing due to extracellular ice crystals was not a factor causing death. At −22°C (−7.6°F) sodium chloride solutions were strongly lethal to these organisms but the addition of glycerol protected against destruction. Dilution of broth cultures favored survival during freezing. It was also found that the survival of organisms in 4.1 or 4.6 *M* sodium chloride held at −22°C (−7.6°F) in the unfrozen state was similar to that obtained in broth at the same temperature when the menstruum was frozen. The concentration of electrolytes in the medium during freezing was believed to be the cause of the death of bacteria under such circumstances.

The cells of *Escherichia coli* in broth were frozen at −22°C and held at this temperature. Survival was determined after freezing and storage (Major, 1953).[19] *Lactobacillus, Micrococcus, Pseudomonas, Chromobacterium, Serratia,* and *Bacillus* species were also investigated. The greatest decrease in viable cells occurred immediately after freezing. During storage the slope of the survival curve flattened out. With *Escherichia coli* and *Serratia marcescens* the lower the concentration of the cell suspension, as frozen, the greater was the percentage destruction. With other types, the percentage of survival was constant and independent of cell concentration.

The cells of various species of bacteria, including *Salmonella,* were frozen in broth containing 15% of glycerol and stored at −10°C (14°F). After a storage period of 5 months the cells of many species, including *Salmonella,* were viable (Howard, 1956).[12]

The cells of *Pasteurella tularensis* were frozen at various rates in 0.14 *M* sodium chloride containing 0.1% of gelatin, defrosted fast and slowly, and survival was determined (Mazur, 1960).[20] Cooling was carried out at rates between 1°C (1.8°F)/minute and 300°C (540°F) per minute; a rate of 50°C (90°F) per minute was considered to be rapid cooling. Rapid warming was carried out at rates between 500°C and 1100°C (900°F and 1980°F) per minute. With cells cooled to −75°C (−103°F) and warmed slowly, the more rapid the cooling rate, the fewer the cells which survived. When cells were cooled only to −30°C (−22°F) or when cells cooled to any temperature were warmed rapidly, the rate of cooling did not greatly influence the degree of survival. It was concluded that injury of cells during cooling and warming was due to the freezing of intracellular water and was not concerned with the formation of extracellular ice crystals.

Frozen Storage

The behavior of microorganisms during storage varies considerably. In one study it has been shown that for *Vibrio costicolus,* survival during a 12-week period was maximal at −25°C (−13°F). The degree of survival of *Escherichia coli* and *Serratia marcescens* was similar. However, *Pseudomonas chlororaphis* survived best at a temperature of −70°C (−94°F) while *Lactobacillus fermenti* survived best at the lowest temperature range used in the study −50° to −70°C (−58 to −94°F). This study also demonstrated that differences in the survival of a given species could also be due to the manner in which the cells were dispensed for storage. Differences in survival were also found when cells from young and old cultures were studied. These observations demonstrate that microbial survival is influenced by a number of factors. During frozen storage the temperature may fluctuate. This fluctuation will cause recrystallization to occur, which may affect viability.

Defrosting

Defrosting effects *per se* have yet to be clearly defined. It is generally assumed that rapid defrosting of a menstruum is a prerequisite for high survival of the microorganisms contained therein. This is probably because rapid defrosting minimizes the amount of ice recrystallization that occurs during defrosting.

Postthawing

Most, if not all, of the microbiological deterioration of foods takes place during the stages of precooling and after thawing. This is because microorganisms do not grow in foods held at temperatures below −12.2 to −10°C (10–14°F). At temperatures between −6.7 and 3.3°C (20 and 38°F) psychrophilic bacteria can grow, but food-poisoning organisms cannot. Therefore, public health problems, as well as microbial deterioration of frozen foods, are chiefly associated with the handling either prior to freezing or after defrosting or both. Factors influencing or controlling microbial growth on or in defrosted products include:

(a) The microflora of the original product, secondary contamination of the product, and the growth of microorganisms during handling.
(b) The duration of prefreezing refrigeration. This step is said to select psychrophiles.
(c) The freezing method. An important aspect here concerns whether the freezing rate is slow enough to allow microbial growth to take place. Contamination of the product by freezing brines has also been reported to be an important problem.
(d) Composition of the food product. Important variables include the effects of food composition on microbial survival at the freezing point, and the suitability and availability of the substrate for supporting microbial growth.
(e) Length of frozen storage.
(f) Defrosting methods.

All of these factors will influence the growth of microorganisms in food products. It should be kept in mind that some bacteria always survive in frozen foods, and the freezing and storage processes cannot be expected to produce a sterile product. As a matter of fact, present literature indicates that there is little or no difference in the spoilage times for frozen foods versus the fresh product. An investigation on minced and eviscerated fresh and frozen poultry supports this fact (Elliott and Straka, 1964).[3] This investigation demonstrated that the growth rate of the psychrophilic bacteria was slightly greater after thawing—presumably due to an increased availability of the substrate. The initial number of psychrophilic bacteria on the frozen thawed products was smaller than that present on the unfrozen food. This reduction was due to the fact that a certain fraction of the psychrophilic bacteria on the frozen product were inactivated by the freezing and thawing process. The result was that spoilage occurred in both the fresh and the frozen and defrosted chicken at about the same time.

The growth of molds may be the cause of deterioration of frozen foods. The psychrophilic molds which grow at temperatures below 0°C (32°F) may cause food spoilage in this case. Poor refrigeration storage conditions occur which lead to fluctuating temperatures, may facilitate the growth of molds in foods. Fluctuating temperatures should not occur but may do so due to improper storage conditions or faulty handling. A rise in temperature may not only allow for microbial growth, but may also lead to a transfer of moisture from one part of a food product to another part. A wide variety of mold species have been isolated from frozen foods. *Pullularia pullulans* is reported to be the most frequently isolated mold from frozen food products; *Botrytis cinerae* and *Geotrichum candidum* have also been isolated frequently. These molds grow at temperatures below 5°C (41°F). There is a correlation between the type of product examined for molds and the species isolated. For instance, *Aleurisma carnis* (reclassified as *Chrysosporium pannorum*) was the most frequently encountered fungal organism when meat products were examined, while *Phoma* species were most frequently encountered when cherry products were investigated.

Bacterial standards have been recommended for various types of frozen foods. The primary purpose of the bacterial count is to check and improve the sanitary and edible quality of foods. It has been stated that if the bacterial count of a frozen vegetable is below 100,000 per gram the product has been handled in the following manner (Tressler and Evers, 1957):[34] (1) A high blanch temperature was used; (2) quick and adequate cooling was used; (3) the packing plant was sanitary; (4) clean packaging material was used; (5) freezing was rapid; (6) no thawing occurred after freezing. Such information obtained from the bacterial count can be used as a valuable tool in quality control.

It has been found that the total plate count did not indicate quality in frozen vegetables (Humphrey, 1950).[14] This is probably true since in vegetables the bacteria which might cause deterioration prior to processing would be destroyed by the blanching process. Also, quality factors in vegetables are often concerned with maturity factors. However, high plate counts in frozen vegetables would indicate a lack of suitable sanitary procedures in the processing plant and it has been shown (Peterson and Gunderson, 1960)[31] that with some bacteria and some foods, high counts, without further increases in number, may result in flavor changes due to the action of bacterial enzymes at temperatures above freezing. This implies that such changes might take place to some extent at freezing temperatures.

In meats, fish, and poultry, since off-flavors and loss of quality are usually due to the growth of bacteria prior to freezing, high bacterial counts after freezing may be indicative of poor quality.

Regarding tests for indicator bacteria in frozen foods, the suitability of such determinations is not clear. It has been shown that the enterococci may survive freezing better than *Escherichia coli,* hence are more apt to be present in foods which have been frozen and defrosted (Larkin et al., 1955).[17] However, enterococci have been shown to be present in vegetable material which has not been contaminated with animal substances (Larkin, 1955).[17] Also, it is known that strains of the enterococci will grow at low temperatures above freezing and considering this they might increase in numbers in food materials during processing. The situation is, therefore, one

Table 5–1
Suggested Maximum Bacterial Counts for Certain Frozen Foods Products

Product	Maximum viable count[a]	Maximum total microscopic counts[a]
Peas	100,000	500,000
Lima beans	100,000	500,000
Cut corn	500,000	1,000,000
Spinach	150,000	500,000
Broccoli	500,000	1,000,000
Cauliflower	200,000	500,000
Green beans	100,000	500,000
Breaded shrimp	300,000	500,000
Fish fillets	100,000	2,000,000
Eviscerated poultry	5,000[b]	500,000
Precooked frozen foods (including poultry products)	100,000	5,000,000
Egg products (whole egg, yolk, egg white)	200,000	
Cut, uncooked meats	10,000[b]	
Hamburger	5,000,000	

[a] Counts are indicated on a per gram basis excepting in the case of eviscerated chicken and cut, uncooked meats.
[b] Counts indicated on a per square centimeter basis.

in which the presence of *Escherichia coli* in frozen foods might be a better index of unsanitary conditions if this organism survived freezing, storage, and defrosting. Since *E. coli* may be destroyed during processes associated with freezing and, since the source of the enterococci might have been other than animal origin, no absolute statement can be made as to the efficacy of using tests for indicator organisms in frozen foods.

It is felt that both plate-count standards and microscope-count standards can, and should, be established for frozen foods. Table 5-1 lists the recommended maximum bacterial counts for certain frozen foods products.

The recommended maximum counts listed in Table 5-1 are believed to allow a workable tolerance for each individual product (Nickerson, 1968).[30] For example, frozen vegetables should contain only a small number of microorganisms picked up from the processing line, since they are normally rapidly frozen after blanching and only a minimal amount of contamination should occur. If higher counts are obtained in frozen vegetables, it is probable that bacterial growth is occurring on the equipment of the processing lines indicating inadequate cleaning and washing procedure. Under these conditions there is also the possibility that after blanching the products were held at comparatively high temperatures for longer periods than would occur under conditions of good manufacturing practices.

References

1. O. P. Chilson, L. A. Costello, and N. O. Kaplan, Effects of freezing on enzymes, *Fed. Proc.* **24**, Suppl. **15**:S-55, (1965).
2. G. L. Eichhorn, P. Clark, and E. Tarien, The interaction of metal ions with polynucleotides and related compounds. The effect of metal ions on the enzymatic degradation of ribonucleic acid by bovine pancreatic ribonuclease and of deoxyribonucleic acid by bovine pancreatic deoxyribonuclease. *J. Biol. Chem.* **244** (1969), 937.
3. R. P. Elliott and H. P. Straka, Rate of microbial deterioration of chicken meat at 2°C after freezing and thawing, *Poultry Sci.* **43** (1964), 81.
4. G. D. Fasman, C. Linblow, and E. Bodenheimer, Conformational studies on synthetic poly-α-amino acids: Factors influencing the stability of the helical conformation of poly-L-glutamic acid and copolymers of L-glutamic and L-leucine, *Biochemistry* **3** (1964), 155.
5. I. Fridovich, Inhibition of acetoacetic decarboxylase by anions. The pH of meister lyotropic series., *J. Biol. Chem.* **238** (1963), 592.
6. R. B. Haines, The effect of freezing on bacteria., *Proc. Roy. Soc. Ser. B.* **124** (1938), 451–463.
7. A. P. Harrison and R. E. Cerroni, Fallacy of "crushing death" in frozen bacterial suspensions. *Proc. Soc. Exp. Biol. Med.* **91** (1956), 577–579.

8. R. J. Heckly, A. W. Anderson, and M. Rockenmacker, Lyophilization of *Pasteurella pestis*. *Appl. Microbiol.* **6** (1958), 255.
9. C. P. Hegarty and O. B. Weeks, Sensitivity of *E. coli* to cold shock during the logarithmic growth phase. *J. Bacteriol.* **39** (1940), 475–484.
10. C. M. Hilliard and M. A. Davis, The germicidal action of freezing temperatures upon bacteria. *J. Bacteriol.* **3** (1918), 423.
11. D. H. Hollander and E. E. Nell, Improved preservation of *Treponema pallidum* and other bacteria by freezing. *Appl. Microbiol.* **2** (1954), 164.
12. D. H. Howard, The preservation of bacteria by freezing in glycerol in broth. *J. Bacteriol.* **71** (1956), 625–626.
13. G. J. Hucker, R. F. Brooks, and A. Emery, The source of bacteria in processing and their significance in frozen vegetables. *Food Technol.* **6** (1952), 147–155.
14. H. J. Humphrey, The bacteriology of frozen foods., *Quick Frozen Foods* **13** (1950), 50.
15. S. C. Keith, Factors influencing the survival of bacteria at temperatures in the vicinity of the freezing point of water. *Science* **37** (1913), 877.
16. P. Kyes and T. S. Potter, The resistance of avian tubercle bacilli to low temperatures with special references to multiple changes in temperature. *J. Infec. Dis.* **64** (1939), 123.
17. E. P. Larkin, W. Litsky, and J. E. Fuller, Fecal streptococci in frozen foods. *App. Microbiol.* **3** (1955), 98–110.
18. G. Lindeberg and A. Lode, Release of ultraviolet-absorbing material from *Escherischia coli* at subzero temperatures. *Can. J. Microbiol.* **9** (1963), 523.
19. C. P. Major, The effect of the initial bacterial concentration upon survival at −20°C. Masters Thesis, Vanderbilt University (1953).
20. P. Mazur, Physical factors implicated in the death of microorganisms at subzero temperatures. *Ann. N.Y. Acad. Sci.* **85** (1960), 610.
21. P. Mazur, Manifestation of injury in yeast cells exposed to subzero temperatures. I. Morphological changes in freeze-substituted and in "frozen-thawed" cells. *J. Bacteriol.* **82** (1961), 662.
22. P. Mazur, Studies on rapidly frozen suspensions of yeast cells by differential thermal analysis and conductometry. *Biophys. J.* **3** (1963), 323.
23. P. Mazur, M. A. Rhian, and B. G. Mahlandt, Survival of *Pasteurella tularensis* in gelatin-saline after cooling and warming at subzero temperatures. *Arch. Biochm. Biophys.* **71** (1957), 131.
24. V. H. McFarlane, Behavior of microorganisms at subfreezing temperatures below freezing, *Comm. Fisheries Rev.* **12** (1940), 28.
25. V. H. McFarlane, Behavior of microorganisms at subfreezing temperature. II. Distribution and survival of microorganisms in frozen cider, frozen syrup-packed raspberries and frozen brine-packed peas, *Food Res.* **5** (1940), 59.
26. H. T. Meryman, Mechanics of freezing in living cells and tissues. *Science* **124** (1956), 515.

27. H. T. Meryman, The mechanisms of freezing in biological systems. "Recent Research in Freezing and Drying," pp. 23–39. Blackwell, Oxford, 1960a.
28. B. Nagg and W. P. Jencks, Depolymerization of F-actin by salts and amides, *Fed. Proc.* **23** (1964), 530.
29. T. Nei, Effects of freezing and freeze-drying on microorganisms. "Recent Research in Freezing and Drying" (A. S. Parkes and A. Smith, Eds.), pp. 78–86. Blackwell, Oxford, 1960.
30. J. T. R. Nickerson, Microbiology of Foods, Chapter 23, Amer. Society Heating Refrigerating & Air-Conditioning Engineers, 1968, p. 275.
31. A. C. Peterson and M. F. Gunderson, Some characteristics of proteolytic enzymes from *Pseudomonas fluorescens, Appl. Microbiol.* **8** (1960), 98–104.
32. J. M. Sherman and G. M. Cameron, Lethal environmental factors within the natural range of growth, *J. Bacteriol.* **27** (1934), 341.
33. A. U. Smith, Biological effects of freezing and supercooling, Monograph No. 9 of the Physiological Society, Williams & Wilkins, Baltimore, Maryland, 1961.
34. D. K. Tressler and C. F. Evers, "The Freezing preservation of Foods, Vol. 1, 3rd ed., Avi Publ. Westport, Conn., 1957.
35. T. B. Turner, The preservation of virulent *Treponema pallidum* and *Treponema perteune* in the frozen state; with a note on the preservation of filterable viruses, *J. Exp. Med.* **67** (1938), 61.
36. T. B. Turner and N. L. Brayton, Factors influencing the survival of spirochetes in the frozen state, *J. Exp. Med.* **70** (1939), 639.
37. L. Van den Berg, The effect of addition of sodium and potassium chloride to the reciprocal system $KH_2PO_4 - Na_2HPO_4–H_2O$ on pH and composition during freezing, *Arch. Biochem. Biophys.* **84** (1959), 305.
38. P. H. Von Hippl and K. Y. Wong, Neutral salts: The generality of their effects on the stability of macromolecular conformations, *Science* **145** (1964), 577.
39. H. H. Weiser, Survival of certain microorganisms in selected frozen foods, *Quick Frozen Foods* **13** (1951), 50.
40. R. S. Weiser and C. O. Hargiss, Studies on the death of bacteria at low temperatures. II. The comparative effects of crystallization, vitromelting and devitrification on the mortality of *Escherichia coli. J. Bacteriol.* **52** (1946), 71–79.
41. R. S. Weiser and C. M. Osterud, Studies on the death of bacteria at low temperatures. I. The influence of the intensity of the freezing temperature, repeated fluctuations of temperature and the period of exposure to freezing temperature on the mortality of *Escherischia coli. J. Bacteriol.* **50** (1945), 413–439.
42. T. H. Wood and A. M. Rosenberg, Freezing in yeast cells, *Biochim. Biophys. Acta* **25** (1957), 78.

CHAPTER 6

Chemical Preservation of Foods

A variety of chemical compounds may be added to foods to prevent their microbial decomposition. In some instances this is done merely to extend the storage life of foods which must also be held at refrigeration temperatures above freezing. Even under these conditions more than one chemical compound may be added. Chemical compounds are sometimes added to foods to stabilize these products indefinitely at ambient or room temperature. When this is done other factors than the chemical compound itself, such as high soluble-solids content or drying to comparatively low moisture content, are used in combination with the addition of chemicals to prevent microbial decomposition.

The salting of foods and the addition of acid to foods are types of chemical preservation. These methods have been included in this Chapter since the chloride ion or the undissociated acid moiety may be an inhibitory factor for microorganisms. When these methods of preservation are used, a chemical has, in fact, been added to the foods.

Salting

Sodium chloride is used to preserve foods in many ways. In some cases limited amounts of salt, usually 2–5%, are present in the final product which, together with refrigerated storage or with the addition of acid and refrigerated storage, is sufficient to prevent the growth of the psychrophilic and psychrotrophic organisms which would otherwise grow and spoil the product. In other instances salt is added to foods in such a manner as to essentially saturate the water phase of the product with sodium chloride. After salting, in this instance, the food may or may not be dried to remove a considerable fraction of the moisture present.

Whether or not salt is to be used as the main method of preservation, such treatment of foods must be carried out under controlled conditions of temperature. If the temperature (natural or artificial) under which the product is held immediately after salting is not below 15.6°C (60°F) (preferably 4.4°C (40°F) or below), then spoilage organisms, including putrefactive bacteria, may invade the tissues and grow before salt has penetrated the inner portions of the food to provide a sufficient concentration to arrest the growth of microorganisms. This applies equally well to the growth of pathogenic bacteria such as *Clostridium botulinum* as to the growth of strictly spoilage types.

Microorganisms vary greatly in their capacity to grow in the presence of sodium chloride. Some of them are not able to grow in materials containing even less than 3% of salt while others grow well even in concentrations higher than 10%. Halophilic bacteria, requiring high concentrations of salt for growth, are fortunately comparatively uncommon in food spoilage.

Inhibition of Microbial Growth by Sodium Chloride

One factor responsible for the inhibition of microbial growth by salting is the removal of available water. Microorganisms grow only in aqueous solutions. If this water is bound to chemical compounds, it may become unavailable to cells of this type. A term, "Water activity," a_w, has been coined to express the degree of availability of water in foods (Scott, 1957).[20] This term is applied to all food, the ordinary fresh-type food having an a_w of about 0.99–0.96 at ambient temperatures. Low water activities, which limit the growth of microorganisms in foods, may be brought about by the addition of salt or sugar, as well as by the removal of water by drying of one type or another. Under such conditions the remaining water has been tied up by chemical compounds added to or concentrated in the food or bound to some food component such as protein.

According to Raoult's Law

$$\frac{p}{p_0} = \frac{N_2}{N_1+N_2},$$

where p is the vapor pressure of the solution, p_0 is the vapor pressure of water; N_1 is the number of moles of solute present and N_2 is the number of moles of solvent present.

Also,

$$\frac{p}{p_0} = a_w = \frac{\text{equilibrium relative humidity}}{100}.$$

A 1-molal solution of a perfect solute in water would reduce the vapor pressure of the solution 1.77% over that of pure water, or the vapor pressure of the solution would be 98.23% of that of pure water, or the a_w would be 0.9823.

In general, molds grow at lower water activities than yeasts; and yeasts grow at lower water activities than bacteria; but there are some exceptions. Bacteria usually requires water activities between 0.99 and 0.96 for growth; but some, which are not halophiles, will grow at water activities as low as 0.94 (some strains of *Pseudomonas* and some bacilli will grow at an a_w as low as 0.90). *Micrococcus halodenitrificans,* a xerophilic (organisms growing at low a_w values) red halophilic organism, will grow at a_w values as low as 0.75, the a_w of a saturated sodium chloride solution (Christian and Waltho, 1962).[7] *Halobacterium salinarium* requires 3 molal sodium chloride for growth when this is the only solute. *Lactobacillus viridescens* grows better at an a_w of 0.975 than at an a_w of 0.992. *Staphylococcus aureus* is known to grow in salt concentrations higher than 10%, but it has been shown that at this concentration enterotoxin formation is greatly reduced (Martins, 1964).[15]

Many yeasts will grow at a_w values as low as 0.90 or lower. *Saccharomyces rouxii,* for instance, will grow on fruit at an a_w of 0.81.

Some molds require low water activities for growth. Thus, *Eremascus albus* and *Aspergillus glaucus* require 40% sugar solutions and *Xeromyces bisporus* 60% sugar solutions for growth. The mold, *Sporendonema epizoum,* grows on salted and dried fish and, while not an obligate halophile, will not grow at a_w values above 0.983.

The effect of salt or sugar added to foods, then, is at least partly to limit the growth of microorganisms by lowering water activity, there being a range of water activities at which microorganisms will or will not grow. There are other factors which may play a part in limiting the growth of microorganisms in foods to which these materials have been added. It has been shown (Baird-Parker and Freame, 1967),[4] for instance, that *Clostridium botulinum* types A and B will grow at water activities only as low as 0.96 (6.1%) when sodium chloride is use to adjust the a_w; but when glycerol is used for the same purpose, these organisms will grow at water activities as low as 0.93. The inhibiting effect of sodium chloride, therefore, is not entirely due to the lowering of water activity although this may be the main reason for the growth-limiting properties of this compound.

There are reports of the growth of *Clostridium botulinum* in foods at higher sodium chloride contents than 6.1%. However, it is probably the case, in these instances, that growth occurred prior to complete diffusion of the salt into all portions of the food. For this reason and because of the growth of spoilage organisms, the salting of foods should be done at ambient temperatures below 15.5°C (60°F), the lower temperature reducing the rate of growth of pathogenic and spoilage organisms and providing for diffusion of salt into the product before significant growth has occurred.

At least part of the inhibiting effect of salt or sugar on microorganisms may be due to osmotic effects in which the cell becomes plasmolyzed or dehydrated, water being drawn from the cell into the solution.

The Addition of Acid

Different species of microorganisms have a pH or hydrogen ion concentration at which growth occurs at the fastest rate and an upper and lower pH at which no growth will occur. Generally, molds and yeasts are considered to grow best in media which are somewhat acid while many bacteria grow best in media at a pH of 7.0 or even on the slightly alkaline side. There are, however, species of bacteria which will grow at hydrogen ion concentrations as low as pH 3.0 and as high as pH 11.0.

It is usually not possible to add acid to foods to lower the pH below that at which microorganisms will grow since this would cause such products to be rejected because of taste. What is done, therefore, when acids are used, is to combine such treatment with other methods of preservation such as refrigeration or heat pasteurization. In some fermented foods, such as sauerkraut, common salt is added and a fermentation by the *Lactobacteriaceae* allowed to take place to produce lactic acid and lower the pH, a factor largely responsible for the preservation of the product.

Acetic acid, as vinegar, is added to pickled fish products along with some sodium chloride, and microbial growth may be inhibited in such products so long as they are held at suitable refrigeration temperatures, preferably below 4.4°C (40°F). In this instance part or most of the inhibiting effect of the acid may be due to the undissociated molecule rather than to lowering of pH. *Clostridium botulinum* has been reported to grow at pH values only as low as 4.8 in artificial media, although it is generally considered that foods with a pH above 4.5 should be sterilized by heating at temperatures above 100°C (212°F) or given a heat process equivalent to what is known as a minimum botulinum kill (sufficient heat treatment of all parts to destroy approximately 1×10^{12} spores of the most heat-resistant strains of *Clostridium botulinum)*.

Vinegar is also added to products such as pickles and ketchup which are then heat pasteurized. The added acid, through lowering the pH, causes the microorganisms to become more heat sensitive so that all vegetative types are destroyed and those spore-forming types not destroyed are inhibited by acid, salt, and possibly also by spices which may be present.

Another instance of the use of acid to prevent spoilage of foods is the addition of citric acid to tomato juice. An organism, *Bacillus coagulans,* is sometimes present in this product, and in some areas in which tomatoes are grown soil conditions are such that the product has a relatively high pH. *Bacillus coagulans,* although not extremely heat resistant, may survive in tomato juice which is processed by heating the cans in boiling water. However, this organism will not grow in tomato juice at a pH of 4.0 or below. Citric acid may, therefore, be added to this product, the pH of which may be 4.5 or higher, to reduce the pH to a point unfavorable to the growth of this organism.

Acetic acid, as vinegar, is used in mayonnaise and doubtless plays a part in stabilizing the product microbiologically. Citric acid is added to jams and jellies where it provides the right acidity, about pH 3.6, for gel formation with pectin but combined with some heating and the removal of some of the oxygen from the headspace of the container, serves as well to prevent the growth of microorganisms, especially of molds.

Fatty Acids and Their Salts

Fatty acids, usually short chain, and their salts are added to bread or bread-type products to inhibit molding or the formation of rope. Ropiness in bread may be caused by the growth of *Bacillus panis, Bacillus mesentericus,* or *Bacillus subtilis* which produce materials such as levan (a polymer of fructose), which has a slimy consistency.

Sodium diacetate (Dykon) is a complex of sodium acetate and acetic acid (42.25% available acetic acid in water) and probably has the formula:

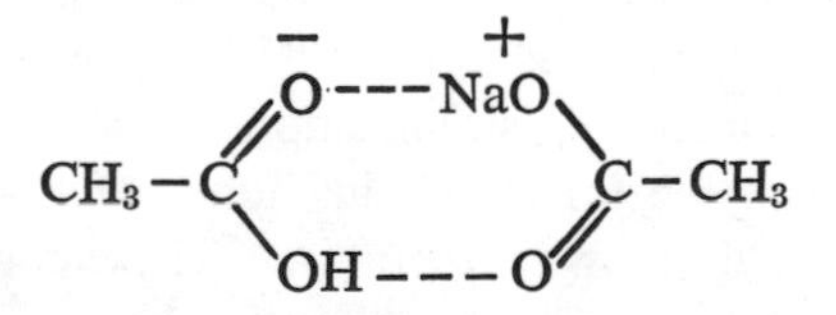

This compound may be added to bread in concentrations up to 0.4 parts per 100 parts of flour. This and other fatty acid compounds are GRAS (generally recognized as safe); but since Standards of Identity have been set up for bread, the amount which may be added is limited

Propionic acid and especially calcium or sodium propionate,

$$CH_3\text{–}CH_2\text{–}\overset{\overset{\displaystyle O}{\|}}{C}ONa$$

(mycoban), may be added to bread as a mold and bacterial inhibitor. These compounds are GRAS but again may be used in bread in concentrations up to 0.32 parts per 100 parts of flour. The propionates are allowed in foods in many European countries, in Scandanavian countries and in Canada.

Caprylic acid (C_8) may be used in cheese as a mold inhibitor in the United States. This compound has been added as a component of the wax, nitrocellulose, or other compound used to coat the paper or plastic material with which cheese is wrapped.

While fatty acids such as acetic or propionic, when added as such, may lower the pH of a food product, their effect in inhibiting microorganisms is considered to be due mainly to the undissociated part of the molecule. The mechanism by which these compounds tend to prevent the growth of microorganisms is not known. It has been speculated that they may destroy cell membranes or inhibit some enzyme systems of the microbial cell concerned with metabolism or reproduction.

The Benzoates and Parabenzoates

In the United States benzoic acid, parahydroxy benzoic acid, and their esters and salts may be added to foods in a maximum concentration of 0.1% by weight. Benzoates are allowed in foods in Canada, European and Scandinavian countries, Japan, and Turkey.

COOH

Benzoic acid, , and its sodium salt; parahydroxybenzoic acid and its methyl, propyl, or butyl esters are used in a number of food products, where, along with other treatments, such as heating or refrigerating, they become a factor in the microbial stabilization of the food. Generally, these compounds are not suitable for stabilizing foods of high moisture content unless the product is held at low temperatures. These compounds, especially the benzoates, are much more effective in foods when the pH is 4.0 or below, since it is the undissociated part of the molecule which is effective. Benzoic acid is not very soluble in water, hence the sodium salt is most often used.

COOH

Parabenzoic acid, , and its esters are more soluble than benzoic

OH

acid and the sodium salts of the parabenzoates are even more soluble. (sodium propyl parabenzoate)

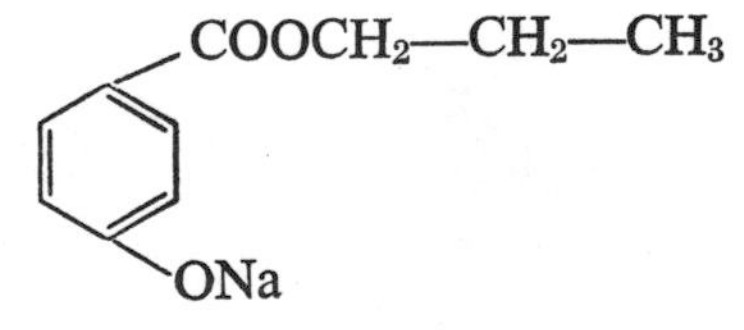

Whereas the parabenzoates are more inhibitive to microorganisms at lower pH values, these compounds are much more effective over a wider range of pH and inhibit a broader spectrum of microorganisms than do the benzoates (Anonymous, 1962).[1] With the exception of France and Turkey, the parabenzoates are used in the countries indicated allowing benzoates in foods.

In the United States and Europe, benzoates and the paraben compounds have been used in fruit juices; fruit and other types of syrup, especially chocolate syrup; candied fruit peel; fruit and other pie fillings; fruit juice concentrates; pickled vegetables; relishes of certain types; and horseradish. Concentrations of 0.1% of the parabenzoates are allowed in foods. It has been suggested that the paraben compounds might be used in such foods as frankfurters and in salad dressings with a cheese base.

Recent investigations (Bosund, 1960)[5] indicate that the benzoates and related compounds inhibit the growth of microorganisms by, in some way, interfering with the utilization of acetate required for the formation of energy-rich compounds which are necessary for cell metabolism.

In another investigation concerned with the inhibitory effect of sodium benzoate on spore-forming bacteria (Gould, 1964)[12] it was found that, if the phases of growth of bacterial spores are considered to be (1) germination (loss of refractility under the phase microscope, increase in heat sensitivity, etc.), (2) rupture and shedding of the spore wall, (3) enlargement and outgrowth into the vegetative cell, and (4) cell division of multiplication, benzoate allowed stages (1) and (2) to take place but prevented stages (3) and (4) or prevented elongation of the vegetative cell and multiplication.

Oxidizing Agents

Hydrogen peroxide, compounds such as sodium or calcium hypochlorite and chloramine T (sodium paratoluene sulfonchloramide)

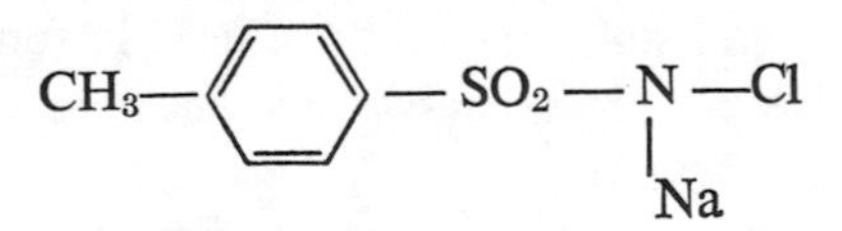

liberating free chlorine, chlorine itself, and compounds such as the iodophors liberating iodine, must often come in contact with foods. The iodophors are chemical complexes of diatomic iodine and synthetic nonionic surfactants. Some of the iodine in such complexes is "available" or titratable and acts in the same manner as elemental iodine in alcohol solution. Hydrogen peroxide may be added to some foods provided that, after addition and allowing time for the compound to act, the residual hydrogen peroxide is eliminated by the further addition of catalase. The addition of hydrogen peroxide followed by catalase is permitted in some cases for milk used for the manufacture of certain dairy products.

Chlorine in the form of hypochlorite or as gaseous chlorine, is added to foods directly; for instance, as when fish fillets are passed through a bath of chlorinated brine. Also, since chlorine is often used as a residual (usually about 5 ppm) in water added to foods, it in many cases becomes a minor component of foods.

Both chlorine and iodine compounds are used for sanitizing equipment, and it is probable that very small amounts of these compounds sometimes get into foods from residues on equipment.

Hydrogen peroxide can be used to reduce the number of viable bacteria in liquid foods such as milk. Chlorine, added to a food as such, would not be an effective bacterial inhibitor or destroy appreciable numbers of bacteria since it reacts with food components, especially proteins. Some work (Ranken et al., 1965),[18] however, indicates that chicken chilled in slush-ice containing 200 ppm of chlorine have an extended refrigerated shelf life as compared to controls. In the past chlorine has generally been used to maintain low total counts in brines and other solutions with which the food was to be treated.

The U.S. Food and Drug Administration has not specified the amounts of any of the oxidizing compounds which may be present in foods since there would be no method of detecting these compounds.

Oxidizing agents are believed to destroy microorganisms by oxidizing sulfhydryl groups in the protein portions of enzymes essential to the metabolic processes of these organisms.

Miscellaneous Compounds

Sulfur Dioxide

Sulfur dioxide or salts of sulfurous acid are used in foods to prevent browning, but they are also used in numerous instances to prevent the growth of undesirable yeasts or bacteria.

One method of preserving fruit juices or fruits, to be used later for the manufacture of juices, is to add sugar together with 350 – 600 ppm of sulfur dioxide or an equivalent amount of the sulfurous acid salt. Sodium benzoate is sometimes added together with the sulfur dioxide, and cool storage is generally combined with sulfur dioxide treatment as a method of preservation.

In wine making and in the manufacture of vinegar sulfur dioxide is used to sanitize equipment and containers for purposes of preventing the growth of molds and undesirable types of yeast in the product. A concentration of 50 – 75 ppm of sulfur dioxide may be added to wine held in bulk storage for the manufacture of vinegar or may be added to the must prior to allowing it to undergo fermentation. Sulfur dioxide suppresses the growth of wild yeasts in the latter application but permits the growth of desirable types of these organisms.

The inhibiting effect of sulfur dioxide may be due to its reaction with the carbonyls of carbohydrates, thus preventing their utilization as a source of energy, or it may be due to the reduction of S-S linkages in enzyme proteins, thus preventing the action of enzymes essential to cell metabolism (Josyln and Braverman, 1954).[13]

It has been shown (Rehm, 1964)[19] that nicotineamide adenine dinucleotide (NAD)-dependent factors concerned with the metabolism of certain carbohydrates by some yeasts and bacteria are inhibited by sulfurous acid, probably due to the formation of addition compounds between the two components. The antimicrobial action of sulfurous acid has been found to be decreased by the formation of sulfonates with carbonyls in foods (Rehm, 1964).[19] However, carbonyl addition compounds formed with acetaldehyde, pyruvate, ketoglutarate, and acetone had an inhibiting effect on yeasts.

In the United States of America, no specification has been made as to how much sulfur dioxide may be used in certain foods; but it may not be used in fresh meats, since it fixes the color of red meat preventing metmyoglobin formation, hence may mask spoilage. With the exception of England, Finland, and Japan, sulfur dioxide is allowed in foods in the countries listed as allowing benzoates.

Sorbic Acid

Sorbic acid, CH_3—CH=CH—CH=CH—COOH, acts to inhibit the growth of molds and yeasts. It has also been shown that sorbic acid and its potassium salt may inhibit the growth of catalase-positive bacteria. It is not effective in preventing the growth of catalase-negative bacteria (Emard and Vaughn, 1952),[10] and thus, the *Lactobacteriuceae* and the *Clostridia* would not be affected. The sorbates have the greatest microbial inhibiting effect in foods at pH values below 5.0.

Sorbic acid and its salts are considered to be comparatively safe as food additives and in the United States no limits as to the amount which may be added have been specified, excepting for those foods for which a standard of identity has been established. When the standards of identity for foods are set up by the Food and Drug Administration, the components of the foods, additives or otherwise, which may be present, are specified. Sorbates are allowed in foods in all countries listed as allowing benzoates.

Sorbic acid or its salts may be used in a number of types of cheeses or in materials used for wrapping certain cheeses. These compounds may also be used in pickled cucumbers or other pickled vegetables, or in relishes. The use of sorbic acid in fresh fruit salads, in fruit juices, in soft drinks, and in cooked sausage products has also been suggested. In Europe sorbates are used to stabilize such foods as tuna fish salad, which are prepared and held at refrigeration temperature, for purposes of distribution.

The solubility of sorbic acid in water at 20°C (68°F) is 0.15% by weight, but the potassium salt is quite soluble. However, when approved for specific foods a maximum of about 0.2% has been allowed.

Sorbic acid has been allowed on the basis that these materials are metabolized by the human in the same manner as are fatty acids, the end products being carbon dioxide and water (Melnick et al., 1954).[16]

Sorbic acid is said to inhibit molds by suppressing the activity of dehydrogenase enzymes (Melnick et al., 1954).[16] Clostridia have an abundance of dehydrogenase enzymes, and this may be the reason why these compounds are not effective in inhibiting this type of organism. It is also stated (Melnick et al., 1954)[16] that sorbic acid is not effective when added to materials containing high concentrations of microorganisms.

In microorganisms on which sorbic acid has an inhibitory effect, this compound has been found to allow germination, shedding of the spore wall and some elongation of the vegetative cell but to prevent division of the vegetative cell (Gould, 1964).[12]

Diethyl Pyrocarbonate

Diethyl pyrocarbonate, $C_2H_5{-}O{-}\overset{\overset{\displaystyle O}{\|}}{C}{-}O{-}\overset{\overset{\displaystyle O}{\|}}{C}{-}O{-}C_2H_5$, has been used, especially in Europe, for stabilizing some fruit juices and carbonated beverages Its use in foods has also been approved by the U.S. Food and Drug Administration. This compound is effective only in materials in which the microorganism concentration is relatively low. In the United States pyrocarbonate may be added to wine or to soft drinks in a concentration of 200 ppm.

It has been proposed (Genth, 1964)[11] that diethylpyrocarbonic acid first hydrolyzes in water to form ethyl carbonate* and that this product, which quickly decomposes to ethyl alcohol and carbon dioxide, is the active form of the compound, as far as the destruction of microorganisms is concerned. It has been further postulated that the ethyl carbonate reacts with intracellular components of microorganisms or with the amine groups on essential proteins.

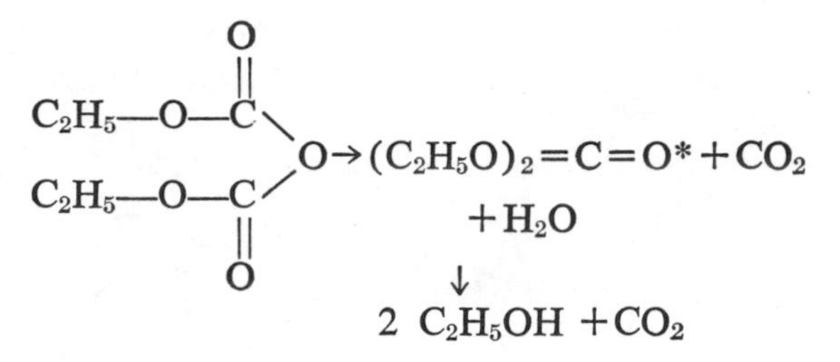

When added to beverages diethyl pyrocarbonate will destroy limited numbers of microorganisms, especially of yeasts. This has been shown to be the case (Anonymous, 1968).[1] It would be expected that the effect of this compound is destructive rather than inhibitive since, when added to beverages, it soon decomposes to water and alcohol, hence would not be present as the pyrocarbonate to act as a growth inhibitor for long periods. Considering the concentration of pyrocarbonate which may be used, the amount of ethyl alcohol formed by its decomposition would have no inhibitive effect on the growth of microorganisms.

Sodium Nitrite

In the United States of America sodium nitrite, $NaNO_2$ (200 ppm), is added to such foods as cured meats for purposes of fixing color. In Canada, it was at one time added to fish fillets in concentrations up to 200 ppm to extend the storage life of these products at refrigerator temperatures above freezing. It has been shown that nitrite is suitable for inhibiting bacteria only in products at pH values below 6.4, and the mechanism of this inhibition is unknown. However, at higher pH values, as are found in fish fillets in which spoilage may be due to the formation of trimethylamine by bacterial reduction of trimethylamine oxide, nitrite will also extend the refrigerated storage life. Although bacterial growth occurs, nitrite is preferentially reduced instead of trimethylamine oxide (Castell, 1949).[6] Hence, while the ordinary spoilage bacteria continue to grow, they do not reduce trimethylamine oxide as long as nitrite is present. Therefore, the bacterial population will reach a much larger number before sufficient end products are formed to cause spoilage odors (Tarr and Sunderland, 1940).[22] Since it takes more time to reach these higher population densities, the storage life at refrigerator temperatures above freezing is extended. Presently, nitrite cannot be added to fish in Canada, and it has been shown (Sen et al., 1970)[21] that nitrosamines (considered to be carcinogenic) are formed when salt water fish are heated in the presence of nitrites.

Nitrite is not allowed to be used in fresh fish in the United States. It may be added to some cured or canned fish products. The reason for adding nitrite to cured fish is to prevent the growth of certain pathogenic bacteria. In canned fish nitrite is added to promote color retention. Nitrite may be added to cured meats in Canada, in Turkey, and in most European countries, including Germany and Italy, as well as in Denmark, Norway, and Sweden.

Hexamethylene Tetramine (Hexamine)

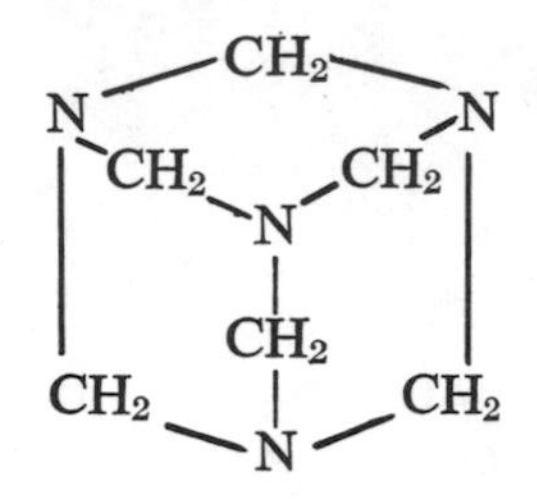

This compound is used to destroy molds which may be present on the peel of citrus fruits. In the United States no amount of this compound is allowed on or in foods. However, citrus fruits may be washed in a solution containing soap and hexamine with or without other antimicrobial agents. Since none of this compound is permitted on or in the food, citrus fruits must be rinsed to remove all residues after treatment of this kind. In some European countries hexamine is allowed in foods in a maximum concentration of 750 ppm. This compound may be used either alone or with sodium benzoate to stabilize pickled fish products which are to be held without refrigeration in some countries.

No explanation of the mechanism by which hexamine destroys microorganisms has been proposed.

Sodium Chlorophenate

Cl

—ONa

This compound, like hexamine, is not allowed to be present in foods in the United States. It is used to wash citrus fruits to prevent rotting. A solution of 2% sodium chlorophenate, 1% hexamine, and 0.5% of soap is used in the wash water for fruit after which thorough rinsing is applied to eliminate residues.

2-4-D and 2-4-5-T

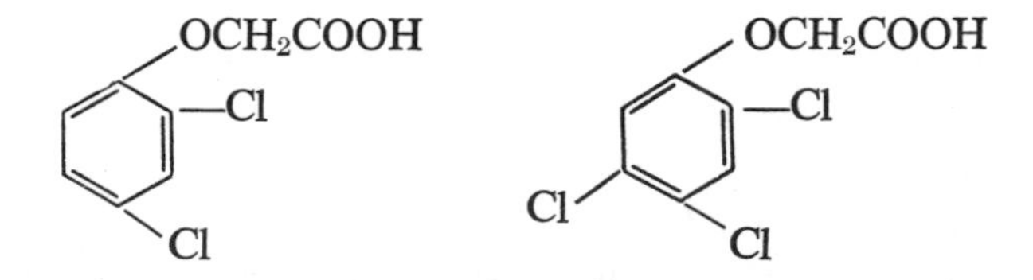

Both 2-4-D, dichlorophenoxy acetic acid, and 2-4-5-T, trichlorophenoxyacetic acid, have been used in wash water for fruits. Dichlorophenoxy acetic acid may also be added to the wax applied to citrus fruits before polishing. The compound 2-4-5-T is not permitted to be used in food processes in the

United States but 2-4-D may be present in a maximum concentration of 5 ppm. Both 2-4-D and 2-4-5-T are used in wash waters for citrus fruits, apples, peas, and quinces. Antimicrobial agents of this type must be removed by subsequent rinsing.

Diphenyl

This compound, , has been used mainly as a constituent of the oiled paper in which citrus fruits may be wrapped, hence may become a constituent of the peel. It is especially a mold inhibitor and in the United States may be present on citron, grapefruit, kumquat, orange, tangerine, tangelo, and other citrus fruit peel in concentrations of 100 ppm.

Boric Acid

Boric acid, H_3BO_3, and its salts are not allowed in or on foods in the United States, but citrus fruits may be washed in an 8% solution of borate then rinsed, prior to waxing and polishing. Also in Norway, France, and the Netherlands this compound may be used on the surface of salt-dried fish as a preservative, or in bread, cake, cheese, butter, and rennet.

Antibiotics and Related Compounds

A number of compounds produced by microorganisms, which are inhibitive to other types of microorganisms, have potential use for food preservation applications. Actually, some of these compounds have formerly been approved by the United States Food and Drug Administration for use in foods, but this approval was withdrawn in 1966.

The Tetracyclines

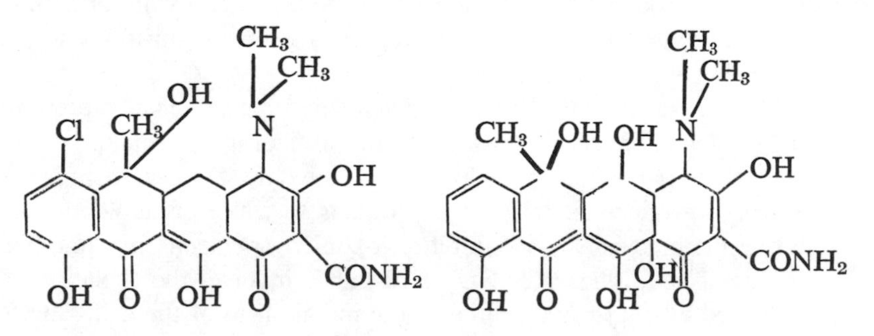

Chlortetracycline and oxytetracycline are wide-spectrum antibiotics which were allowed in fresh poultry, scallops, shrimp, and eviscerated but not filleted fish, in concentrations of 5 – 7 ppm. Because of the fact that small residues, about 0.5 ppm, may be left after cooking and the effect which such residues might have on the intestinal flora of the human, permission to use such compounds in foods was withdrawn.

The tetracyclines are still used in Argentina, Canada, England, and Japan for the treatment of fresh meat, fish, or poultry where it has been found that the storage life of the product at refrigerator temperatures above freezing may be extended for several days, enabling the shipment and distribution of fresh foods to inland points (Anonymous, 1968).[1]

At one time there was some controversy as to the efficacy of using oxytetracycline as compared to chlortetracycline. It was later found that in alkaline waters oxytetracycline was not stable; but if tartaric acid was added to the water used for application of the antibiotic, the results with either tetracycline were similar.

It has been shown (Davis et al., 1967)[8] that in treating fish with the tetracyclines there may be a reduction of the bacterial flora; or the mean generation time of the flora might be lengthened for a time, this being restored to normal after a few days of storage. In some cases the bacterial flora eventually causing spoilage might be the normal types, nonpigmented pseudomonads; in other instances it was found to consist of a much greater proportion of pigmented pseudomonads; and in still other instances *Achromobacter* constituted an important segment of the spoilage flora.

It is evident that there are species or strains of both *Pseudomonas* and *Achromobacter* which are resistant to the tetracyclines and that rigorous sanitation and application of suitably low refrigerator temperatures above freezing must be used together with these antibiotics, to extend the storage life of fresh foods. Under less satisfactory conditions a resistant flora will be established in the processing plant, and the resistant bacteria will not be inhibited by these compounds. The use of the tetracyclines is no substitute for the application of suitably low temperatures.

In those cases in which *Achromobacter* become the chief representatives of the spoilage flora of foods, an extension of storage life might be expected since, in comparison to pseudomonads, they must grow to much higher numbers to produce sufficient end products to cause organoleptic changes.

While the mechanism of inhibition of microorganisms has not been exactly established, it has been suggested that it involves the chelation of metals required by enzymes essential to the metabolism of the cell. The formation of metal chelate compounds such as the following are known:

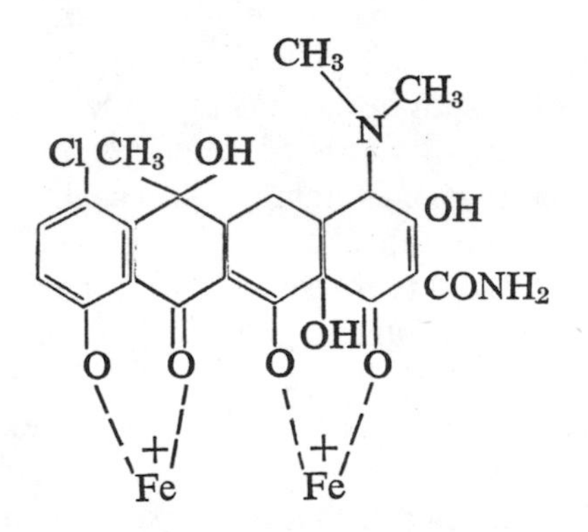

It is also known that certain species of *pseudomonas* require small concentrations of iron; and it is entirely probable that other metals which might be required for enzyme action, essential to bacterial growth, would be chelated by the tetracyclines.

Another explanation (Davis et al., 1967)[8] of the antimicrobial action of the tetracyclines is that they interfere with the binding of charged tRNA by ribosomes, hence, interfere with protein synthesis.

Sodium α-Acyldipeptides

The sodium salt of a fatty acid combined with the amide, methyl, or ethyl ester of a lysine dipeptide has been shown to have microbial inhibiting properties. Na palmitoyl-L lysyl-L lysine-ethylester, (R-1):

$$\begin{array}{l} CH_3—(CH_2)_{14}—CO—NH—\underset{\underset{\underset{NH_2}{|}}{\underset{(CH_2)_4}{|}}}{CH}—NH—\underset{\underset{\underset{NH_2}{|}}{\underset{(CH_2)_4}{|}}}{CH}—COOCH_2—CH_3 \end{array}$$

has been found to have the most suitable properties as a bacterial inhibitor (Molin, 1964).[17] Since these compounds in foods may be eventually broken down into their constituents by proteolytic enzymes, no problem with wholesomeness is anticipated, although in the United Stated the use of sodium α acyldipeptides has not been approved.

The compound R-1 was found to have a strong antimicrobial effect on gram-negative and gram-positive bacteria, certain yeasts and certain molds. *Clostridium botulinum* type E was found not to be inhibited by this compound. Both the long-chain fatty acid and the dipeptide (or longer-chain amino acid residue) were found to be essential to the inhibitory properties of this type of compound.

D-amino acids or L-amino acids in the acyldipeptides were equally effective and optimal activity occurred at a pH of about 8.

On repeated transfers to media containing R-1, sensitive bacteria became less sensitive and even resistant to this compound but recovered their original sensitivity on repeated transfers to media not containing R-1. Ferrous and ferric compounds tended to neutralize the inhibitive effect of R-1.

The ordinary spoilage organisms, *Pseudomonas, Achromabacter, Lactobacillus,* and *Enterococci* have been found to be sensitive to R-1, as well as certain bacilli, the indicator organism *Escherichia coli,* certain yeasts, and molds. It can, therefore, be concluded that this type of compound might be of use in extending the storage life of flesh-type foods held under refrigeration temperatures.

It was shown that the effect of R-1 on bacteria was an exponential killing of the organisms during that phase of destruction in which 90 – 99.9% of the organisms are destroyed. There was, however, in any bacterial population a certain fraction of resistant cells. It is known that when R-1 is added to media in which bacteria are present, R-1 accumulates in the cell and interferes with the uptake of other amino acids. This may be, at least in part, the mechanism by which microorganisms are inhibited or destroyed by this compound. It has also been suggested that the surface activity of such compounds may cause cell wall destruction or cause the cells to undergo leakage or a loss of essential compounds to the surrounding medium.

Nisin and Tylosin

Nisin and tylosin are polypeptide antibiotics (formulae not known). Nisin is produced by a special strain of *Streptococcus lactis.* Tylosin is produced by *Streptomycetes fridiae.* Nisin is effective against thermophiles such as *Bacillus sterothermophilus* when present in concentrations of at least 2 ppm but is ineffective in inhibiting the growth of the *Lactobacteriacea.* It is said that heat-damaged cells are more susceptible to the inhibitive effect of nisin than are unheated cells, although this has been disputed (Tramer, 1964).[23] It has been proposed, therefore, that nisin be added to canned foods to be given a heat process sufficient to reduce the number of *Clostridium botulinum* type A spores from 60×10^9 to 0.1. Under these conditions the danger of botulism would be ruled out, and spoilage due to the germination and growth of the more heat-resistant thermophilic spores would be prevented. Since it has been shown that about 75% of nisin is destroyed during pressure cooking, some allowance for this would have to be made by adding greater quantities than that which should be effective.

It has also been suggested that tylosin be used for canned food products since, even in concentrations of 2.5, it appears to prevent the outgrowth and toxin production by *Clostridium botulinum,* even after very mild heat treatment (Malm and Greenberg, 1964).[14] The difficulty is, however, that some spoilage organisms are able to grow in the presence of this compound. It would appear, therefore, that the best application of tylosin might be in canned foods which are given a mild heat treatment, then held under refrigeration or to irradiated foods which are handled in the same manner.

Neither nisin nor tylosin may be added to foods in the United States of America. In Australia, Belgium, Czechoslovakia, England, France, Germany, Israel, Italy, Mexico, and Sweden the addition of nisin to some foods is allowed.

Regarding the mechanism by which nisin and tylosin prevent bacterial growth, the general observation has been that they prevent outgrowth of spores. It has been found that nisin allows germination but prevents the shedding of the spore wall while tylosin allows germination, shedding of the spore wall, and some enlargement of the cell to take place, but prevents division of the vegetative cell.

References

1. Anonymous, "Handbook of Food Additives," The Chemical Rubber Co., Ohio, 1968.
2. Anonymous, Parabens inhibit food spoilage molds, yeasts and bacteria at any pH, *Food Process.* **23** (1962), 64.
3. Anonymous, "The Merck Index," Merck, Rahway, N.J., 1968.
4. A. C. Baird-Parker and F. Freame, Combined effect of water activity, pH and temperature on the growth of *Clostridium botulinum* from spore and vegetative cell inocula, *J. Appl. Bacteriol.* **30** (1967), 420.
5. J. Bosund, The bacteriostatic action of benzoic and salicylic acids, *Acta Chem. Scand.* **14** (1960), 111.
6. C. H. Castell, Nitrite-reducing bacteria on cod fillets, *J. Fish. Res. Bd. Canada* **7** (1949), 528.
7. J. H. B. Christian and J. A. Waltho, The water relations of staphylococci and micrococci. *J. Appl. Bacteriol.* **25** (1962), 369.
8. B. D. Davis, R. Dulbecco, H. N. Eisen, H. S. Ginsberg, and W. B. Wood, "Microbiology," Harper, New York, 1967.
9. N. N. DeSilva, Changes in the spoilage flora during preservation of fish with antibiotics, "Microbial Inhibitors in Food," pp. 245–259. Almquist and Wiksell, Stockholm, Sweden, 1964.
10. L. O. Emard and R. H. Vaughn, Selectivity of sorbic acid media for the catalase negative lactic acid bacteria and clostridia, *J. Bacteriol.* **63** (1952), 487.

11. H. Genth, On the action of diethylpyrocarbonate on microorganisms, "Microbial Inhibitors in Food," pp. 77–85. Almquist and Wiksell, Stockholm, Sweden, 1964.
12. G. W. Gould, Effect of food preservatives on the growth of bacteria from spores, "Microbial Inhibitors in Food," pp. 17–24. Almquist and Wiksell, Stockholm, Sweden.
13. M. A. Josyln and J. B. S. Braverman, The chemistry and technology of the pretreatment of fruit and vegetable products with sulfur dioxide and sulfites, *Advan. Food Res.* **5** (1954), 97.
14. B. Malm and R. A. Greenberg, The effect of tylosin on spoilage patterns of inoculated and noninoculated cream style corn, "Microbial Inhibitors in Food," pp. 86–95. Almquist and Wiksell, Stockholm, Sweden, 1964.
15. S. T. Martins, A study of staphylococcal enterotoxins, Ph.D. Thesis, Massachusetts Institute of Technology, 1964.
16. D. Melnick, F. H. Luckmann, and C. M. Gooding, Sorbic acid as a fungistatic agent for foods. VI. Metabolic degradation of sorbic acid in cheese by molds and the mechanism of mold inhibition, *Food Res.* **19** (1954), 44.
17. N. Molin, Antimicrobial effects of N_2–acyldipeptides, II. The influence of molecular configuration on antimicrobial activity, *Physiol. Plantarum* **17** (1964), 456.
18. M. D. Ranken, G. Clewlow, D. H. Shrimpton, and B. J. H. Stevens, Chlorination in poultry processing, *Brit. Poultry Sci.* **6** (1965), 331.
19. H. J. Rehm, The antimicrobial action of sulfurous acid, "Microbial Inhibitors in Food," pp. 105–115. Almquist and Wiksell, Stockholm, Sweden, 1964.
20. W. D. Scott, Water relations of food spoilage microorganisms, *Advan. Food Res.* **7** (1957), 83.
21. N. P. Sen, D. C. Smith, L. Schwinghamer, and B. Howsam, Formation of nitrosamines in nitrate-treated fish, *Can. Inst. Food Technol. J.* **3** (1970), 66.
22. H. L. A. Tarr and P. A. Sunderland, Conditions under which nitrites inhibit bacterial spoilage of fish. Progr. report **44**, 16, Pacific Fisheries Exp. Stat., *Fish. Res.* Bd., Canada, 1940.
23. J. Tramer, The Inhibitory action of nisin on *Bacillus stearothermophilus,* "Microbial Inhibitors in Food," pp. 25–33. Almquist and Wiksell, Stockholm, Sweden, 1964.

CHAPTER 7

Microbiology of Flesh-Type Foods and Eggs

Spoilage of food products of animal origin is usually caused by the growth of microorganisms, resulting in changes in appearance, odor, and taste. Chemical changes may also occur involving discoloration; for instance, oxidation of the myoglobin in meat giving rise to darkening or a change from a bright-red color to brown or gray. Other chemical changes may cause the development of unacceptable odors and flavors in the product, as, for instance, the rancidification of fats, due to oxidation of unsaturated fatty acids. Enzyme action, chiefly changes due to proteolysis, constitutes another type of chemical change which may be involved in spoilage of flesh-type foods. Ordinarily enzyme changes causing spoilage are confined to fish held uneviscerated. Thus, uneviscerated mackerel held under refrigeration may undergo a sufficient degree of proteolysis (autolysis) to cause the tissues of the abdominal cavity to be completely liquefied in certain areas, resulting in exposure of the entrails in parts of the abdominal (belly) region.

While chemical changes are important factors in the deterioration of foods, methods of controlling such changes, including fast handling during processing and distribution, have mostly eliminated spoilage due to oxidation of myoglobin or rancidification and autolytic changes are chiefly confined to a few species of fish under conditions in which fast distribution is not possible. The chief cause of spoilage in fresh meats, poultry, fish, and eggs, as well as in some processed products manufactured from these foods, is the growth of microorganisms. Generally, the organisms which spoil flesh-type foods and eggs are bacteria; but there are instances in which molds or yeasts are responsible for the spoilage of these products.

Fresh Meats

Source of Contamination with Spoilage Bacteria

The bacteria which cause spoilage of fresh meats held too long at refrigeration temperatures or for shorter periods at higher temperatures are soil and water types. One may ask the question, "How does the animal carcass, or portion of the carcass, become contaminated with spoilage bacteria?."

The main source of the spoilage bacteria with which fresh meat becomes contaminated is the hide of cattle or the skin of hogs. It has been shown (Jensen and Hess, 1941)[21] that cattle hides and hog skin, in the area where the stick knife is inserted, may contain millions of bacteria (anaerobes and aerobes) per square centimeter of surface. In cattle these bacteria are transferred to the surface of the meat by direct contact of the outer hide with the meat, in case of careless skinning methods; by the skinning knife used to cut through the hide before skinning can begin and through contact of the workers clothing and hands, first with the hide and then with the skinned carcass. It has been found that under the conditions of humidity and temperature encountered on the slaughtering and dressing floor, bacteria on the clothing and hands of workers may grow and increase in numbers.

Implements, such as saws, used for the dressing of cattle may become a source of contamination. The contents of the bung (colon) and of the rumen may contaminate meat surfaces through expurgation or regurgitation if not tied off. Also, the sawdust used in refrigerated rooms may become a source of airborne contamination of the carcass. In most cases, however, the hide is the chief source of spoilage bacteria on beef carcasses (Empey and Scott, 1939).[15]

During those periods in which long-time curing methods were used in the United States of America, the stick knife and the scald water were the chief source of spoilage bacteria in pork products. In passing through the skin of the hog, the knife picks up bacteria which are transferred to the blood and through the blood into the muscle tissues and even into the marrow of the bone. The heart of the hog sometimes continues to beat even after the carcass enters the scalding water, and this tends to pump some of the water through the carcass. The temperature of the scald water is high enough to destroy some of the spoilage bacterial types, but due to the continuous entry of unscalded hogs into the bath, even vegetative types may be present in the bath and may enter the tissues of the carcass. In pork products fast cures have eliminated most of the spoilage due to bacteria present in the tissues, but satisfactory pasteurization of refrigerated ham products and commercial sterilization of such products as small hams and luncheon meat depends upon low levels of bacterial contamination of the tissues.

Surface contamination of hogs may be of somewhat less importance than in cattle since retail cuts, if not distributed in short periods of time, are ordinarily frozen and defrosted for retail sales. When spoilage of fresh cuts does occur, the source of the spoilage bacteria is probably most often the brushes of the dehairing machine, the hands and clothing of workers, and saws and knives used in the dressing operations.

Organoleptic Indications of Fresh Meat Spoilage

When there is sufficient growth of bacteria on the surface of meat carcasses or on cut surfaces, taint or malodor and slime formation will occur. In beef carcasses the effect of bacterial growth is first evident in the areas of the deep fissures of the neck muscle where the head has been severed and in the folds of flesh under the forelegs, since these are the areas of the carcass where sufficient moisture for good bacterial growth is likely to be present.

The actual compounds produced by bacteria and causing off odors and flavors are not known but, in some instances, the formation of hydrogen sulfide may be involved. Spoilage odors have been described as "tainted," "sour," "putrid," and "stale." At the time of slime formation on meat the bacterial count is between 50–100 $\times$ 10^6 cells per square centimeter (Ayres, 1955).[3]

Cut meat generally has a shorter storage life than sides or quarters, and ground meat has a shorter storage life than cut meat when these products are held at refrigerator temperatures above freezing. The reason for this is that the surfaces of sides or quarters are relatively dry and tend to be covered with fat while cut surfaces are moist, providing a better growth media for bacteria. Ground meat is an especially good growth medium because of the extensive surface area provided by the grinding and because of the fact that, whereas on the uncut meat the bacteria would be present almost entirely on the outer surfaces, during grinding these organisms are distributed throughout the product.

The Bacteria Causing Spoilage of Fresh Meat

Under the ordinary conditions in which fresh meat is held, the pseudomonads are the organisms which eventually cause spoilage as indicated by off-odor or off-flavor development. There may be instances in which the pigmented pseudomonads would be chiefly involved in spoilage, but such a situation would be unusual. It is the nonpigmented pseudomonads which are ordinarily responsible for the spoilage of fresh meat.

In meat products spoilage may be caused, in some cases, by *Achromabacter* species, especially if the meat has been subjected to some type of processing.

Under some conditions which are used in the handling of fresh meat, such as might be brought about by the packaging of cut or ground meat under vacuum (Thornley, 1967),[34] the *Lactobacteriaceae* species may be the eventual cause of spoilage. These organisms are considered not to be included in the normal flora of fresh meat but enter the product through contamination from plant equipment or from the workers who handle such products.

When processing of some types is employed, a short-chained coccal type, *Microbacterium thermosphactum,* may cause spoilage. This organism is a facultative anaerobe as are the *Lactobacteriaceae* and was, at one time, included with this group; but, in contrast to members of the family *Lactobacteriaceae* which are catalase-negative, *Microbacterium thermosphactum* is weakly catalase-positive.

Under conditions in which sides or quarters of beef are held for comparatively long periods of time at temperatures near 0°C (32°F) for purposes of aging, the surface may become covered with molds. Generally these organisms are species of *Cladosporium, Thamnidium,* or *Mucor.* The purpose of this aging process is to promote desirable tenderization of the meat and the production of special "aged" flavors. No investigation seems to have been carried out to determine that molding, per se, produces a more tender or more flavorful meat. Molding, of course, may cause an increased trim loss when the product is prepared for cooking.

Factors Affecting the Spoilage of Fresh Meat

Under ordinary circumstances the rate of bacterial growth will determine the period after which fresh meat will have become spoiled. The factors governing the rate of bacterial growth are: the pH of the meat, the moisture conditions on the meat surfaces, and the temperature of the meat during holding.

The pH of beef varies between 5.1 and 6.2; that of pork between 5.3 and 6.9; and that of lamb between 5.4 and 6.7. The pseudomonads, the ordinary meat spoilage bacteria, grow best at pH values near 7.0 or slightly on the alkaline side. When a pH range on the acid side is reached, for instance at pH 6.0, even a slight decrease in pH tends to reduce the growth rate of the pseudomonads. From the standpoint of spoilage, therefore, in preparing animals for slaughter it is important that they not be excited or fatigued, since such a condition will bring about a depletion of glycogen in the muscle. This will mean that during and after rigor (through anaerobic glycolysis) (Wismer-Pedersen, 1960),[39] only small amounts of lactic acid will be produced, resulting in a meat of comparatively high pH which is especially suitable for the growth of spoilage bacteria. It is even possible that high glycogen content provides a reserve source of lactic acid, so that when, through alkaline products produced by bacterial metabolism, there is a trend for the pH to shift toward higher levels, this may be neutralized by acid in the muscle.

As has been previously pointed out, bacteria will not grow without moisture; and in flesh-type foods the higher the moisture content, the faster the growth of bacteria. For this reason the surface of cut or ground meat always provides a better medium for bacterial growth than do sides or quarters. In holding quarters or sides, therefore, the objective is to keep the surfaces relatively dry. This is usually possible for such portions. Meat spoilage due to the growth of bacteria is generally a surface phenomenon. The spoilage bacteria are strictly aerobic or grow much more slowly under conditions of low oxygen tension, hence there is little if any invasion of the deep tissues excepting in the fissures of the neck muscle where the head was severed.

The bacteria causing spoilage of meat are psychrophiles or psychrotrophs. Psychrophilic bacteria may grow at temperatures well below 0°C (32°F) if formation of ice does not occur. The psychrotropic bacteria will not grow at the lower temperatures but will grow at some temperatures in the refrigeration range. For both types of bacteria growing at temperatures below 10°C (50°F), the lower the temperature, the longer the lag phase before growth begins, and the slower the growth rate once the exponential growth phase has been reached.

The Control of Bacterial Spoilage of Fresh Meat

Considering the factors favoring the growth of meat spoilage bacteria, these products should be handled in such a manner as to provide a low normal pH, dry surfaces and comparatively low temperature in the product.

All attempts should be made to dress and handle carcasses in such a manner as to have as little contamination with bacteria as possible since, as indicated in Chapter 3, while the size of the initial contamination will neither affect the duration of the lag phase nor the exponential growth rate, it will affect the spoilage time, since the larger the number of bacteria present at the start the shorter the time required to reach a particular final population.

It has been shown that if cattle are passed through a chlorinated water bath, shortly prior to slaughter, many of the bacteria present in the hide may be destroyed; and this results in eventual lessened contamination of the carcass (Empey and Scott, 1939).[15] Such procedures are not ordinarily used in the United States.

The moisture content of surfaces must be kept as low as is practicable to limit bacterial growth on sides or quarters of meat. This is done by maintaining suitable relative humidities and air speeds over the meat held in coolers and especially in storage rooms.

Generally in the United States of America relative humidities between 90 and 95% are maintained in storages. A major problem in meat storage is maintaining a high enough relative humidity to prevent excessive weight loss through drying. In order to dry surfaces and to maintain a suitable dry condition in rooms 115,000–120,000 cu ft in size, an air circulation of about 135,000 cu ft per minute may be used in coolers and 40,000 cu ft per minute in storage rooms.

Little can be done to regulate the moisture content of cut portions of

FIG. 7–1. a. Boning operation in meat plant. b. Boning operation in meat plant. (Courtesy of Oscar Mayer & Co.)

FIG. 7-1. Continued

meat since moisture from the inside will diffuse to the cut surface. However, in areas of the slaughterhouse where meat is cut, the ambient or room temperature is generally higher than that of the meat itself, freshly removed from the cooler or storage room. This situation generally exists since better efficiency of workmanship is attained at the higher temperature of cutting rooms (about 10°C (50°F)). The relative humidity of cutting rooms should, therefore, be kept low enough to prevent condensation of moisture from the air on meat surfaces (sweating), since such condensed moisture will provide an excellent growth medium for bacteria. Meat should be held at such higher temperatures for cutting for only short periods of time.

The temperatures of rooms used for the holding of fresh beef should be maintained as low as is possible without freezing the meat. The freezing point of beef is between −2.2 and −1.1°C (28 and 30°F). However, if an attempt is made to maintain this range of temperatures some alternate freezing and thawing may occur which would cause a deterioration of quality, especially from the standpoint of excessive drip. Considering the variation in temperatures which occurs in the ordinary commercial refrigeration system, it is probably better to maintain storage and handling temperatures between −1.1 and +1.7°C (30 and 35°F).

Pork, due to the higher fat content of the tissues, is less affected by lower temperatures near 0°C (32°F) than is beef. Actually pork portions, such as loins and shoulders, may be frozen and defrosted for retail sales (see Figs. 7-1a and 7-1b). It is recommended that pork which is not to be shipped the same day that it is cut should be held at −1.1–0°C (30 to 32°F) if it is not to be frozen.

Proper low temperatures when used in cooler and storage rooms and, when applied during transportation and retail display, may entirely prevent the growth of psychrotrophic bacteria and will greatly retard the growth of the psychrophilic spoilage bacteria.

In some coolers where beef is held as sides or quarters, special ultraviolet ray lamps may be used which may enable the holding of these products for 3 days or longer at temperatures as high as 15.6°C (60°F) for special tenderization (American Meat Institute Foundation, 1960).[2] Ultraviolet rays would destroy microorganisms only on those surfaces contacted by the rays. They do not penetrate and, hence, would not affect such areas as fissures in the neck muscle or the folds of tissues under the forelegs. However, some ozone is produced by the ultraviolet lamps used; and this would have an added bactericidal effect, possibly even in areas not penetrable by the rays. Both ultraviolet rays and ozone may bring about some oxidative chemical changes in the fats of meat.

Sides of meat are rarely entirely spoiled by bacterial growth due to the fact that in the United States ordinarily the product is consumed within 10–12 days after slaughter and, in any case, since spoilage is confined to surfaces, the major portions could always be salvaged after considerable trim loss. Various times have been specified for the storage life of sides or quarters, but this varies greatly with the degree of contamination at the start, the conditions of handling, etc. It is evident that under ideal conditions of processing and handling fresh carcasses may be held for approximately 15–20 days without excessive trim loss. It has recently been shown that cut portions of beef, steaks, roasts, etc. may have a bacteriological storage life as long as 21 days if handled in a sanitary manner, packaged, and held at −1.1°C (30°F) (Urbain, 1968).[36]

Cooked and Cured Meat Products

Cooked Sausage Products

There are a variety of sausage types which are both cooked and cured. Most important among these would be frankfurters and bologna. Both of these sausage types are heated to the point that the pseudomonads, the fresh meat spoilage bacteria, are destroyed. Spoilage due to the growth of other bacterial species may occur.

One type of spoilage in cooked sausage products is greening. These products contain nitrite which reacts with the natural pigment myoglobin to form nitrosomyoglobin which upon heating becomes nitrosohemachrome or denatured nitrosomyoglobin. The latter compound has a pink color and the iron in the porphyrin ring of this compound is in the ferrous state. Under conditions of bacterial oxidation of this compound the porphyrin ring structure in the nitrosohemachrome compound may be changed chemically to form verdoheme or choleglobin which are green in color.

Bacterial greening of cured cooked meats is of three types: irregular green spots on surfaces of the product, green cores, or narrow green rings within the product.

In surface and core greening, oxidation is brought about by hydrogen peroxide produced by the growth of members of the *Lactobacteriaceae*. When these organisms grow in the presence of air, they produce hydrogen peroxide. Since they do not have the enzyme catalase, this compound tends to accumulate in the substrate or to react with the ingredients of the substrate. Specifically, *Lactobacillus viridescens* has been found to be the organism causing surface or core greening; but it would seem that members of the streptococci (enterococci), the pediococci, or *Leuconostoc* species might also cause this type of change under the proper conditions. It is known that *Streptococcus faecium* and *Streptococcus faecalis* may cause greening of heat-pasteurized hams in some cases (Riemann, 1957).[30] It is also known that *Leuconostoc* and *Pediococcus* species some times cause greening (American Meat Institute Foundation, 1960).[2]

Surface greening is due to the growth of *Lactobacteriaceae* on surfaces, contamination having occurred after the product was smoked and cooked, during such operations as skinning and packaging (Niven, 1951).[26]

In core greening, which most often occurs in large sausages, such as bologna, the organisms first grow under anaerobic conditions producing no hydrogen peroxide. When the sausage is cut, after the organisms have grown to large numbers, greening occurs in an hour or two owing to the production of hydrogen peroxide because of access to oxygen. It has been shown that the core-greening strains of *Lactobacillus* are considerably more heat resistant than those strains which cause surface greening (Niven et al., 1954).[27]

Green rings are formed a few millimeters from the surface of cooked sausage, usually in frankfurters, because of oxidation-reduction potentials suitable for the conversion of hemachrome to verdoheme. The actual cause of this change is not known, but it is known that this is likely to occur when meat with a high bacterial count has been used as raw material. It is believed that in this case during processing (smoking and cooking) the bacteria are killed; but enzymes are not entirely inactivated. Thereafter, the oxidizing enzymes remain active and provide the conditions suitable for the conversion of hemachrome to verdoheme.

Sliming of Cooked Sausage

Cooked sausage products sometimes spoil because of surface slime produced by the growth of microorganisms. The organisms causing slime formation are yeasts, *Leuconostoc, Lactobacillus, Streptococcus* species, and *Microbacterium thermosphactum* (American Meat Institute Foundation, 1960)[2]; (Licciardello et al., 1967).[23] It might be expected that streptococcal species and especially members of the genus *Leuconostoc* would be most often the cause of slime formation on cooked sausage, since the members of these groups are known to produce levan or dextran and have been shown to be the cause of ropy brines used for curing fish and meat products.

Fresh Sausage Products

Fresh or uncooked sausage products are sometimes spoiled by bacterial growth. When this occurs the product either becomes putrid or it becomes sour. It appears that *Microbacterium thermosphactum, Streptococcus faeceum,* or even *Streptococcus faecalis* may be the cause of souring of fresh sausage. This type of spoilage occurs infrequently. When putrefaction is the cause of spoilage, more often the case, the pseudomonads and *Achromobacter (Acinetobacter)* species are present in large numbers.

Cured Meats

Meat products, such as hams and bacon, are handled in several forms: as the product without further processing; as the sliced and packaged product; or as the canned, heat-pasteurized product. Hams may also be boiled, sliced, and packaged. All of these processed meats, except the commercially sterile type of small canned hams and luncheon meat, should be held at refrigerator temperatures above freezing.

Sliced packaged ham and bacon products rarely spoil in the United States of America since curing times are short, and they are distributed and consumed within a reasonable time. Greening and souring of such sliced meats occurs rarely. When sliced meats spoil by souring, it appears that the *Lactobacteriaceae,* particularly *Streptococcus faecium* (original source from the intestine of the hog); and even *Streptococcus faecalis* (original source from the human intestine) may be the causative organisms; or spoilage of this type may be caused by *Microbacterium thermosphactum.*

In the processing of canned refrigerated-type hams, the product is heated in boiling water to an internal temperature of only 71°C (160°F) (Because of the large size and low thermal conductivity of canned hams, thermal processing to attain high temperatures in all parts cannot be used for these products since such treatment would cause color and flavor changes in hams of this size.). This temperature–time treatment is not sufficient to always destroy all of the relatively heat-resistant streptococci which may be present in this product, and certainly will not destroy all bacilli or clostridia, some of which may usually be found therein. The streptococci, of the type indicated, can grow at temperatures below 10°C (50°F), while most bacilli and clostridial species will not. Under adequate conditions of refrigeration, therefore, when spoilage occurs, it is generally due to the enterococcal streptococci or other heat-resistant lactic acid types.

Canned commercially sterile-type hams or luncheon meat products are given a heat process equivalent to an F_0 of only 0.3–0.6. This extent of heating would not be sufficient to prevent many types of canned heat-processed foods from spoiling. It is believed that the following factors account for the lack of spoilage of canned meat products of this type: (1) comparatively low level contamination with spore-forming bacilli and clostridia, (2) sensitization of spores to heat by contact with meat fluids prior to processing, (3) sensitization of spores to heat by the sodium chloride present in the product, (4) sensitization of spores to heat by the nitrite present in the product, and (5) the unsuitability of heated meat media containing salt and nitrite for the growth of small numbers of surviving spore-forming organisms.

Some Characteristics of Cured or Cured and Cooked Meat Products Spoilage Bacteria

It seems to be the case that more often than not, when spoilage occurs in cured or cured and cooked meat products or sausage, the organisms causing spoilage are members of the family *Lactobacteriaceae*. These organisms are of the type which seem to survive comparatively well on processing plant equipment, having been brought into the plant on the raw material *(Streptococcus faecium)* or through human-borne contamination *(Streptococcus faecalis* and some species of *Lactobacillus)*. These organisms are relatively heat resistant for vegetative types, hence survive when ordinary spoilage types *(Pseudomonas* and *Achromobacter)* are killed during processing involving heating. This eliminates the types of organism which might grow faster than the *Lactobacteriaceae* at refrigerator temperatures above freezing. The lactic organisms are relatively salt tolerant, and cured products contain sufficient salt to also cause prevention of growth of some *Pseudomonas* and *Achromobacter* species which otherwise might grow and cause spoilage before a sufficient growth of the *Lactobacteriaceae* occurred to cause deterioration of the product. The lactic bacteria are facultative anaerobes, permitting growth under conditions in which the *Pseudomonas* or *Achromobacter* species would not proliferate or would grow at a reduced rate. This may be a factor in core greening. Many species or strains of the *Lactobacteriaceae* grow at temperatures below 7.2°C (45°F). This characteristic would not eliminate ordinary fresh meat-spoilage bacteria, but it would eliminate many other types which otherwise might grow and compete with the lactic bacteria.

Dried or Summer Sausage

Certain types of sausage are prepared and then held at temperatures of 7.2°C (45°F) or higher in rooms with low humidity for a period of 40 days or more. During the holding or curing of these sausages, a lactic acid fermentation takes place, and a considerable dehydration of the product occurs. The salt content which is increased by the drying (lowering the water activity of the product), together with the acid produced by the growth of species of *Lactobacteriaceae,* tend to preserve the product after curing; and the lactic acid also gives this type of sausage its particular taste or tang.

At one time dried sausage could only be made from meat (usually trimmings) which was old or high in bacterial count. The reason for this was that lactic acid organisms are present in relatively low concentrations in fresh meat and tend to establish themselves only when the growth of aerobic types have provided suitable conditions of anaerobiosis. It was at one time necessary, therefore, to use old trimmings for the production of dried sausage, in order to have sufficient numbers of the lactic bacteria to provide a suitable fermentation. This use of old meat did not always produce a good-tasting sausage because of other odors and flavors produced in the trimmings due to the growth of ordinary spoilage types of bacteria or because of chemical changes in the meat such as the oxidation of fats.

Today fresh meat or trimmings are used in the manufacturing of dried sausage; and, when prepared, the product is inoculated with a known culture of the *Lactobacteriaceae.* The organisms used in this case are *Pediococcus cerevisiae* or several species of *Lactobacillus.* Also, either very small amounts of nitrite or nitrate are added to these sausages; and it is desirable to have the pink color of nitrosomyoglobin in the interior of the finished sausage. This means that during the curing there must be some reduction of nitrate to nitrite if no nitrite has been added. For this reason a culture of *Micrococcus aurantiacus* can be added to accomplish the reduction of nitrate to nitrite (American Meat Institute Foundation, 1960).[2]

Since nitrite or nitrosomyoglobin is present in dried sausage and the lactic acid organisms which grow therein to large numbers produce no catalase, it might be expected that, when this type of product is cut and exposed to air for a short time, hydrogen peroxide would be produced and the product would develop a green core. This does not occur. The reason that green-core formation does not take place appears to be that: (1) During growth of the *Lactobacteriaceae* under anaerobic conditions within the sausage no hydrogen peroxide is produced. (2) By the time that curing is completed the organisms have all died, probably because of acid production and high salt content. At the time that the product is cut the acid and salt content of the product is high enough to inhibit or inactivate the enzymes which form hydrogen peroxide in the presence of air, should they be present. (3) Sugars may be depleted by the time that curing is complete in which case no hydrogen peroxide would be formed.

Control of Spoilage of Cured and Cooked Meats

Certain procedures may be employed to limit the spoilage of processed meats: (1) All meats to be used for curing or for sausage manufacture should be promptly cooled after slaughter to an internal temperature of 1.7°C (35°F). (2) Products such as hams or bacon should be pumped with pickle solution the same day they are cut or should be held at −3.3 to −2.2°C (26–28°F) if they are to be processed the next day. Otherwise, they should be frozen and held in this state until defrosted (at 4.4°C (40°F) or below) for further processing. (3) Fresh cuts, to be cured as such or to be processed as cured cooked sausage products, should not be brought from refrigerator temperatures to areas of higher temperature and high relative humidities since this situation usually results in condensation of moisture on the product. (4) Meat used for the manufacture of sausage should not be spoiled or slimy and should be of comparatively low bacterial count. (5) Pickle solution used for the pumping or covering of meat portions to be cured should not be reused for short-cure procedures. (6) Fresh or cooked sausage products should be promptly refrigerated after stuffing or after cooking and water cooling. (7) Product returned to the manufacturer from retail outlets should not be reused or reprocessed. Such procedures serve only to establish in the plant a bacterial flora of the types which cause spoilage of product. (8) Meats which have been comminuted should not be held in this condition for long periods of time prior to stuffing, smoking, cooking, etc., but should be promptly processed. (9) During the manufacture of cooked sausage, such as frankfurters and bologna, the internal temperature of the product should be brought to 71.1°C (160°F). (10) Good plant sanitation (adequate cleaning and sanitizing of floors and walls of processing areas, of all equipment contacting food products or used for processing of products, and of the room where product materials or finished product is held) should be practiced at all times. (11) Clean outer clothing should be provided to workers daily, and workers should be required to practice good personal sanitation. (12) Finished product should be quickly distributed to retail outlets and should not be held for long periods prior to such distribution.

Poultry

Although some types of poultry are frozen in considerable quantities, especially turkeys, by far the largest volume is handled in the fresh, eviscerated state and is distributed while holding at refrigerator temperatures above freezing, usually at temperatures below 7.2°C (45°F).

The source of the bacteria causing spoilage of refrigerated poultry is from soil and water, especially the former. After killing and bleeding, the birds are given a semiscald treatment during which they are passed through water at 54.5–51.1°C (130–124°F), the carcass remaining in the water for 30–60 sec. This facilitates the removal of feathers. During scalding some of the spoilage types of bacteria are destroyed. The extent of this destruction is not known, but it would appear that many organisms may survive. Some authors have found that there is a buildup of bacteria on the skin of chickens during dressing and evisceration (Drewniak et al., 1954).[13] Others (Walker and Ayres, 1956)[37] have found little difference in the number of bacteria present on the skin of poultry before and after dressing. Some authors have found that procedures after dressing, which would include washing with pressurized sprays of water, actually decrease the number of bacteria present on the skin of poultry, as compared to the number present after rough picking (Drewniak et al., 1954).[13]

Regardless of whether most of the bacteria present on poultry surfaces consist of flora already present prior to killing or whether they are mainly picked up during such procedures as defeathering (from machines), pinning (from workers hands and from pinning knives), eviscerating (from intestinal contents) or from cooling (cross contamination from one carcass to another), numerous spoilage bacteria may be found on freshly dressed and eviscerated poultry. Investigations have shown that under good sanitary conditions the total count on the skin of freshly eviscerated poultry will be in the order of 10^2–10^3 per square centimeter while under conditions of poor plant sanitation the count may be as high as 10^5 per square centimeter (Farrel and Barnes, 1964).[17]

Experiments have been carried out to determine the effects of adding chlorine to the slush ice water used for chilling chickens (Barnes, 1965).[7] Colony counts made on the skin were not significantly different where 0, 5, 50, or 100 ppm of chlorine was used in the ice water. The addition of 200 ppm of chlorine did reduce the total count on poultry after cooling but did not affect the numbers of coliforms or fecal streptococci present. Chlorination of cooling water at a level of 200 ppm slightly extended the storage life of the product at 1°C (33.8°F) and, of course, would prevent cross-contamination during cooling.

During the holding of eviscerated poultry, bacteria grow on both the skin and flesh portions producing slime and causing off odors which have been described as tainted, acid, or sour. There is no doubt that hydrogen sulfide, produced by bacterial action, also sometimes contributes to the off-odor of these products at the time of spoilage.

It has been found that off-odors are evident in eviscerated poultry when the skin count reaches approximately 10^7 per cm^2 and that slime formation occurs when counts are in the order of 10^8 per cm^2 (Ayres et al., 1950).[6]

The bacteria on the skin of poultry immediately after processing have been found to belong primarily to *Pseudomonas* species. Also, at the time of spoilage during holding at refrigerator temperatures above freezing the bacterial flora is made up chiefly of the pseudomonads. Mostly the nonpigmented pseudomonads are present at the time of spoilage. It has been suggested that poultry spoilage pseudomonads be classified as: *Pseudomonas fragi, Pseudomonas putida,* and *Pseudomonas ambigua,* some pigmented strains such as *Pseudomonas fluorescens* being present in the spoilage flora and also some species of *Xanthomonas* (Ayres, 1960).[4]

The storage life of eviscerated poultry depends on the extent of contamination after eviscerating and cooling and the temperature at which the finished product is stored. Under the best conditions of storage and sanitation a storage life of approximately 16 days or possibly slightly longer may be expected. Under poor conditions of sanitation a storage life of approximately 8 days at 0°C (32°F), 4 days at 5°C (41°F), and 3 days at 10°C (50°F) has been reported.

Since, considering the manner in which they are handled, surfaces of poultry cannot be kept free of moisture, the conditions for preventing spoilage of eviscerated poultry handled in the unfrozen state seem to be concerned with good plant sanitation and the application of low temperatures including quick and adequate cooling and low storage temperatures. Adequate cleaning of equipment such as defeathering machines, knives, and eviscerating tables (when used), thorough washing of eviscerated products and the maintenance of low-count ice slush for cooling will help to extend the storage life of processed poultry. Temperatures applied during the storage of eviscerated poultry during preshipment, transportation, and retail display should be as near to the freezing point of the product as is practicable. Eviscerated poultry which is not frozen should be promptly shipped to retail outlets and not held for appreciable lengths of time prior to shipment.

Fresh Fish

Fish which are handled in the eviscerated state (chiefly cod, haddock, halibut, and some salmon) and some which are not eviscerated (chiefly small flounders and ocean perch) are subject to spoilage by bacterial growth. Pelagic fish, such as herring and mackerel, and most salmon, are handled without evisceration. Due to the short time that they are held in the fresh state prior to processing, pelagic fish and most salmon are usually not subject to extensive bacterial deterioration. Also, with herring and mackerel, enzyme action often causes significant deteriorative changes prior to noticeable spoilage of bacterial origin. While considerable information is available regarding the bacteriology of certain fish species from the North Atlantic, little data have been reported that is related to the bacteriology of fish or shellfish from warmer waters such as the Gulf of Mexico. Even though a number of investigations dealing with the microbiology of some cold water marine species are available, this type of information is limited to only certain species and the picture of how bacterial spoilage proceeds is not entirely clear.

Sources of Contamination

It might be expected that the chief source of the bacteria causing spoilage of fresh marine fish is the ocean. This appears not to be the case. It is true that at the time when fish are removed from the ocean the bacteria present on surfaces, gills, and in the intestine are of marine origin. Also, when marine fish are eviscerated, the lining of the body cavity is generally contaminated by sea water or through spillage of intestinal contents. However, fish are held in fresh water ice aboard fishing boats. This ice contacts the fish directly and also the walls of the holding pen, which contact the fish in places. Generally, therefore, the organisms causing spoilage of fresh eviscerated marine fish are psychrophilic and psychrotrophic bacteria of fresh water origin (Shewań and Hobbs, 1967).[33]

When fish are further processed, as by filleting or steaking, the organisms which become predominant on surfaces are again of fresh water origin, these organisms having become established on equipment through contact with fish, ice, or the water used for cleanup.

While fresh water psychrophilic and psychrotrophic bacteria may become established on holding pen walls, equipment, etc., it must also be the case that when first stored in ice, at least as many of the organisms present on fish surfaces are of marine origin as are from fresh water. A situation must, therefore, exist in which conditions for growth are more favorable to the fresh water types of bacteria; or the fresh water types must grow faster at low temperatures and crowd out the marine species. It may be a situation in which marine species do not grow well because of the dilution of salts (sodium chloride and other salts) brought about by contact with the melting ice.

Organoleptic Changes in Fish Due to Bacterial Growth

When eviscerated fish or cut portions of fish spoil aerobically, due to bacterial growth, the organoleptic changes which occur are entirely different from those taking place when fish are held on boats under conditions which are essentially anaerobic. The latter type of spoilage is less frequent but occurs often enough to warrent some description.

Aerobic spoilage has been reported (Shaw and Shewan, 1968)[32] to bring about the organoleptic changes listed in Table 7–1.

It is known that many of the bacteria which may be involved in the spoilage of fish will reduce trimethylamine oxide [$(CH_3)_3{\equiv}N{=}O$] to trimethylamine [$(CH_3)_3{\equiv}N$]. Trimethylamine has an odor usually described as fishy. It is also known that certain bacteria produce an enzyme, triamineoxidase, which activates trimethylamine oxide to cause it to be reduced by the many dehydrogenase enzymes of various bacteria.

The tissues of marine fish contain trimethylamine oxide, this material having accumulated as an end product of their nitrogenous metabolism or as a component of the food ingested by fish.

It is said that the reduction of trimethylamine may take place in the following manner:

$$2(CH_3)_3{\equiv}N{=}O + CH_3{-}\underset{\displaystyle OH}{\underset{|}{CH}}{-}COOH \rightarrow 2(CH_3)_3{\equiv}N + CH_3{-}COOH + CO_2 + H_2O.$$

TABLE 7–1
Organoleptic Changes in Fish During Refrigerated Storage

Strong off-odors	Slight off-odors	Other off-odors
Sour	Slight ammonia	Slight musty
Fecal	Slight hydrogen sulfide	Sweet
Brine-like	Slight sour	Milky
Indole	Slight amine-like	Soapy
Burned matches	Slight indole	Yeasty
Decaying vegetables	Slight fruity	

In eviscerated fish trimethylamine may play a small part in causing organoleptic changes. This is probably not the case when cut portions held under refrigeration undergo bacterial spoilage. Under the latter conditions the trimethylamine content increases as the number of bacteria increase, and fillets eventually take on a trimethylamine-like odor prior to the development of ammoniacal and putrid odors.

When fish are held aboard boats under anaerobic conditions an odor develops which has been described as bilgy. Bilginess is an especially offensive odor usually reminiscent of some hydrogen sulfide mixed with other odors. The situation of anaerobiosis occurs aboard boats when fish become pressed tightly against the pen walls or even against one another. In this condition there probably is no significant increase in bacteria present on surfaces. The organisms producing hydrogen sulfide and other compounds are already present on the surface of the pen walls, on the surface of the fish, or both. These bacteria normally produce hydrogen sulfide, but in air this is oxidized to sulfur which has no organoleptic effect. Under anaerobic conditions then, bacterial enzymes, already present, produce hydrogen sulfide which now cannot be oxidized to sulfur since no oxygen is present. The hydrogen sulfide and other compounds then diffuse into the tissues to cause the organoleptic change known as bilginess.

The Bacteria Causing Spoilage of Fresh Fish

The flora of fresh-caught fish from northern waters are usually gram-negative rods which include pseudomonads and *Moraxella*-like species *(Achromobacter or Acinetobacter)* with some coryneforms, *Flavobacterium, Micrococcus,* and *Cytophaga* species also being present (Shewan and Hobbs, 1967).[33] Also, it has been stated that among the pseudomonads 40–50% are *Pseudomonas pellucidium,* 20–30% *Pseudomonas geniculatum,* 10–20% *Pseudomonas pavonacea* and *Pseudomonas nigrificans,* and 5–10% *Pseudonomas schuylkilliensis* and *Pseudomonas fluorescens.* Species related to *Pseudomonas ovalis, Pseudomonas fragi,* and *Pseudomonas multistriatum* are also present in small numbers (Shewan and Hobbs, 1967).[33]

It would appear that on fish from warmer waters such as the Indian Ocean or the African coastal waters, coryneforms and micrococci may comprise the major constituents of the flora of certain fish when removed from the water.

Regardless of the flora originally present on fish freshly removed from the water, it is considered that fresh-water bacterial types will take over and eventually become the predominating flora due to the manner in which fish are handled.

The spoilage flora of iced northern fish consists almost entirely of *Pseudo-*

monas and *Achromobacter* species, and usually 20–50% of the pseudomonads will be of the *Pseudomonas fragi* type. It has been found that *Pseudomonas* organisms from Groups 2, 3, and 4 [produce no diffusible pigment but may oxidize carbohydrates (Group 2); do not act on carbohydrates but produce alkali in Hugh and Leifson's medium (Group 3); or do not act on carbohydrate and produce no change in Hugh–Leifson's medium (Group 4)] bring about the greatest organoleptic changes in fish press juice or in previously sterilized fish flesh (Shaw and Shewan, 1968).[32] Volatile bases are produced during the growth of spoilage bacteria in fish, but trimethylamine is not necessarily a significant component of such bases when spoilage occurs. The *Achromobacter,* though present as a part of the spoilage flora, would appear to play a minor role in the production of organoleptic changes.

Investigations on cod and haddock (Castell et al., 1949; Chai et al., 1968)[10,11] have indicated that *Pseudomonas putrefaciens* (produces H_2S from proteins) is especially active in causing spoilage of whitefish from the North Atlantic.

A recent finding on haddock (Laycock and Regier, 1970)[22] indicated that, in fish iced 2 days (fresh), *Achromobacter* was the predominating flora, some *Pseudomonas* and *Flavobacterium* species also being present. As the fish was held (5 days, borderline quality) the *Achromobacter* further increased, the proportion of pseudomonads remaining approximately constant while *Flavobacterium* species dropped out. In the 9-day iced fish (spoiled), Group I and II *Pseudomonas* and *Pseudomonas putrefaciens* increased and the *Achromobacter* species decreased to the point where the latter species was present in only slightly greater proportions than the other two species combined. In fillets cut from fish iced 2, 5, or 9 days at spoilage Group I and II pseudomonads always predominated (present in roughly equal proportions) while *Achromobacter* made up approximately one fifth of the flora.

While *Achromobacter* were present in the greatest numbers in iced haddock, their contribution to organoleptic spoilage was probably not so great as that of the combined pseudomonads since concentrations of 10^8–10^9 per gram of *Achromobacter* species would be required to bring about noticeable organoleptic changes while the pseudomonads would produce such changes at concentrations of 10^7–10^8 per gram.

Group I pseudomonads (oxidize carbohydrates and produce a diffusible pigment) have, in the past, not been considered to play an important part in fish spoilage except in special cases, but more recently findings indicate that their contribution to organoleptic changes in white-fleshed fish is much greater than formerly supposed.

At one time it was considered that the spoilage of fresh fish is entirely a surface phenomenon, in that the bacteria grow on skin and body cavity lining surfaces and their end products diffuse into the tissues to cause organoleptic changes. Some groups still consider this to be the case. On the other hand, at the time of advanced spoilage flesh counts as high as 10^7–10^8 per gram may be obtained; and there may be obvious invasion of bacteria into the tissues, through the skin, and even through the arterial system of the gills.

Clostridium botulinum, type E, may be present in the bacterial flora of fresh water or marine northern fish. Fortunately, unless grossly mishandled, this organism does not increase in numbers to any extent by the time that fish have spoiled.

Factors Affecting the Spoilage of Fish and the Control of Spoilage

Even under the best conditions of handling [good sanitary surrounding and temperatures just below 0°C (32°F)] fish such as cod or haddock will have a storage life not longer than 15 days. Halibut may have a storage life of 19–24 days under the best of conditions. The reason for this may be that it has a relatively low pH (5.8–6.8 with an average of about 6.3 as compared to cod and haddock which have a pH of 6.5–7.0 with an average near to 6.75) which would cause this fish to be less suitable as a substrate for the growth of spoilage bacteria (Patashnik, 1966).[28]

As fillets or steak, most white-fleshed fish under the best conditions of sanitation (in cutting) and at low holding temperatures above freezing may be expected to have a storage life of approximately 12 days.

The factors governing the spoilage time of fish held aboard boats are the sanitary conditions under which they are held and the adequacy of icing. Regarding sanitary conditions, fish hold pens are generally not cleaned well and even the water used for cleaning may come from the harbor, which is usually high in bacterial count. It is preferable to hold fish in boxes stored in the hold of the boat. Fish boxes are of wooden, plastic, or metal construction, and the latter types are preferable since it is difficult to clean and sanitize wooden boxes sufficiently to remove most of the bacteria from the wood. The removal of bacteria from plastic and especially from metal boxes can be readily accomplished.

Fish held aboard boats should be iced in such a manner that there is a layer of ice between the sides and bottom of the container in which they are held and the fish. Such a condition prevents the development of bilginess. Also, icing should be sufficiently extensive to provide a temperature in the fish of about 0.6°C (33°F). The use of fish boxes provides a distinct advantage in that the fish held in bottom layers are not subjected to physical degradation due to high pressures exerted by the weight of deep layers of fish and ice. Also, the bottom layers of fish in boxes are not continuously contaminated with a stream of water (high in bacterial count) from melted ice.

The cutting of fish into fillets or steaks should be done under the best conditions of sanitation since the higher the bacterial population on cut surfaces at the start, the shorter the storage life. It has been shown (Shewan and Hobbs, 1967)[33] that prior to filleting, fish can be washed so as to remove more than 90% of the bacteria on skin surfaces. It has been shown that although fillets tend to become grossly contaminated from cutting boards, this can be avoided with suitable techniques (Castell et al., 1949).[10] Fast processing (prevention of pile-up of cut fillets) and the use of clean water for washing or rinsing also contribute to longer storage life. After packaging, fillets or steaks should be held at temperatures as near 0°C (32°F) as can be applied without freezing the fish.

In some countries freshly cut fillets may be immersed in solutions of one of the tetracyclines to provide a concentration of about 7 ppm of these antibiotics in the product. Treatment in this manner, together with good plant sanitation and suitable holding temperatures, may extend the storage life of such products for several days beyond that attainable without antibiotic treatment.

Cured Fish

Various combinations of salting, cold smoking, hot smoking, drying, or the addition of vinegar or acetic acid may be used in the curing of fish. These various treatments will have different effects on the storage life, refrigerated or otherwise, of the different products.

Lightly salted and lightly smoked fish, such as kippers and smoked haddock or smoked fillets, are quite perishable and must be held under refrigeration. Although these products usually contain only 2–2.5% of added salt, the surface flora may be changed by this treatment. Also, the smoking, though cold, 20–23.9°C (68–75°F), contributes phenolic compounds which have some bacteriostatic effect. While little information on the subject is available, it has been indicated that the spoilage flora of lightly salted and smoked fish products may be pseudomonads or micrococci.

Heavily salted and heavily smoked fish (low-temperature smoking) is not subject to bacterial deterioration due to the high salt content (10–12%) and inhibiting compounds in the smoke. Molds may cause deterioration of such products on occasion.

Hot smoked fish are [final temperature near 82.2°C (180°F)] lightly salted and lightly smoked and are subject to bacterial decomposition and should be held at temperatures of 4.4°C (40°F) or below at all times. It is to be expected that during the processing of this product the ordinary spoilage bacteria (*Achromobacter* and especially the *Pseudomonas*) would be destroyed. Coccal types, including micrococci, could survive this treatment. The spoilage flora of hot-smoked fish seems not be have been investigated, but it is known that clostridia may sometimes be present in comparatively large numbers.

Some fish (herring) are merely salted and held in this condition or salt and sugar is added at the start and later on they are canned with special sauces (Swedish tidbits). The salt content of the first type eventually reaches approximate saturation with sodium chloride while tidbits have a salt content above 7%. Cured at temperatures below 15°C (59°F) and held at these temperatures salted herring do not spoil bacteriologically although there may be some growth of halotolerant micrococci and yeasts. When herring tidbits are spoiled by holding at high temperatures, it has been shown (Erichsen, 1967) that the organisms involved are heterofermentative *Lactobacteriaceae*. The pediococci (especially *Pediococcus halophilus* and *Pediococcus cerevisiae*), first grow and produce some acid but do not cause spoilage. *Leuconostoc mesenteroides* may cause slime formation and spoilage by converting sucrose to dextran.

Fish such as cod may be lightly salted and dried or heavily salted and dried. The former type of dried fish develops odor and flavor during curing which is said to be due to the growth of certain micrococci. When properly cured, heavily salted and dried fish do not undergo bacterial spoilage; but during curing a deterioration known as "pink" may develop causing slime, softening of the fish, and eventual falling apart of the fish. The organisms causing this type of spoilage are present in the salt used for curing and consist of halotolerant bacteria of pleomorphic forms. No absolute identification of these bacteria has been made, and a number of different genera may be involved. Some of these organisms require sodium chloride for growth.

A type of spoilage known as dun may occur in salt dry fish especially in light or medium cures. Small black, brown, or fawn-colored spots develop on the fish surfaces in this case. This type of spoilage is due to the growth of molds of the genera *Sporendonema* and *Oöspora*. These organisms require 5–10% of sodium chloride in the medium for growth, and some strains will tolerate 20% of sodium chloride. The source of these organisms is the salt used in curing.

Curing Brines

Curing brines consist of about 10–22% of sodium chloride; they may or may not contain sugar (sucrose or dextrose) in concentrations of 3.5–10%; sodium or potassium nitrate is often present, especially in meat-curing brines, in concentrations of 0.4–0.7%; and especially in meat-curing brines, sodium or potassium nitrite may be added in a concentration of about 0.25%. In modern meat cures, ascorbate or related compounds and lactic acid may be added to curing brines at a level of about 0.25%, and one or more of several phosphate compounds may be added in concentrations up to 5%.

In the short cures used for meat in the United States, nitrite is present in the brine for purposes of combining with the myoglobin of the meat to form the pink nitrosomyoglobin. Also, the brine is injected into the meat by artery pumping or by stitch pumping. Curing in this case rarely requires more than 5 days; and since curing is carried out at 7.2°C (45°F) or below, there is little chance for the significant development of bacterial populations in the brine.

It has been shown (Deibel and Niven, 1958)[12] that in short-cure brines the bacterial flora consists mainly of the *Lactobacteriaceae,* although some micrococci and yeasts may be present in surface layers. Species related to *Lactobacillus plantarum* and *Lactobacillus casei* were found to represent the lactic organisms present, these strains being motile. These organisms are generally not present in sufficient numbers to contribute special flavors or off-flavors to meat products.

When long-curing methods are applied to meat, as is sometimes done in parts of Europe, the meat is covered with the curing brine and allowed to stand for periods as long as 60–90 days. Also, in this type of curing, the ambient temperatures under which the product is held may be higher than those ordinarily used for short cures. Nitrate is added to such brines in many cases, but there is no addition of nitrite. Bacteria, therefore, may play an important part since they convert nitrate to nitrite and thus provide for the formation of nitrosomyoglobin in the meat.

In converting nitrate to nitrite bacteria use the oxygen in the nitrate to oxidize lactic acid to acetic acid at the same time reducing nitrate to nitrite.

$$\underset{\substack{|\\ OH}}{CH_3—CH}—COOH + 2NO_3^- \rightarrow 2NO_2^- + CH_3—COOH + CO_2 + H_2O$$

Lactic Acid *Nitrate* *Nitrite* *Acetic Acid*

The specific organisms providing for the reduction of nitrate to nitrite in long-cure brines have not been finally identified, but they appear to belong to the two genera *Vibrio* and *Achromobacter.*

At times the wrong flora may be established in long-cure brines, micrococci or other types which do not reduce nitrate becoming the principal flora. Also, under some conditions a too active growth of the desirable types of bacteria may take place in the brine. Under such conditions the oxidation-reduction potential of the brine falls below normal; and the oxygen of the nitrite is used to oxidize lactate anaerobically, this reaction resulting in the release of nitrogen gas.

$$\underset{\text{Lactic Acid}}{CH_3{-}\underset{\displaystyle OH}{\underset{|}{CH}}{-}COOH} + \underset{\text{Nitrite}}{4NO_2^-} \rightarrow 2N_2\uparrow + 3CO_2 + 4OH^- + H_2O$$

When this happens the brine becomes frothy due to the nitrogen gas which is produced and becomes alkaline, since potassium or sodium carbonate is formed from the carbonic acid present, the carbon dioxide having dissolved in the brine. Such conditions cause the nitrate to be depleted in a few days, hence nitrite is not available for color formation in the meat. Also, as the oxidation-reduction potential falls to low levels, amino acids present in the brine undergo bacterial decomposition resulting in foul odors.

There are instances, especially in the curing of herring or sprats (about 10% of sodium chloride and 10% of sucrose are added to the ungutted fish in barrels) in which the brine becomes ropy or slimy leading to spoilage of the product. This product would normally be held in the salt–sugar brine for some months and then filleted and packed in tins in special sauces.

It is known that *Leuconostoc,* which form dextran (a polymer of dextrose) from sucrose, or *Achromobacter* species, which form levan (a polymer of fructose) from sucrose, may be the cause of ropy brines under these conditions. Neither of these organism types can form the polymer (dextran or levan) from dextrose or fructose but can only form the polymer from sucrose.

It has been found (Lindeberg, 1954)[25] that the *Achromobacter* species causing ropiness in brine produced maximum growth in salt concentrations of 5–10%. These organisms did not utilize carbohydrates as a source of energy but could utilize a number of organic acids for this purpose. Native proteins were not utilized, but ammonium salts, amino acids, or polypeptides were utilized.

The levan-forming enzyme produced by the *Achromobacter* species was found to be activated by sodium, and the enzyme was most active in salt concentrations of 5–6%.

Bacteriological standards have been proposed for curing brines (Proceedings Second International Symposium on Food Microbiology, 1958).[29]

These proposed standards relate to the number of salmonellae, coagulase-positive staphylococci, enterococci, *Escherichia coli,* and *Clostridium botulinum* in curing brines. The proposed standards were more lenient, with respect to numbers, for cover-pickle brines than for pump-pickle brines. This is understandable on the basis of what may be attainable, but it would seem that products in long cure might tend to be more of a public health hazard than those in short cure, and for this reason more rigid sanitary requirements might be expected for cover-pickle brines.

Shell Eggs

Source of Contamination

It is probable that most hen's eggs when laid contain no viable bacteria. Actually, considering the manner in which the egg is formed, the yolk passing along the oviduct and the white being added before the whole becomes covered with a shell, there is some chance of contamination of the egg with bacteria. Tests made to determine viable bacteria in fresh-laid eggs would indicate that as much as 10% may contain bacteria which are able to grow on artificial media. However, considering the difficulty of sampling eggs in such a manner that the contents are not contaminated with bacteria from the outside, the estimation that 10% of eggs contain viable bacteria when laid is probably high. It has been shown that less than 1% of naturally clean eggs become spoiled bacteriologically, which would indicate that most rot-producing bacteria get into the egg from the outside. This does not necessarily indicate that less than 1% of freshly laid eggs contain viable bacteria when laid since bacteria when present might be of the type which were not able to overcome the antimicrobial factors present in eggs and thus might be unable to grow in the egg contents. It has been found, for instance, that bacteria of the family *Lactobacteriaceae* are sometimes present in freshly laid eggs; and these organisms do not cause spoilage of eggs.

The source of the bacteria or molds causing spoilage of eggs is the outside of the shell. When the egg is laid, the shell becomes contaminated with bacteria from the nest; and if the nest is not clean, the outside of the shell becomes contaminated with hen droppings which greatly increases the bacterial load. For this reason roll-away nests in which the egg rolls away from the laying area once the hen leaves, produces cleaner eggs which are less subject to spoilage than do ordinary nests. It has been shown that the total number of bacteria on the outside of the shell of an egg will vary with the egg and numbers between 10^2 and 10^8 per egg have been found (Board, 1964).[8]

Organoleptic Changes in Eggs Spoiled by the Growth of Microorganisms

Typical egg changes when spoilage is due to the growth of microorganisms have been listed by various investigators (Romanoff and Romanoff, 1949; Alford et al., 1949)[1,31] as follows:

Green rots. The white takes on a greenish appearance, but the yolk is not affected in early stages. In advanced stages some encrustation may appear on the yolk, or the yolk may disintegrate and mix with the white. Odors are frequently sweetish or fruity. The white fluoresces strongly under uv light.

Colorless rots. No color is produced in the egg, but the yolk becomes encrusted and may disintegrate. There may be no off odor, or there may be a fruity odor.

Red or pink rots. The white may be liquefied or still viscous and may contain spots of brown, pink, or red. Also, a pinkish coagulum may appear on the yolk.

Black rots. The white becomes liquefied, the yolk becomes coagulated and turns black. The egg has a fecal, putrid, or hydrogen sulfide odor.

Miscellaneous rots. Various changes may take place, including changes in the color of the albumin, liquefaction of the albumin, etc. It is probable that this type of change is the same, in part, as those listed under colorless rots.

Fungal rots. Molds grow on the surface of the egg shell, penetrate the shell, and grow on the inside causing coagulation or liquefaction of various parts of the egg and also causing musty flavors and odors.

The Bacteria Causing Spoilage of Shell Eggs

It is known that green rots are caused by the growth of types of *Pseudomonas fluorescens.* Since the pigment products by these organisms fluoresces strongly under ultraviolet light, this type of spoilage can be detected by candling under ultraviolet light.

Colorless rots are said to be caused by the growth of species of *Pseudomonas* and *Achromobacter.*

Pink or red rots are known to be caused by species of *Pseudomonas* which produce pigments not of the fluorescent type.

Black rots are known to be caused by the genus *Proteus (Proteus melanovogenes* has been cited specifically as causing this change), and it has been indicated that species of *Pseudomonas* could be involved.

Miscellaneous rots could be caused by various species of bacteria including *Pseudomonas, Serratia, Alcaligenes,* and others.

Fungal rots are usually due to the growth of *Penicillium, Mucor, Cladisporium,* or *Sporotrichum.*

In one investigation of egg spoilage in Australia (Alford et al., 1950)[1] green rots were most frequently followed by black rots and colorless rots. Pink rots and other types were relatively infrequent. Other investigations (Trussel, 1955)[35] indicate that green rots are most frequent since it has been found that only 18.5% of shell eggs showing bacterial decomposition did not fluoresce under ultraviolet light.

All investigations indicate that the *Pseudomonas* species are by far the most important agent of spoilage of shell eggs. Also, since it is indicated that about 80% of spoiled eggs contain pyoverdine, the water-soluble fluorescent pigment, it appears that especially *Pseudomonas fluorescens* and *Pseudomonas ovalis,* and also *Pseudomonas convexa* and *pseudomonas aeruginosa* are the important species causing spoilage of fresh eggs. These organisms produce pyoverdine and have also been associated with the bacterial decomposition of shell eggs. *Pseudomonas putrefaciens* and *Pseudomonas geniculatum,* which do not produce water-soluble fluorescent pigments, have also been found to be the cause of rot in shell eggs in some instances.

Proteus, Alcaligenes, and *Serratia* species have been reported as sometimes causing the spoilage of shell eggs; but it is evident that these genera are of much less importance in this respect than the *Pseudomonas* species, especially those types which produce water-soluble fluorescent pigments.

Factors Affecting the Bacterial Spoilage of Shell Eggs

The resistance of shell eggs to bacterial spoilage depends on a number of properties of the egg and its anatomical components. Among these are the cuticle, the shell itself, the outer and inner shell membranes, and the various properties of the albumin or egg white.

The cuticle is deposited over the surface of the shell of the egg as it is laid. This consists of a thin layer of mucin, a conjugated protein containing glucose. The cuticle effectively seals the pores in the egg shell against the entry of invading microorganisms, as long as it is present and retains its elastic properties. However, the cuticle is easily removed by moisture, even by moist droppings which may contact the shell; and after the egg is about 3 weeks old, the cuticle becomes brittle and is easily chipped away. The anti-contamination effect of the cuticle is, therefore, not permanent.

The egg shell itself acts to some extent as a mechanical barrier against the invasion of microorganisms. However, there are numerous pores in the shell, about 110 per square centimeter, and many of these are large enough to allow bacteria and molds to pass through. About 10% of the pores are quite large. The shell itself is, therefore, limited in its effectiveness as an obstruction to the entrance of microorganisms into the egg contents.

Directly under the egg shell are the inner and outer shell membranes. These are composed of keratin (proteinaceous material similar to that present in horn), but they contain large pores. While it has been found (Lifshitz et al., 1964)[24] that the inner shell membrane is more efficient than either the shell or the outer shell membrane in preventing the entrance of bacteria into the egg, this investigation was done in such a manner as to negate the effect of the cuticle. Present opinion would appear to consider that the shell membranes are not an effective barrier to the entrance of microorganisms into the egg.

The albumin or egg white has several factors which tend to inhibit the growth of microorganisms. At one time it was believed that egg white, being a native protein, would not support the growth of bacteria. It has been shown (Board, 1966; Garibaldi, 1960)[9,19] that there are nitrogenous materials in egg white which will support bacterial growth.

There is an enzyme, lysozyme, present in egg white which lyses gram-positive bacteria. This is one of the reasons that shell egg are rarely, if ever, decomposed by gram-positive bacteria. Also, avidin, a substance present in egg white, sequesters biotin and causes it to be unavailable to those bacteria which require it for growth. The proteins in egg albumin also sequester riboflavin and vitamin B_6, also required by some bacteria for growth (Eliott and Michener, 1965).[14]

Probably the most effective inhibitor of bacterial growth in eggs is the conalbumin (that portion of the albumin which cannot be crystallized), which sequesters iron. Small amounts of iron are required for the growth of *Pseudomonas* species, the group of bacteria mostly responsible for egg spoilage; and, while the egg is comparatively fresh, the albumin effectively ties up iron so that it is unavailable to bacteria. When the egg has been held in storage for some time, the conalbumin is no longer as active in sequestering iron; and it is known that old eggs are more subject to bacterial decomposition than fresh eggs. Also, if the yolk is mixed with white as in liquid whole egg magma, the yolk supplied enough iron, biotin, and vitamins so that the albumin is no longer effective in sequestering compounds necessary, in small amounts, for bacterial growth. It is interesting to note that when the egg white is thin enough to allow the yolk to move and contact the shell membrane, the egg becomes much more subject to bacterial decomposition since the yolk is a much better substrate for bacterial growth than is the albumin.

The washing of shell eggs is looked upon with disfavor in the trade since bacteria are apt to penetrate the shell during washing. At one time it was believed that bacterial penetration during washing was due to the fact that water colder than the egg itself was often used. This colder water, it was thought, created a slight vacuum within the egg due to contraction of contents because of cooling, and thus sucked bacteria in the wash water into the egg (Forsythe, 1952).[18] It has since been shown that bacteria may enter the egg during washing even if there is no temperature difference between the egg and the bacterial suspension (Ayres, 1967).[5] It is evident that, if eggs are to be washed, the wash water should contain a disinfectant. The quarternary ammonium compounds have shown promising results as disinfectants for egg washing.

If eggs are to be broken out, they should be washed prior to breaking since fewer bacteria will enter the product during breaking. This is especially the case if dirty eggs are being broken.

Control of Spoilage of Shell Eggs

One of the best means of controlling spoilage of shell eggs is to store only clean eggs. This means that nests which will prevent eggs from being dirtied when laid should be used by the producer. Stored eggs should be held at −1.1–0°C (30–32°F) under relative humidities of 80–85%. Higher humidities facilitate molding.

Shell eggs may be thermostabilized by dipping in hot water or oil for periods long enough to coagulate a thin layer of the albumin next to the shell. This is effective in preventing bacterial spoilage during holding; but there is, of course, some loss of albumin; and the functional characteristics of the egg proteins may be degraded to some extent by this method of treatment.

Liquid Egg Products

Liquid egg products which are to be dried or frozen ordinarily contain a few hundred thousand bacteria per milliliter. Today such products are pasteurized at 60–62.8°C (140–145°F), for 1–4 min. This is actually done to destroy salmonellae, but it also lowers the bacterial count.

In an investigation to determine the bacterial flora of liquid whole egg products, the principal genera before and after freezing were *Flavobacterium, Alcaligenes, Proteus, Pseudomonas,* and *Escherichia* although other genera were present (Wrinkle et al., 1949).[40]

In the manufacture of liquid whole-egg products the magma should be quickly mixed, screened, pasteurized, and cooled to 4.4°C (40°F) prior to placing in cans for freezing or prior to drying. Prompt handling and cooling will greatly assist in the manufacture of egg products of suitably low bacterial counts. Care should be taken to eliminate all spoiled eggs from the batch since even one bad egg will greatly increase the bacterial count of the batch.

The defrosting of frozen egg products presents somewhat of a problem, especially, if to accomplish melting, the product is held at room temperature. Significant increases in bacterial counts are not obtained if defrosting is carried out in air at 12.8°C (55°F) (requires 72 hr for a 30-lb tin) or in running water at 14.4°C (58°F) (requires 24 hr) (Winter and Wrinkle, 1949).[38]

References

1. L. R. Alford, N. E. Holmes, W. J. Scott, and J. R. Vickery, Studies on the preservation of shell eggs. 1. The nature of wastage in australian export eggs, *Aust. J. Appl. Sci.* **1** (1950), 208.
2. American Meat Institute Foundation, "The Science of Meat and Meat Products," Freeman, San Francisco, Calif., 1960.
3. J. C. Ayres, Microbiology of meat animals, *Advan. Food Res.* **6** (1955), 109.
4. J. C. Ayres, The relationship of organisms of the genus pseudomonas to the spoilage of meats, poultry and eggs, *J. Appl. Bacteriol.* **23** (1960), 471.
5. J. C. Ayres, Sanitation practices in egg handling, *J. Appl. Bacteriol.* **30** (1967), 106.
6. J. C. Ayres, W. S. Ogilvy, and G. F. Stewart, Post mortem changes in stored meats. Part 1, Microorganisms associated with development of slime on eviscerated, cut-up poultry, *J. Food Technol.* **4** (1950), 199.
7. E. M. Barnes, The effect of chlorinating chill tanks on the bacteriological condition of processed chickens. *Froid Suppl.* (1965), 219.
8. R. G. Board, The growth of gram negative bacteria in the hen's egg, *J. Appl. Bacteriol.* **27** (1964), 350.
9. R. G. Board, The course of microbial infection of the hen's egg. *J. Appl. Bacteriol.* **29** (1966), 319.
10. C. H. Castell, J. F. Richards, and I Wilmot, *Pseudomonas putrefaciens* from cod fillets, *J. Fish Res. Board Can.* **7** (1949), 430.
11. T. Chai, C. Chen, A. Rosen, and R. E. Levin, Detection and incidence of specific species of spoilage bacteria on fish. II. Relative incidence of *Pseudomonas putrefaciens* and fluorescent pseudomonads on haddock fillets, *Appl. Microbiol.* **16** (1968), 1738.
12. R. H. Deibel and C. F. Niven, Microbiology of meat curing. I. The occurrence and significance of a motile microorganism of the genus lactobacillus in ham curing brines, *Appl. Microbiol.* **6** (1958), 323.
13. E. E. Drewniak, M. A. Howe, H. E. Goresline, and E. R. Baush, Studies on sanitizing methods for use in poultry processing, U.S. Dep. Agr. Circ. 930, 1954.
14. R. P. Elliott and H. D. Michener, Factors affecting the growth of psychrophilic microorganisms in foods. *U.S. Dep. Agri. Tech. Bull.,* No. 1320, 1965.

15. W. A. Empey and W. J. Scott, Investigations on chilled beef. 1. Microbial contamination acquired in the meatworks. Commonwealth of Australia, Council for Scientific and Industrial Research, Bulletin No. 126, 1939.
16. I. Erichsen, The microflora of semi-preserved fish products. 111. Principal groups of bacteria occurring in tidbits. *Antonie Van Leewwenhoek J.* **33** (1967), 107.
17. A. J. Farrel and E. M. Barnes, The bacteriology of chilling procedures used in poultry processing plants, *Brit. Poultry Sci.* **5** (1964), 89.
18. R. H. Forsythe, The effect of cleaning on the flavor and interior quality of shell eggs, *Food Technol.* **6** (1952), 55.
19. J. A. Garibaldi, Factors in egg white which control growth of bacteria, *Food Res.* **25** (1960), 337.
20. M. Jaye, R. S. Kittaka, and Z. J. Ordal. The effect of temperature and packaging material on the storage life and bacterial flora of ground beef, *Food Technol.* **16** (1962), 95.
21. L. B. Jensen and W. R. Hess, A study of ham souring, *Food Res.* **6** (1941), 273.
22. R. A. Laycock and L. W. Regier, *Pseudomonads* and *Archromobacter* in the spoilage of irradiated haddock of different pre-irradiation quality, *Appl. Microbiol.* **20** (1970), 333.
23. J. J. Licciardello, L. J. Ronsivalli, and J. W. Slavin, Effect of oxygen tension on the spoilage microflora of irradiated and nonirradiated haddock *(Melanogrammus aeglefinus)* fillets, *J. Appl. Bacteriol.* **30** (1967), 239.
24. A. Lifshitz, R. C. Baker, and H. B. Nayler, The relative importance of chicken egg exterior structures in resisting bacterial penetration, *J. Food Sci.* **29** (1964), 94.
25. G. Lindeberg, Experimental view-points on ropiness of fish brine. *Proc. Symp. Cured and Frozen Fish Technol.,* Goteborg, Sweden, 1954.
26. C. F. Niven, Jr., Sausage discolorations of bacterial origin, *Amer. Meat Inst. Found. Bull.* No. 13, 1951.
27. C. F. Niven, Jr., L. G. Buettner, and J. B. Evans, Thermal tolerance studies on the heterofermentative lactobacilli that cause greening of cured meat products, *Appl. Microbiol.* **2** (1954), 26.
28. M. Patashnik, New approaches to quality changes in fresh chilled halibut, *Comm. Fish. Rev.* **28** (1966), 1.
29. Proceedings Second International Symposium on Food Microbiology. The Microbiology of Fish and Meat Curing Brines, Her Majesty's Stationary Office, London, England, 1958.
30. H. Riemann, The Survival of Some Group D *Streptococci* in Ham Curing Brines. Publication No. 8, Danish Meat Res. Inst. 263–269, 1957.
31. A. L. Romanoff and A. J. Romanoff, "The Avian Egg," Wiley, New York, 1949.
32. B. G. Shaw and J. M. Shewan, Psychophilic spoilage bacteria of fish, *J. Appl. Bacteriol.* **31** (1968), 89.

33. J. M. Shewan and G. Hobbs, The bacteriology of fish spoilage and preservation, *Progr. Ind. Microbiol.* **6** (1967), 171.
34. M. J. Thornley, A taxonomic study of *Acinetobacter* and related genera, *J. Gen. Microbiol.* **49** (1967), 211.
35. P. C. Trussel, Bacterial spoilage of shell eggs. 1. Bacteriology. *Food Technol.* **9** (1955), 126.
36. W. M. Urbain, Radiation pasteurization of fresh meat and poultry. 8th Annual A.E.C. Food Irradiation Contractors Meeting, pp. 161–166, 1968.
37. H. W. Walker and J. C. Ayres, Incidence and kinds of microorganisms associated with commercially dressed poultry, *J. Appl. Microbiol.* **4** (1956), 345.
38. A. R. Winter and C. Wrinkle, Proper defrosting methods keep bacterial counts low in frozen egg products, *U.S. Egg Poultry* **55** (1949) 28.
39. J. Wismer-Pedersen and H. Riemann, Preslaughter treatment of pigs as it influences meat quality and stability, Found. Univ. of Chicago, *Proc. Res. Conf. Amer. Meat Inst.* 12th pp. 89–106, 1949.
40. C. Wrinkle, H. H. Weiser and A. R. Winter, Bacterial flora of Frozen egg products, *Food Res.* **15** (1949), 91.

CHAPTER 8

Fruits and Vegetables

All fresh, unprocessed, plant foods go through various stages of ripening known as senescence (Duckworth, 1966).[1] This phenomenon of ripening is a normal physiological response which may result in many physiological disorders affecting the quality of the product. However, the greatest single cause of reduced quality and quantity of fruits and vegetables is microbial spoilage which occurs or is initiated prior to harvesting.

Microbial spoilage can occur during any stage of development up to final harvesting. Food processors are more concerned with microbial spoilage during harvesting, transportation, and storage of fruits and vegetables than during the growing of these products since microbial spoilage prior to harvesting is usually the responsibility of the farmer.

Certain microorganisms cause various plant diseases. Some types of microorganisms are true pathogens in the sense that they are able to invade perfectly healthy tissues in order to develop at the expense of the host. As the tissues become more senescent, there is a deterioration of tissue integrity and the over-all resistance to invasion by microorganisms is lost. This type of invasion of microorganisms is primarily saprophytic in nature, rather than parasitic; and although differences in the characteristics of the fruit or vegetable may be selective as far as the invading species are concerned, a more general type of spoilage is now observed. Parasitic types of spoilage can be controlled by development of disease-resistant plants. Also, the use of certain sprays during development of the fruit or vegetable can be quite effective in controlling disease and spoilage.

The second facet of the microbiology of fruits and vegetables occurs during the postharvesting handling of these products. Any type of mechanical handling or bruising which softens, ruptures, or in any way breaks down natural barriers to microbial invasion increases the probability of microbial spoilage.

Another important factor in determining the types of organism which can grow and cause losses of fruits and vegetables is the pH of the particular specimen. For most fruits the pH is quite low ranging from about 2.3 for lemons to around 5 for bananas. This fact reduces the probability that bacteria will be the spoilage agent of fruit and increases the probability that molds will be the etiological agent. The pH range for vegetables is somewhat higher and is normally between 5.0 and 7.0. Therefore, vegetables can be expected to be more susceptible to bacterial spoilage than are fruits. In fact, it has been reported that while rotting of fruit as a result of bacterial action is negligible, up to 36% of the total losses of vegetables due to microbiological spoilage is caused by bacteria.

The environment to which fruits and vegetables are normally exposed is such that these products come in contact with large numbers of different types of microorganisms. This is especially true of fruits and vegetables which are in close contact with the soil. Fruits that are tree-borne would be expected to be primarily exposed to air-borne spores of molds. When fruits and vegetables are contaminated with microorganisms, those organisms contacting surfaces may or may not enter the product in question. This is because of the fact that layers of the product concerned, when intact, tend to prevent the penetrating of microorganisms. Microorganisms may enter the internal tissues of fruits or vegetables by passing through natural orifices, such as the stomata and lenticles. The cuticle, or outer covering, may also be penetrated by certain invading species, especially if it is thin. Other organisms responsible for stem-end rots in certain fruits normally enter through the calyx during the flowering stage (Zdenka et al., 1959).[2]

Microorganisms Primarily Responsible for Spoilage of Fruits and Vegetables

Certain general types of microbial spoilage are commonly found in most fruits and vegetables. The nature of the decomposition and the type of organism causing it are somewhat characteristic of the particular fruit or vegetable (Tomkins, 1951; Charley, 1959).[3,4] Some general types of spoilage follow.

Bacterial Soft Rot

This form of spoilage is normally caused by *Erwinia carotovora* and related species, as well as by *Pseudomonas* species. Practically all vegetables may be attacked by these bacteria, but carrots, celery, and potatoes are the commodities most often affected. A common feature of these spoilage organisms is that they have the ability to secrete pectin-decomposing enzymes which soften and breakdown the plant tissues. This facilitates spread of the organism through the host and provides for the entrance of other organisms devoid of the ability to produce pectin-decomposing enzymes.

Gray Mold Rot

Gray mold rot is caused by species of *Botrytis* commonly *Botrytis cinerea* (This mold has a gray-colored mycelium.), and high humidity and warm temperatures favor its growth. Most fruits and vegetables can be attacked by this organism.

Rhizopus Soft Rot

This type of rotting is caused by molds of the genus *Rhizopus*. A cottony growth of mycelia with small, black dots of sporangia, sometimes covering the food, is characteristic of this type of spoilage. Most fruits, except citrus species, can be attacked by *Rhizopus*.

Alternaria Rot

This fungus causes a firm black rot in the later stages of development. In the earlier stages the rot is greenish-brown in color. Citrus, pome fruits, tomatoes and root vegetables are the commodities most commonly attacked by this fungus.

Blue Mold Rot

Various *Penicillium* species cause this type of rot. The bluish-green color of the rot is the result of the blue–green spores produced by this fungus. The various species of *Penicillium* are probably the most destructive fungi for fruits and vegetables. It has been reported that 30% of all fruit decay is caused by *Penicillium italicum* (a blue mold), *Penicillium digitatum* (a green mold), and *Penicillium expansum. Penicillium italicum,* and *Penicillium digitatum* are commonly found on citrus fruits while *Penicillium italicum* is also associated with pome fruits.

Downy and Powdery Mildews

Species of *Phytophtora* and *Peronospora* are the primary causative agents of downy mildew. These molds grow as white woolly masses. Lemons, limes, strawberries, melons, tomatoes, leafy vegetables, potatoes, and carrots are the commodities most often spoiled by *Phytophtora* species, while onions are more frequently spoiled by *Pernospora*.

Sclerotinia Rots

Molds of the genus *Sclerotinia* cause several types of spoilage. For instance, these organisms usually cause a brown rot in fruits and a watery soft rot in vegetables. Fruits commonly attacked include pome fruits, lemons, and limes, while among the vegetables endives, lettuce,. legumes, onions, and root vegetables are subject to rotting because of the growth of these fungi.

Stem-End Rots

Some species of molds characteristically grow and invade the stem ends of fruits. *Fusarium, Diplodia, Phomopis,* and others are the genera involved in this type of spoilage. Bananas, citrus fruits, sweet potatoes, and egg plant are sometimes spoiled by this type of rot.

Black Mold Rot

Aspergillus is the genus of mold which is commonly associated with black mold rots and the species most frequently involved is *Aspergillus niger*. The dark-brown to black masses of spores causes the black appearance of the mold. Fruits, including bananas, figs, dates, and grapes are often spoiled by the growth of this mold.

Other fungi may cause a black rot of fruits or vegetables, and these are species of *Alternia, Physalospora,* and *Ceratostomella. Alternaria* species usually cause rotting in citrus fruits, pome fruits, tomatoes, and root vegetables. *Physalospora* species usually infect the pome fruits and *Ceratostomella* species more often affect sweet potatoes.

Anthracnose Rot

A spotting of the leaves or the edible portion of fruits or the seed pods of vegetables is caused by *Colletotrichum lindemuthianum*. Citrus fruits, bananas, avocados, tomatoes, and beans are examples of commodities most commonly invaded by this organism.

Other

Other types of spoilage that are caused by fungi include pink mold rot caused by *Trichothesium roseum* and green mold rot caused by *Cladisporium*. A variety of fruits and vegetables are affected.

Control of the Microbial Spoilage of Fruits and Vegetables

There are several aspects to reduction of spoilage of fruits and vegetables. It is desirable to keep the initial microbial contamination as low as possible. Also, these commodities should be handled and stored in such a manner that further contamination with infective organisms is limited and the conditions for the proliferation of such organisms are unfavorable. In addition, it is important that these commodities are handled in such a manner that mechanical disruption of tissues does not occur.

Control of Initial Contamination

Contamination with spoilage organisms may take place either during or after harvesting. Much of the initial microbial load of fruits and vegetables is due to contamination with soil-borne organisms. Soil-borne contamination should be minimized. The collection boxes and field machinery used for harvesting, as well as for grading of the various products, should be kept as clean as possible. The containers used to hold fruit and vegetables during storage must be kept scrupulously clean and the floors and walls of the storage areas should be free from soil and infective organisms.

Control of Contaminating Spoilage Organisms

The prevention of contamination of various products with microorganisms involves the removal, inhibition, and destruction of these agents.

The number of microorganisms contaminating a given product depends upon such factors as the locality where the crop was grown, the weather during the growth season, the species or infective agent, and the degree of exposure to contamination of various parts of the product in question.

The simplest and most commonly used method of removal of contaminating organisms is by washing. However, certain precautions and limitations of this method should be pointed out. First of all, washing can cause the microorganisms to spread from an infected area to an uninfected one. Also, the water film left on the surface of a product may encourage the growth of microorganisms. Furthermore, the water itself may be a source of contamination. Consequently, the use of a germicide in the water is advisable. Chlorinated water at a concentration of 50–125 ppm of available chlorine is quite effective in decontaminating surfaces and preventing the spread of microorganisms from one part of the product to another.

Dehydroacetic acid,

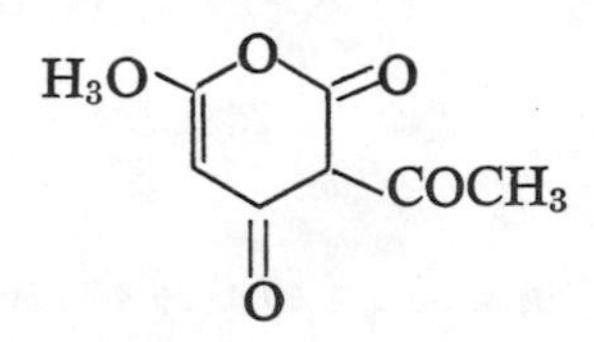

in water at concentrations between 0.5 and 1.5% has been shown to be quite effective in lowering the incidence of rotting in peaches, strawberries, raspberries, cherries, and blackberries (Smith, 1962; VonSchelhorn, 1951).[5,6] Other acids, such as sorbic and peracetic acid, $CH_3—\overset{O}{\overset{\|}{C}}—O—OH$, also have been shown to reduce spoilage of various products when applied in water solution in the form of postharvest washes, dips, and sprays.

The inhibition of growth of microorganisms on contaminating fruits and vegetables involves the same physical principles as has been indicated for various other food products. These include storing the product at a temperature that is low enough to slow down the growth of microorganisms but high enough to be compatible with the physiology of the commodity in question and to prevent freezing. Furthermore, controlling the storage atmosphere by use of "inert" gases, such as nitrogen and carbon diozide, not only retards various physiological aspects of ripening or other biochemical changes that normally occur when a product is stored, but also will tend to inhibit the growth of many microorganisms.

As discussed earlier, the use of germicides in the wash water is an effective method of reducing spoilage or rotting of various fruit and vegetables. Other chemicals that destroy or inhibit the growth have been impregnated in wrapping papers, box liners, and so on, as well as being sprayed on or dipped onto the product in question. Sodium orthophenylphenate has been effectively used at a concentration of 2% or less. This compound when used as a dip is said to be quite effective in combating decay in several products including peaches, pears, mangoes, and sweet potatoes.

Impregnating wrapping papers and box liners with diphenyl has proved highly effective against *Penicillium* and the organisms causing stem-end rots in citrus fruits (Smith, 1962; VonSchelhorn, 1951).[5,6] Diphenyl is fungistatic rather than fungicidal; and, since the compound volatizes slowly from the paper, rotting may, therefore, occur if the product is removed from the source of the inhibiting compound.

Since the bruising of fruit and vegetables disrupts tissues and provides an avenue of entrance for spoilage microorganisms, it is evident that these products should be handled in such a manner that mechanical injury does not occur. Bruising also liberates cellular enzymes which may now accelerate physiological changes, as well as further facilitate the entrance of spoilage organisms.

References

1. R. B. Duckworth, "Fruits and Vegetables," pp. 95–118. Pergamon, New York, 1966.
2. S. R. Zdenka, Etinger-Tulczynska and M. Buck, The microflora within the tissue of fruits and vegetables, *J. Food Sci.* **28** (1959), 259.

3. R. G. Tomkins, The microbiological problems in the preservation of fresh fruits and vegetables, *J. Sci. Food Agri.* **2** (1951), 381.
4. V. L. S. Charley, The prevention of microbiological spoilage in fresh fruit, *J. Sci. Food Agr.* **10** (1959), 349.
5. W. L. Smith, Chemical treatment to reduce post-harvest spoilage of fruits and vegetables, *Botan. Rev.* **28** (1962), 411.
6. M. VonSchelhorn, Control of microorganisms causing spoilage in fruit and vegetable products, *Advan. Food Res.* **3** (1951), 431.

CHAPTER 9

Microbiology of Dairy Products

No attempt will be made in this chapter to cover the microbiology of all types of dairy products or to discuss all of the different precautions and procedures significant to the dairy industry. More complete coverage of the microbiology of dairy products should be and has been taken up by separate volumes on the subject.

The most commonly used dairy products have been included in this volume since they represent a very significant proportion of the food of the United States and other countries. Some consideration of the most important dairy products is, therefore, pertinent in understanding the microbiology of foods.

Fluid Milk and Cream

Bacteria Present

Even milk drawn aseptically has been found not to be sterile (Wayne and Macey, 1933),[34] although the number of bacteria in milk removed from the udder of the cow in this manner has been found to be in the range of only 10^2–10^3 per milliliter. *Micrococcus caseolyticus* and *Micrococcus freudenreichii* and *Streptococcus liquefaciens* have been the organisms most frequently found in aseptically drawn milk (Knowles, 1936).[22]

Actually if no infection of the animal is involved, the microorganisms present in the milk prior to drawing have little significance, since external sources will contribute greater numbers of bacteria; and these types will be better suited to growth in milk, under the conditions in which it is held, than are those already present as the milk is drawn.

The equipment used in milking and in the holding of milk, the coat of the animal, and the air in the barn or stable, are the chief sources of the organisms added to fluid milk during drawing, cooling, etc. Those workers involved in the milking operation will also contribute some bacteria to the product from hands, clothing, and by other means. After milk has been drawn, it will ordinarily contain a few thousand bacteria per milliliter, depending upon the existing conditions of sanitation and the precautions used in handling the product. Various proportions, again influenced by sanitary conditions, of *Streptococcus lactis* types, *Pseudomonas, Bacillus, Aerobacter,* and *Escherichia* species are present in freshly drawn milk although it should not be interpreted that only these species may be present.

Spoilage of Raw Milk

If milk is adequately and promptly cooled on the farm and transported to the bottling plant under suitable conditions, it will not become spoiled prior to pasteurization or other processing because of the growth of microorganisms. Under unsuitable conditions of handling, raw milk may undergo various types of microbial deterioration although today in the United States spoilage of raw milk is infrequent. Various changes in milk, of microbial origin, have been described.

Streptococcus lactis grows well in milk, and, if its growth is not arrested either by pasteurization or by the application of low temperatures [growth is slow at 10°C (50°F)] will produce enough acid (about 0.25%, mostly lactic but also small amounts of acetic and propionic) to cause the milk to sour. Also, the *maltigenes* variety of *Streptococcus lactis* may produce off-flavors which have been characterized as malty (Hammer and Babel, 1957)[13]; the *hollandicus* variety of this organism may grow and cause the milk to become ropy or slimy (Hammer and Babel, 1957)[13]; and the *anoxyphilus* and *tardus* varieties may produce enzymes which cause coagulation of the milk before souring occurs (Haniman and Hammer, 1931).[15]

Raw milk may undergo what is called a gassy fermentation due to the growth of the coliform group, certain clostridia (Hammer and Babel, 1957)[13] or yeasts (Hunter, 1918).[16] *Candida* and *Torula* species of yeasts have been cited as causing this type of change in raw milk.

Micrococcus (Prouty, 1943),[26] *Alcaligenes* (Gainor and Wegemer, 1954),[11] *Lactobacillus* (Haniman and Hammer, 1931),[15] *Streptococcus* (Hammer and Babel, 1957),[13] and *Escherichia–Aerobacter* (Kelly, 1932)[21] species have been found to cause ropiness in raw milk.

Certain bacilli and streptococci have been found to grow in raw milk and cause curdling without souring in certain instances.

Milk contains fats, in part composed of short-chain fatty acids such as butyric acid. This acid and others have a definite flavor and odor which is undesirable in milk or in a number of other dairy products. *Pseudomonas fluorescens* or *Pseudomonas fragi,* which produce lipase, have been found to cause butyric off-flavors in raw milk (Collins and Hammer, 1934).[6] Some *Achromobacter* species have also been found to cause this type of change. The growth of coliform bacteria in raw milk may be the cause of unclean flavors, and organisms of the *Actinomyces* group may grow and cause moldy flavors (Fellers, 1922).[9]

Red or blue or yellow colors may be produced in milk (Hammer and Babel, 1957).[13] Blue colors may be due to the growth of *Pseudomonas syncyanea; Serratia marcescens* or the yeast *Rhodotorula glutinis have* been found to cause red colors in milk. *Pseudomonas synxantha* may cause a yellowing in milk together with off-flavors.

In the usual circumstance growth to large numbers of organisms, around 1.0×10^7 per milliliter or higher, are required to cause color, flavor, or consistency changes in milk, but some organisms may cause sweet curdling at cell concentrations between 1.3 and 4.9×10^6 per ml (Hammer and Babel, 1957)[13]; and the growth of *Pseudomonas graveolens* has been reported to cause off-flavors at cell concentrations of 5.5×10^6 per milliliter (Gould and Jensen, 1944).[12]

Pasteurized Milk

Heat pasteurization of milk was started in the United States in the early part of the 20th century, not as a public health measure but for the purpose of extending the storage life of milk at refrigerator temperatures above freezing (Shrader, 1939).[30] As time has gone on pasteurization has become a necessity for the prevention of the transmission of such diseases as typhoid, salmonellosis, tuberculosis, and brucellosis.

In pasteurization treatment by the batch method milk or cream must be heated to 62.8°C (145°F) and held at this temperature for at least 30 min before cooling. By the high-temperature–short-time method of pasteurization milk or cream is flash-heated to 71.7°C (161°F), held for at least 15 sec and then cooled. It is only in the last few years that the heating temperature requirements have been raised from 60 to 62.8°C (140 to 145°F) and from 71.1 to 71.7°C (160 to 161°F), respectively. This has been done to prevent the transmission of Q fever, a rickettsial disease, the causative agent of which was apparently not necessarily destroyed by the former methods of pasteurization.

Specifications for the time–temperature requirements for pasteurized milk seem to have been established on a somewhat arbitrary basis. It does not appear to be the case that survival curves for the various organisms pathogenic to man, and transmittable through milk, have been determined for various temperatures in milk and related to the destruction of some number of organisms by present time–temperature requirements. The adequacy of pasteurization treatment seems to have been judged on the basis of whether disease outbreaks could be traced to pasteurized milk, and this is apparently the reason for recent increases in the temperature requirements for pasteurization procedures.

Thermophilic bacteria, both vegetative *(Lactobacillus thermophilus)* and spore-forming types *(Bacillus calidolactis)* have been isolated from pasteurized milk. Various thermoduric organisms have also been isolated from pasteurized milk (Hammer and Babel).[13] The latter include *Micrococcus, Streptococcus, Microbacterium,* and *Sarcina* species. *Streptococcus thermophilus, Streptococcus durans, Streptococcus bovis, Streptococcus faecalis,* and *Streptococcus paracitrovorus* are probably of greatest importance among thermoduric strains in milk since some of them may be concerned with desirable changes taking place in the development of the normal flavor or texture of certain dairy products, and others may grow at refrigerator temperatures above freezing and thus may be concerned with the spoilage of pasteurized milk.

Spoilage of Pasteurized Milk

When milk, after pasteurization, is mishandled by holding at high temperature, spoilage may occur due to acid formation and souring caused by the growth of lactic streptococci. However, sweet curdling may occur before souring. At low temperatures some of the lactic streptococci may grow, but their growth is slow, and souring or acid curdling usually does not take place (Haniman and Hammer, 1931).[15] Sweet curdling or curdling with the development of foul odors sometimes occurs when milk is held at refrigerator temperatures above freezing for long periods. This is said to be the type of change brought about by the growth of psychrophilic bacteria such as *Pseudomonas fragi* (Burgwald and Josephson, 1947).[5] However, when *Pseudomonas* species cause the spoilage of pasteurized milk, these organisms must get into the product through recontamination, since the heat treatment applied during pasteurization is sufficiently severe in time–temperature effect to destroy large numbers of vegetative cells with survival characteristics of the *Pseudomonas.*

Standards for the bacterial content of milk and cream have been established by the U.S. Public Health Service. These standards are listed in Table 9–1. No limits are provided for pasteurized grade C milk and cream.

It should be noted that since standards of milk were first established in the early part of the 20th century, the requirements for bacterial count standards for some dairy products have several times been made more restrictive and the number of viable cells allowed for a particular product has been reduced.

Control of Spoilage of Fluid Milk

Good sanitary procedures on the farm, prompt cooling to a suitably low temperature, and prompt delivery to the bottling plant are the factors which

TABLE 9–1

Bacterial Standards for Milk and Cream
(U.S. Public Health Service, 1953)

Product	Maximum count/ml		Maximum coliform count/ml	
	Raw	Pasteurized	Raw	Pasteurized
Certified milk	10,000	500	10	1
Raw Grade A milk	50,000	—	—	—
Pasteurized Grade A milk	200,000	30,000	—	10
Pasteurized Grade B milk	1,000,000	50,000	—	10
Pasteurized Grade A cream	400,000	60,000	—	10
Pasteurized Grade B cream	2,000,000	100,000	—	10

provide for the prevention of spoilage of raw milk. Regarding sanitation, the air of the stable or barn should be kept as free of dust as possible. To facilitate this, milking should be done before feeding. The coat of milk cows should be kept clean, and the udder should be cleaned before the cow is milked. The clothes and hands of the milker should be kept clean. All equipment contacting milk, including milking machines (all parts), pails and milk cans, or bulk cooling tanks should be thoroughly cleaned and treated with chlorine solution or other suitable sanitizing agent after they have been used. If not cleaned and sanitized just prior to using, these utensils should be stored in such a manner that no puddles of liquid remain therein; and they should be stored under conditions in which they are not subject to air-borne contamination.

Milk should be promptly cooled soon after it is drawn. Today cold-wall bulk-cooling milk tanks are in use on most farms. The milk in such tanks can be gently agitated to facilitate cooling. The larger cooling surfaces and the agitation, which provides contact of more milk with cooling surfaces, allows much faster cooling in bulk tanks than is possible by placing milk in cans and immersing the cans in refrigerated water. It is preferable to cool milk to a temperature below 1.7–4.4°C (35–40°F) promptly after drawing and to hold it at this temperature until delivered.

High-count pasteurized milk may result from improper control of pasteurization, poor plant sanitation, or holding at high temperatures after pasteurization. The controls on pasteurizing equipment should be such that proper time–temperature treatment is given to all parts of the milk (including foam in batch-type pasteurizing methods). All pipelines should be cleaned and sanitized periodically so that there is no bacterial build-up in such areas. All equipment, including holding tanks, pipe lines, joints in pipe lines, pumps, homogenizers, filling machines, etc., should be of sanitary design. Also, this equipment should be cleaned and sanitized periodically. At one time the washing and sanitizing of milk bottles was a very important factor in the prevention of spoilage of pasteurized milk. This is still a factor when bottles are used. Today retail milk is usually packaged in waxed cartons, and it has been shown that such containers contribute insignificant numbers of bacteria to fluid milk.

After pasteurizing milk should be cooled to about 1.7°C (35°F) since it will warm up slightly during filling into cartons or other containers. Filled cartons should be held at temperatures between 1.7 and 4.4°C (35–40°F) until delivered to retail outlets.

Butter

Source of Organisms Present

Butter may be manufactured from cream which has been pasteurized, inoculated with known species of bacteria, and incubated, or the cream may be allowed to develop some acid from the growth of bacteria already present. In cultured cream the bacteria present, hence those present in the butter,

FIG. 9–1. Technician preparing starter culture. Courtesy of Kraftco.

will consist primarily of the organisms inoculated; but some organisms present in the cream will have survived the pasteurization treatment, and during the manufacturing procedures some bacteria, yeasts, and molds will be picked up from piping, pumps, cooling vats, the churn, the water used for washing the butter, and the employees who carry out the various procedures. Butter from uncultured cream would contain, as well as the organisms responsible for souring the cream, some of those organisms, including vegetative types, which were picked up at various points between drawing the milk and churning. Ordinarily, the lactic acid streptococci would have grown to greatest numbers in cream allowed to sour naturally and especially in cultured cream.

Deteriorative changes in butter may be caused by vegetative types of bacteria. These organisms, either present in butter in small numbers at the start because the cream was not pasteurized prior to culturing and churning, or present in small numbers because of contamination during manufacturing procedures, may be very important to the eventual quality of the product.

FIG. 9–2. Bulk holding tanks for raw material. Courtesy of Kraftco.

The Process of Flavor Development in Butter Manufacture

In order to have suitable flavor development in the cream used for manufacturing butter, there must first be production of lactic acid to provide the correct pH, after which the flavor compounds, acetylmethylcarbinol and especially diacetyl, are formed by fermentation (see Fig. 9-1).

Streptococcus lactis or *Streptococcus cremoris are* the organisms which grow and produce lactic acid to lower the pH to suitable levels. *Streptococcus citrovorus* and *Streptococcus paracitrovorus* metabolize compounds in the cream to acetylmethylcarbinol and diacetyl, and these organisms are ordinarily used to culture cream after pasteurization if the cream is to be used for butter manufacture (Hammer and Babel, 1957)[13] (see Fig. 9-2).

Streptococcus lactis and *Streptococcus cremoris* are homofermentative producing chiefly lactic acid, and the manner in which this acid is produced from lactose in milk or cream is well known. This is a long and complex series of chemical reactions; conversion of lactose to lactic acid is said to take place by an adaption of the "Embden–Myerhof System" (Jenness and Patton, 1959).[17] Other systems for the fermentation of lactose to lactic acid may be possible.

One outstanding feature of the fermentation of lactose to lactic acid is that the reaction is one which generates the energy-rich ATP molecule as a source of energy for the microorganisms involved in its formation. Two molecules of ATP are utilized for the phosphorylation of sugars in the early part of the reaction scheme, but four molecules of ATP are produced from ADP and phosphate with the formation of pyruvic acid from diphosphoglyceric acid. Thus, there is an overall gain of two molecules of ATP (Stanier et al., 1971)[32] per mole of hexose metabolized.

The real flavor components of butter are acetylmethylcarbinol and especially diacetyl. There are at least two pathways which may be used by bacteria for the formation of these compounds. One of these pathways involves the formation of pyruvic acid during the conversion of lactose to lactic acid. In this case, however, the pyruvic acid is converted to diacetyl by one of the following sequences:

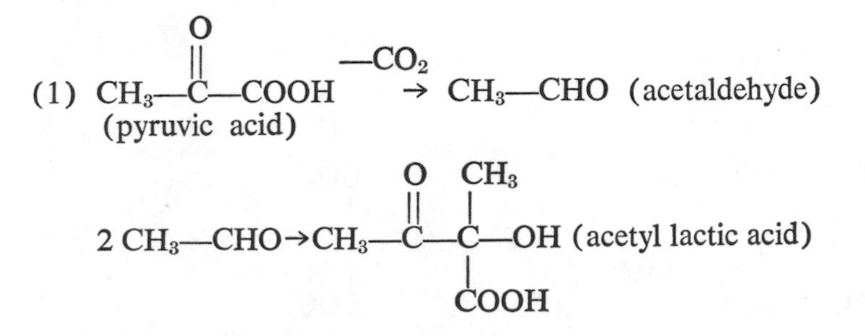

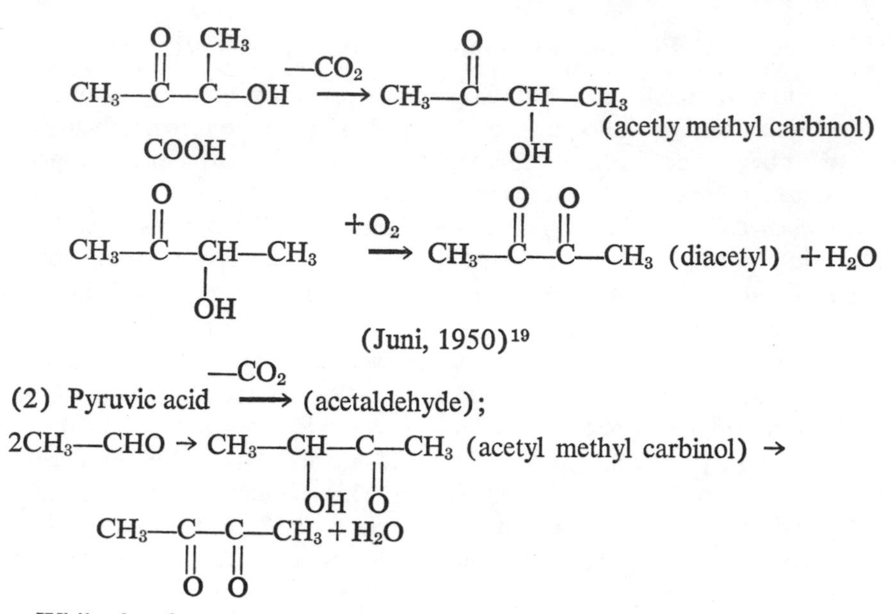

While the above mechanisms may be involved in flavor compound formation in cream cultures for butter production, it is generally considered that these flavor compounds are formed from citric acid present in the cream. It is known that in the udder of the cow citric acid is formed from pyruvic acid and that the concentration in milk may be as high as 0.2% (Stanier et al., 1971).[32] It is also known that increases in acetylmethylcarbinol formation may be obtained in cultured dairy products by the addition of citric acid in concentrations up to 0.15%.

The proposed mechanisms for the formation of diacetyl from citrates are as follows:

(1)

$$\underset{\text{(citric acid)}}{HO{-}C(COOH)(CH_2{-}COOH)_2} \rightarrow \underset{\text{(oxal acetic acid)}}{HOOC{-}C({=}O){-}CH_2{-}COOH} + CH_3{-}COOH \xrightarrow{-CO_2} \underset{\text{(pyruvic acid)}}{CH_3{-}C({=}O){-}COOH} \rightarrow \text{To diacetyl as previously indicated}$$

(Singer and Pensky, 1951)[31]

(2)

$$HO{-}C(COOH)(CH_2{-}COOH)_2 \rightarrow \underset{\text{(acetic)}}{CH_3{-}COOH} + CO_2 + CH_3{-}C({=}O){-}COOH \rightarrow \text{diacetyl}$$

(Hammer and Babel, 1957)[13]

The *Streptococcus lactis* or *Streptococcus cremoris* organisms, therefore, produce lactic acid from lactose while the *Streptococcus citrovorus* and *Streptococcus paracitravorus* organisms produce acetylmethylcarbinol and diacetyl from citric acid. The main difference between the latter two organisms is that *Streptococcus citrovorus* produces no lactic acid, while *Streptococcus paracitrovorus* produces variable amounts of D(−) lactic acid (Hammer and Babel, 1957).[13]

The question may be raised as to why, in starter cultures for butter, the acid-producing streptococci, which produce lactic acid but no acetylmethylcarbinol or diacetyl, are used. The reason is that the other organisms of the culture produce suitable amounts of these flavorful compounds only under conditions which provide a suitable range of pH. pH values of 4.7 and below furnishes the best conditions for diacetyl formation. The use of acid-producing streptococci in starter cultures, therefore, assures the production of sufficient acid to provide a suitable pH in the cream for adequate production of diacetyl.

In cultures in which acid is not formed or in a menstruum less acid than the general range of pH indicated, acetic acid and ethyl alcohol are formed instead of acetylmethylcarbinol and diacetyl.

Concentrations of diacetyl said to be desirable in dairy products are: in milk less than 1 ppm, in buttermilk 1–2 ppm, and in butter 2–4 ppm (Stanier et al., 1971).[32]

Spoilage of Butter

Some bacteria which have been associated with rancidity in butter are: *Pseudomonas fragi, Pseudomonas fluorescens, Serratia marcescens,* and *Achromobacter lipidis.* Molds of the genera *Geotrichum* and *Cladisporium* are also said to cause hydrolytic rancidity in butter (Haniman and Hammer, 1931).[15]

Whereas various off-flavors, especially those due to oxidative rancidity of the fat or due to excess acidity at the time of churning, may cause deterioration of butter, by far the most frequent cause of loss of quality is hydrolytic rancidity. Butter fat contains short-chain fatty acids which are odorful. This is especially true of butyric acid. These steam-volatile short-chain fatty acids are present in butter fat as glycerides; and while they affect the texture of butter, they are the cause of no particular flavor in the product. When the fat is hydrolyzed, however, liberated free fatty acids, butyric and other steam-volatile fatty acids, cause a definite off-flavor known as "strong" or rancid butter.

Hydrolysis of butter fat is caused by the enzyme lipase, and this enzyme is ordinarily produced in butter because of the growth of bacteria. The hydrolysis takes place as indicated in the following:

$$\begin{array}{llll} CH_2OOCR_1 & & CH_2OH & R_1COOH \\ | & \text{lipase} & | & + \\ CHOOCR_2 & + 3H_2O \rightarrow & CHOH & + R_2COOH \\ | & & | & + \\ CH_2OOCR_3 & & CH_2OH & R_3COOH \\ \text{(fat triglyceride)} & & \text{(glycerol} & \text{(fatty acids)} \end{array}$$

When cream is churned the bacterial count of this material is usually more than 1 × 10^8 bacteria per milliliter. However, during churning and washing most of the bacteria are removed in the buttermilk and wash water. The result is that butter manufactured under good conditions will have a count of a few thousand organisms per milliliter while that manufactured under unsuitable conditions of sanitation may have a count as high as 1 × 10^6 per milliliter or somewhat higher (Shrader, 1939).[30]

Butter is a water-in-oil emulsion, and it is said that there are approximately 1–1.8 × 10^{10} water droplets per gram (Rahn and Boysen, 1928).[27] If this is the case, then considering the bacterial count of the finished product, most of the water droplets must contain no bacteria. It is known that under ordinary circumstances bacteria in butter do not migrate from one water droplet to another, but some of their end products may diffuse into the oil phase of the product (Hammer and Babel, 1957).[13] The growth of bacteria in butter must take place in the water droplets; and when lipase is produced therein, it may cause hydrolysis of the surrounding fats.

Control of Hydrolytic Rancidity in Butter

One method of decreasing the probability that butter will become strong or rancid is to add salt. Concentrations of 1–3% by weight of sodium chloride are added to salted butter. Since butter contains about 18% of water, and the salt dissolves only in the water phase, this means that the salt concentration in the water phase is somewhere between 5.5 and 16.7%. These salt concentrations are sufficient to inhibit the growth of some of the bacteria producing lipase, especially some of the *Pseudomonas* species, but will not inhibit all types of lipase-producing bacteria as evidenced by the fact that salted butter may undergo the development of hydrolytic rancidity.

Since salt tends to preserve butter, sweet or unsalted butter is much more subject to hydrolyic rancidity development and other types of microbial deterioration than is salted butter.

The holding of butter in frozen storage, if the product is to be stored, and at adequately low refrigerator temperatures above freezing during distribution, retail display, and in the home, are factors which tend to discourage the growth of microorganisms in butter.

The application of good plant sanitation and sanitary procedures in the manufacture of butter is the most important means of controlling spoilage. These procedures include the proper pasteurization and culturing of cream, the provision of a suitably clean air supply in the plant, and the proper cleaning and sanitation of vats, tanks, churns, and molding and extruding equipment. Of great importance is the water supply. It has been shown that many potable waters may contain butter-spoilage organisms (Anonymous, 1960).[3] The best control of water supplies to be used for cleaning equipment and washing butter is to use water which has been subjected to breakpoint chlorination (sufficient chlorine added to oxidize all of the reactable organic matter present therein and to provide a slight residual of active chlorine).

Standards for good quality butter have been established (Davis, 1965).[7] These standards state that the yeast and mold count should not exceed 20 per milliliter and the coliform count should not exceed 10 per milliliter (butter melted at low temperatures for purposes of making counts). The U.S. Dept. of Agriculture also has set a standard of 50 per gram as the maximum proteolytic count. Proteolytic bacteria may cause various off-flavors in butter during storage since they may decompose proteins present in the water phase of the product.

Cultured Fluid Dairy Products

Soured cream, buttermilk, and yogurt are the chief cultured milk products used in the United States of America, the first-named product probably representing that of the greatest volume.

Soured cream and cultured buttermilk are prepared in much the same manner with some exceptions. With soured cream the butter fat content is usually regulated to 18%, while with buttermilk the raw material is reconstituted skim milk or fluid skim milk to which 2% of butterfat has been added. The raw material is first pasteurized, the temperature being brought to 82.2°C (180°F) and held for 30 min or more. After pasteurization the fluid material is cooled to 21.1–22.2°C (70–72°F), inoculated with a culture of *Lactobacteriaceae,* and held at the appropriate temperature until sufficient acid has been produced (Judkins and Keener, 1960).[18]

The culture used for soured cream or buttermilk is the same as that used in culturing cream for butter, namely, a mixture of *Streptococcus lactis, Streptococcus paracitrovorus,* and *Streptococcus citrovorus. Streptococcus cremories* may be used instead of *Streptococcus lactis* to produce acid (Davis, 1965).[7]

In producing soured cream the product is held at suitable growth temperatures until approximately 0.6% of acid is produced. Also, small amounts of the enzyme rennet may be added to this product. In producing cultured buttermilk the product is incubated at suitable temperatures until 0.9–1.0% of titratable acidity is produced (Judkins and Keener, 1960).[18]

If held for too long a time or at high temperatures after manufacture, soured cream and cultured buttermilk may be spoiled by the growth of molds or yeasts. These products, after manufacture, should be promptly cooled to 4.4°C (40°F) or below, packaged, and held and distributed at that temperature.

It has been stated that in good-quality cultured milks the coliform count should not exceed 10 per milliliter and that mold and yeasts counts should not exceed 10 per milliliter (Davis, 1965).[7] Higher counts for these types of microorganisms indicate improper pasteurization of the raw material, poor cleaning and sanitizing procedures for equipment, heavy air contamination or other inadequacies of manufacturing practices.

In the manufacture of yogurt the milk is boiled for purposes of concentrating, or milk solids are added to raise the solids content, and this mixture is homogenized and pasteurized and then cooled. This material is inoculated with a mixture usually consisting of *Lactobacillus* (usually *bulgaricus*) and *Streptococcus thermophilus*. The inoculated material is incubated at temperatures between 40 and 46.1°C (104 and 115°F) until a titratable acidity of 0.84–0.90% is reached. To obtain a suitable product it is considered that the proportion of lactobacilli to streptococci after fermentation should be approximately 1 to 1.

Yogurt is subject to the same types of deterioration as are cultured buttermilk and soured cream. The same microbiological standards are applied to yogurt as to the other cultured milks.

Cheeses

Cottage Cheese

In the manufacture of cottage cheese coagulation of the milk or curdling may be brought about by acid produced by bacterial growth or by a combination of acid and the action of a clotting enzyme (Judkins and Keener, 1960).[18]

Generally skim milk is first pasteurized (regular or high–short method), cooled to 31.1–32.2°C (88–90°F) and inoculated with a starter culture, with or without renin, then incubated at ambient temperatures until sufficient acid has been produced to cause coagulation. As much as 5% of culture may have to be added for a coagulation to take place in less than 5 hr while with longer sets (up to 16 hr) less culture may be used (Davis, 1965).[7]

The curd is cut and cooked at 46.1–54.4°C (115–130°F), the temperature depending upon the type of curd desired. The curd is then washed about three times starting with water at 21.1°C (70°F), the final wash water having a temperature of 0–4.4°C (32–40°F) which cools the product. The cheese is eventually milled or cut into small pieces at which time approximately 1–1.5% of salt is added. Cream to provide a butter-fat content of 4% may also be added to produce creamed cottage cheese.

According to the literature, either *Streptococcus lactis* or *Streptococcus cremoris* is used as starter cultures for cottage cheese. However, the statement to the effect that some manufacturers prefer to use *Streptococcus* cultures, as starters, which produce diacetyl (since this compound contributes to the flavor of cottage cheese) would indicate that *Streptococcus citrovorus* or *Streptococcus paracitrovorus*-like organisms may some times be included in starter cultures. Usually about 0.5–0.6% of lactic acid is produced prior to coagulation or curd formation for acid-coagulated cheese (Hammer and Trout, 1928).[14]

Spoilage of Cottage Cheese

Cottage cheese may become spoiled due to the growth of molds, yeasts, or bacteria. A variety of species cause moldiness, the most common being *Geotrichum candidum* and *Penicillium* species (Haniman and Hammer, 1931).[15] Yeasts which grow in this product and cause deterioration seem not to have been well classified, but is known that lactose-fermenting types may be involved and that some varieties are pigmented and may cause a pink discoloration of the product.

Sliminess, fruity odors, and putrid odors and flavors in cottage cheese may be caused by the growth of bacteria. The growth of *Alcaligenes metalcaligenes, Pseudomonas viscosa,* and *Pseudomonas fragi* is often responsible for these changes (Parker and Elliker, 1953; Parker et al., 1951).[24,25] Other genera such as *Proteus, Aerobacter,* and *Achromobacter* may cause sliminess (Parker et al., 1951).[25] All of these organisms, even when present in moderate concentrations in the raw materials, should be destroyed by adequate pasteurization treatment. Spoilage of cottage cheese by the growth of these organisms must, therefore, result from recontamination of the product during the manufacturing process prior to packaging.

The control of spoilage of cottage cheese would appear to involve suitable pasteurization and cooling of fluid raw materials, good sanitation practices in the plant, including cleaning and sanitation of equipment, and especially adequate cooling of the curd during the application of the last wash water. The temperature of the water used for cooling should be such as to lower the temperature of the curd to a point below 4.4°C (40°F). After milling and packaging of the product it should be placed in storage at temperatures below 4.4°C (40°F) and held under such conditions during distribution.

Standards for cottage cheese (Davis, 1965)[7] specify that the coliform, yeast, and mold counts for a product which is less than 24 hr old should be not more than 10 per gram for each type of organism.

Cheddar Cheese

Manufacturing procedures for cheddar cheese are the same as for cottage cheese up to the point of cutting, heating, and draining of the whey except that whole milk is used as raw material. Cheddar cheese is then "cheddared," or the curd is again cut into larger portions and piled, allowing more of the whey to be pressed out. After cheddaring about 2% of salt is added, and the curd is milled. The milled curd is now filled into cheesecloth or plastic-lined metal hoops which are placed in a cheese press in which more of the whey is pressed out over a period of about 18 hr. The cheese is now stored for purposes of maturing and the development of suitable texture and flavor. Starters consisting of *Streptococcus lactis* or *Streptococcus cremoris* are ordinarily used for cheddar cheese. When renin is also used for forming the curd, less acid is usually produced at the time of coagulation (Hammer and Trout, 1928),[14] about 0.2% less than is the case with cottage cheese.

A United States Food and Drug Administration specification requires that the milk used for cheddar cheese either be pasteurized prior to inoculation with starter cultures or that the cheese be aged for at least 60 days at a temperature lower than 1.7°C (35°F) prior to distribution for sale in order to insure that the products are free from pathogenic bacteria.

Cheddar cheese is ordinarily cured or held for some time prior to distribution in order to develop desirable texture and flavor qualities. Holding times during curing vary with the holding temperature and can be 12–18 months at 0–1.1°C (32–34°F), 8–10 months at 3.3–4.4°C (38–40°F) or 60 days at 7.2–12.8°C (45–55°F).

The lactic streptococci usually predominate in cheddar cheese for some time after pressing. Maximum concentrations of these organisms in cheddar cheese may reach 2.5×10^9 per gram after about 2 weeks of curing. *Lactobacteriaceae* of the *Lactobacillus casei* type then take over and eventually become the predominant organism, although strains present in cheese made from pasteurized milk seem to differ somewhat from those present in cheese made from raw milk (Haniman and Hammer, 1931).[15] Streptococci other than those of the *Streptococcus lactis* or *Streptococcus cremoris* types are commonly present in cheddar cheese, and micrococci are also commonly found. It is considered that *Streptococcus lactis* and *Lactobacillus casei* both contribute to texture changes in cheddar cheese and that they may also contribute to flavor, although strains of *Lactobacillus casei* are said to produce a typical cheddar cheese flavor (Allen and Knowles, 1934).[1] Strains of *Micrococcus freudenreichii* may play a part in the development of typical flavor in cheddar cheese (Baribo and Foster, 1952).[4]

Spoilage of cheddar cheese. Considering the manner in which cheddar cheese is handled today, the cake being cut into retail size portions and packaged at some central point prior to retail distribution, deteriorative changes may be separated into those which usually occur during curing and those occurring during retail distribution. It is known that if there is a significant growth of coliform bacteria, either *Escherichia* or *Aerobacter,* in cheddar cheese, the lactose will be fermented with the production of carbon dioxide which produce gas holes in the cake (Haniman and Hammer, 1931).[15] Such a situation, of course can be prevented by either pasteurizing the milk prior to culturing or by curing the cakes at temperatures below 4.4°C (40°F) at which temperatures these organisms will not grow.

Openness or angular open areas are produced in cheese by certain lactobacilli or streptococci which cause a fast fermentation of lactose with formation of carbon dioxide. This occurs most frequently at high curing temperatures (Sherwood, 1934).[29]

It is known that *Streptococcus liquefaciens* and yeasts *(Torula amara)* may produce bitter flavors, and the microorganisms causing rancidity of butter may cause the same type of change in cheddar cheese if proliferation of these bacteria takes place in the product. Such changes can be controlled by the pasteurization of the milk used prior to culturing and by suitable sanitation procedures for the plant and the equipment used in manufacturing the product.

In retail outlets and in the home the type of deterioration occurring most often in cheddar cheese is probably the molding of surfaces. Apparently various species of mold may grow on this product. If the cheese cake is properly waxed or, after cutting, is packaged so as to exclude air, molds will not grow on cheese since the species which are involved in this type of deterioration are aerobic. Mold inhibitors such as sorbic acid or propionates may also be added to packaging materials to prevent the growth of molds on cheese surfaces. The holding of cheese cakes or packaged product at temperatures below 4.4°C (40°F), during curing or retail display, while not absolutely eliminating mold growth, will tend to prevent molding because of the slower growth rates of the organism at these temperatures.

Much of the cheese used in the United States is "process" cheese. This is produced from cheddar cheese cakes of various ages with sodium citrate or phosphate added as an emulsifying agent. The mixture is melted to pasteurize, blended to obtain a smooth texture, then cooled, extruded, and packaged, or sliced and packaged. "Process" cheese may also be made from types of cheese other than cheddar. The pasteurization treatment, good plant sanitation, and packaging so as to exclude air and prevent the growth of molds, seem to have eliminated most microbiological problems associated with these products.

Swiss of Emmenthaler Cheese

In the manufacture of Swiss cheese the culture is added to the raw whole milk which is then warmed to 30°C (86°F); or the milk may be pasteurized, cooled to 30°C (86°F), and the culture added. Renin is added at this point, and the curd coagulates in 20–30 min. The curd is cut into small pieces and cooked. The cooking temperature is high, 48.9–57.2°C (120–135°F). The curd is then stirred in such a manner that it collects at the center of the vat, and it is taken up in a cloth by placing the cloth under the curd then gathering to remove the curd in a single mass. The curd, 90–220 lb in weight, is pressed over a period of 24 hr after which the wheels of cheese are immersed in salt solution for 1–4 days. This is done to develop a thick rind to support the weight of the curd. In recent years a plastic covering may be used to support the mass instead of cheesecloth. Curing of the cheese is carried out at 12.8–18.3°C (55–65°F) over a period of 3–6 months.

Starter cultures for Swiss cheese contain three organisms, *Streptococcus thermophilus, Lactobacillus helveticus,* and *Propionibacterium shermanii.* While most other bacteria present are destroyed by the high-temperature cooking, these organisms have a significantly higher heat resistance and are able to survive.

At first, during processing, *Streptococcus lactis* may grow to some extent along with some growth of *Streptococcus thermophilus;* the latter organism, however, seems to increase at the greatest rate during the first several hours of processing including cooking (Haniman and Hammer, 1931).[15] The lactobacilli are believed to increase during pressing and to some extent during ageing or curing (Sherman, 1921).[28] The *Propionibacterium* species increase primarily during the ripening or curing period (Frazier et al., 1935).[10]

The function of the streptococci in Swiss cheese may involve some modification of texture, through enzymes causing protein breakdown, and there may be some components produced by these organisms which contribute to flavor. One of their main functions, however, is to produce lactic acid which the *Propionibacterium* species convert to propionic acid and carbon dioxide although other compounds can be metabolized to yield these compounds. Lactic acid is apparently broken down in the following manner:

$$3CH_3—\underset{\displaystyle OH}{\underset{|}{CH}}—COOH \rightarrow 2CH_3—CH_2—COOH + CH_3—COOH + CO_2 + H_2O$$

The propionic acid in particular, and probably the acetic acid to some extent, are the components responsible for the chief flavor characteristics of Swiss cheese while carbon dioxide produces holes or "eyes" in the cheese, 0.3–1.0 in. in diameter and 1–3 in. apart, contributing to the particular texture of the product (Sherman, 1921).[28] Amino acids and their decomposition products (small amounts of NH_3 and H_2S) formed primarily by the *Propionibacterium* species are also considered to contribute to the flavor of Swiss cheese (Haniman and Hammer, 1931).[15]

Spoilage of Swiss Cheese. Lactose-fermenting yeasts *(Candida pseudotropicalis)* and strains of *Clostridium perfringens* (Weiser, 1942)[35] have been found to be responsible for the production of excessive gas with consequent excessively large and numerous gas holes in Swiss cheese. The control of this type of deterioration involves the use of a good grade of raw milk and good sanitation of plant and equipment during manufacturing procedures.

Abnormal flavors in Swiss cheese have been attributed to the growth of *Bacterium proteolyticum* and *Bacillus putrificus* (Hammer and Babel, 1957).[13] The use of raw milk of suitable quality would appear to be the best method of controlling this particular type of defect.

Slow Fermentations in Cultured Dairy Products

Insufficient growth of the cultures in fermented dairy products causes slow or insufficient acid production and may result in products of poor quality or even in spoiled products. Generally, two factors have been responsible for defects of this kind: the presence of antibiotics in the milk and contamination of the product with bacteriophage to which the acid-producing bacteria are sensitive.

Cows, producing milk, may sometimes be treated with antibiotics as a cure for infections of one type or another. Sufficient quantities of antibiotics may get into a batch of milk, through the milk of a few cows, to cause inhibition of the bacteria involved in typical fermentations in dairy products.

It has been shown that as little as 0.1–0.15 Iu of penicillin per milliliter of milk will cause inhibition of lactic cultures (Haniman and Hammer, 1931),[15] and greater amounts will completely prevent growth of these organisms. Quantities of chlortetracycline greater than 0.12 μg per milliliter will also cause inhibition of growth in lactic cultures (Katznelson and Hood, 1949).[20]

The prevention of slow fermentations in cultured dairy products due to antibiotics in the milk involves a number of precautionary measures. If possible, which may often not be the case, the dairy products manufacturer may be able to set up a sufficiently good communication with his suppliers of milk to have the milk of any antibiotic-treated cow excluded from products sold to the dairy. If the antibiotic which may be present in the milk is penicillin, then the enzyme penicillinase may be added to the batch of milk being processed. This will decompose the penicillin and prevent any inhibitory effect. The use of excessive amounts of culture will tend to obviate the effect of the presence of antibiotics in milk which is to be manufactured into dairy products through a bacterial fermentation. However, the effectiveness of such a procedure would seem to depend upon several factors and to involve some uncertainties. Finally, if reconstituted milk, instead of raw milk, is used, the problem of antibiotics may be obviated. Apparently as dried milk is held in storage antibiotics are decomposed so that their concentration in reconstituted milks is not sufficient to cause inhibition of lactic cultures.

Bacteriophages are viruses which infect bacteria and may cause them to lyse and be destroyed, thus the typical fermentation will not be produced. The subject of viruses has been treated elsewhere in this volume, and only the sources and control of bacteriophage affecting starter culture will be considered in this chapter.

The cultures which are used for inoculating milk for fermentation may contain bacteriophages to which the particular culture, or other strains of these organisms, are sensitive. Since, like bacteria, bacteriophages are widespread, those phages to which the culture may be sensitive may be present in raw milk. Because, as a rule, at some period during the manufacture of dairy products bacteriophages which affect a particular culture may have multiplied, they are generally present to some extent on utensils and other equipment and especially in the air of the dairy plant.

Since phages affecting bacteria multiply under the same general conditions as those favoring the growth of the bacteria themselves, it is generally not possible to destroy them by physical or chemical treatment without also destroying the bacteria or without adversely affecting the chemical or physical state of the raw material. Ordinary milk-pasteurizing temperatures and times, for instance, are not considered adequate to destroy bacteriophages affecting lactic cultures.

Control of the lysis of cultures used in fermenting dairy products may be brought about in several ways. To begin with, it is often advantegeous to inoculate milk with several strains of the same organism, thus bacterial strains not sensitive to the phages, which might be present in the milk from one source or another, would be included in the culture and would be able to grow in the milk. Also, the cultures used for inoculum should be obtained from several laboratories and rotated every day or, at least, every few days.

In this manner strains of lactic cultures not sensitive to the particular phages will be present in adequate numbers in the inoculum.

Probably one of the best methods of obviating the effect of bacteriophage on the growth of starter cultures in milk products is to transfer and incubate both the mother and the bulk starter cultures in a room well separated from the dairy manufacturing section of the plant. All starter cultures, of course, should be checked periodically for acid production and the viability of the culture.

References

1. L. A. Allen and N. R. Knowles, Ripening of cheddar cheese. IV. The bacterial flora of cheese made from milk of very low bacterial count. V. The Influence of the rennet flora of the cheese. VI. The influence of the starter on the ripening process., *J. Dairy Res.* **5** (1934), 185.
2. Anonymous, Milk Ordinance and Code, U.S. Dept. of Health, Education, and Welfare, U.S. Public Health Service, 1953.
3. Anonymous, "Standard Methods for the Examination of Dairy Products," 11th ed., Amer. Public Health Ass., New York, 1960.
4. I. E. Baribo and E. M. Foster, The intracellular proteinases of certain organisms from cheese and their relationship to the proteinases in cheese., *J. Dairy Sci.* **35** (1952), 149.
5. L. H. Burgwald and D. V. Josephson, The effect of refrigerator storage on the keeping qualities of pasteurized milk, *J. Dairy Sci.* **30** (1947), 371.
6. M. A. Collins and B. W. Hammer, Types of lipolysis brought about by bacteria as shown by nile blue sulfate, *J. Bacteriol.* **27** (1934), 487.
7. J. G. Davis, "Cheese," Vol. 1, Churchill, London, 1965.
8. R. P. Elliker, Retention of mold fragments by butter, butter milk and wash-water during manufacture of butter, *J. Dairy Sci.* **27** (1944), 563.
9. C. R. Fellers, Actinomyces in milk with special reference to the production of undesirable odors and flavors. *J. Dairy Sci.* **5** (1922), 485.
10. W. C. Frazier, L. A. Burkey, A. J. Boyer, G. P. Sanders, and K. J. Matheson, The bacteriology of swiss cheese, II. Bacteriology of the cheese in the press., *J. Dairy Sci.* **18** (1935), 373.
11. G. Gainor and D. E. Wegemer, Studies on a psychrophilic bacterium causing ropiness in milk. I. Morphological and physiological considerations., *Appl. Microbiol.* **2** (1954), 95.
12. I. A. Gould and J. M. Jensen, Lactic acid in dairy products. II. Relation to flavor, acidity measurements, bacterial count and methylene blue reduction., *J. Dairy Sci.* **27** (1944), 753.
13. B. W. Hammer and F. J. Babel, "Dairy Bacteriology," 4th ed., Wiley, New York, 1957,
14. B. W. Hammer and G. M. Trout, A study of the yellow cocci which survive pasteurization, *J. Dairy Sci.* **11** (1928), 18.

15. L. A. Haniman and B. W. Hammer, Variations in the coagulation and proteolysis of milk by *Streptococcus lactis.*, *J. Dairy Sci.* **14** (1931), 40.
16. O. W. Hunter, A lactose fermenting yeast producing foamy cream. *J. Bacteriol.* **3** (1918), 293.
17. R. Jenness and S. Patton, "Principles of Dairy Chemistry," Wiley, New York, 1959.
18. H. F. Judkins and H. A. Keener, "Milk Products and Processing," Wiley, New York, 1960.
19. E. Juni, α-Acetolactic acid as an intermediate in acetyl-methyl carbinal formation, *Fed. Proc.* **9** (1950), 396.
20. H. Katznelson and E. G. Hood, Influence of penicillin and other antibiotics on lactic streptococci starter cultures used in cheddar cheese making, *J. Dairy Sci.* **32** (1949), 961.
21. C. D. Kelly, The rate of change in milk brought about by certain lactic acid streptococci, *J. Bacteriol.* **23** (1932), 59.
22. N. R. Knowles, Bacterial flora of foremilk and of rennit., *J. Dairy Res.* **7** (1936), 63.
23. H. B. Monison and B. W. Hammer, Distribution of *Pseudomonas fragi*, *J. Dairy Sci.* **24** (1941), 9.
24. R. B. Parker and P. R. Elliker, Effect of spoilage bacteria on biacetyl content and flavor of cottage cheese., *J. Dairy* Sci. (1953), 843.
25. P. B. Parker, V. W. Smith, and P. R. Elliker, Bacteria associated with a gelatinous or slimy curd defect of cottage cheese., *J. Dairy Sci.* **34** (1951), 887.
26. C. C. Prouty, A ropy milk outbreak caused by a thermoduric *Micrococcus.*, *J. Milk Technol.* **6** (1943), 263.
27. O. Rahn and H. H. Boysen, Distribution and growth of bacteria in butter., *J. Dairy Sci.* **11** (1928), 446.
28. J. M. Sherman, The cause of eyes and characteristic flavor in emmental or Swiss cheese, *J. Bacteriol.* 6 (1921), 379.
29. I. R. Sherwood, The relation of certain lactic acid bacteria to open texture in cheddar cheese,
30. J. H. Shrader, "Food Control—Its Public Health Aspects," p. 112, Wiley, New York, 1939.
31. T. P. Singer and J. Pensky, Acetoin synthesis by highly purified α-carboxylase, *Arch. Biochem.* **31** (1951), 457.
32. R. Y. Stanier, M. Doudoroff, and E. S. Adelberg, "The microbial world," Prentice–Hall, Englewood Cliffs, N.J., 1971.
33. J. Van Beynum and J. W. Petter, Lactic acid streptococci in the formation of aroma in starters., *J. Dairy Res.* **10** (1939), 250.
34. R. Wayne and H. Macey, The effect of various methods for drying up cows on the bacterial and cell content of the milk, *J. Dairy Sci.* **16** (1933), 79.
35. H. H. Weiser, Microorganisms associated with gassy swiss cheese, *J. Bacteriol.* **43** (1942), 46.

CHAPTER 10

Food Infections

All food-borne diseases are classified as food infections or food intoxications. Although these classifications are somewhat arbitrary, food infections are those in which the bacteria present in the food at the time of eating grow in the host and cause disease. Food intoxications are those diseases in which bacteria grow in a food, producing a substance therein which is toxic to humans and other warm-blooded animals. In the latter case, the bacteria grow in the food and in ordinary circumstances do not grow in the host; and it is during the growth of the bacteria in the food that the poisonous substances or toxins are produced. The poisonous materials, in this case, are already present in the foods when they are eaten.

Salmonellosis

Salmonellosis is caused through the ingestion of living bacteria of the *Salmonella* group. These bacteria invade body tissues, grow, and cause typical symptoms in the host.

While salmonellosis is a notifiable disease (attending physicians must report all cases to the U.S. Public Health Service), it is recognized that cases of this infection are often not reported to public health authorities. Actually it is considered that less than 1% of the 7–10 million acute digestive illnesses which occur annually in the United States are investigated to the extent that either the actual cause, or the means by which the disease was transmitted, can be determined.

Some years ago typhoid fever, caused by *Salmonella typhi,* was a fairly common and sometimes fatal disease, whereas the ordinary *Salmonella* infection was rarely recognized. Today there are comparatively few cases of typhoid fever but the incidence in the United States of salmonellosis (excluding typhoid fever) has increased from 723 cases reported in 1946 to 18,120 cases reported in 1967 (Annual Supplement, Morbidity and Mortality Weekly Report, 1968). Even with this increase in outbreaks recorded, it is believed that many cases of this disease are still not reported.

While antibiotics are available for the treatment of both typhoid fever and salmonellosis, there are still some deaths attributed yearly to both diseases. Thus, in 1966 there were 73 deaths reported as caused by salmonellosis and 15 deaths attributed to typhoid fever (only 396 cases of the latter diseases were reported).

Even though the organism causing typhoid fever belongs to the genus of bacteria known as *Salmonella,* the disease is usually separated from salmonellosis epidemiologically because of its severe symptoms, because the organism seems to infect only the human, and because of the fact that only a few cells of *Salmonella typhi* need be ingested by the normal adult to cause the malady.

Salmonellosis—the Disease

The ordinary symptoms of salmonellosis are: abdominal pain, diarrhea, chills, frequent vomiting, and prostration. However, when a severe case of this disease is encountered much more severe symptoms may be involved. Such symptoms may include: septicemia with leukopenia (a reduced white blood cell count); endocarditis or pericarditis; meningitis, and other types of illness, including osteomyelitis.

Severe symptoms are not often encountered in the ordinary case of salmonellosis, the average adult having a gastroenteritis, although the very young and the aged may be afflicted to a greater degree and may be infected by the ingestion of fewer cells than is the case with adults.

Typhoid, more frequently a severe disease than is the case with salmonellosis, in some instances causes very high fever and ulceration of the small intestine (Peyer's patches). Death may follow because of fever or because of perforation of the intestine resulting in peritonitis.

While salmonellosis is ordinarily thought of as an intestinal disease, actual growth of the bacteria in the intestine has been difficult to demonstrate. Infection ordinarily takes place through the ingestion of milk, food, or water, but apparently the organisms then pass into the tissues through the lymph system, eventually infecting the spleen and liver and infecting the gallbladder through the liver, bile being a suitable medium for growth of these organisms. Eventually the organisms may be found in the blood, and there may be a secondary occurrence of the organism in the intestine via the contaminated bile which empties into the intestine. Fever is said to be caused by or to occur at the time of destruction of these bacteria in the blood by white blood cells (Cecil and Loeb, 1959).[8]

The incubation period for ordinary salmonellosis, is from 7 to 72 hr. With typhoid fever the incubation period is from 7 to 21 days.

Carriers

People having had salmonellosis often become carriers of the organism for a period of time after the symptoms have gone. Such carriers excrete *Salmonella* bacteria in their feces. Carriers are a public health hazard because when they handle food which will be eaten by others, they may contaminate it with the particular disease-causing bacteria and thus may be responsible for transmitting the disease to others.

The carrier stage with ordinary salmonellosis usually does not persist for periods longer than 12 weeks. With typhoid fever, however, the carrier stage may last for periods much longer than this and at least one case is on record in which a woman (the so-called typhoid Mary) was the cause of transmitting this disease to others over a period of years.

Use of Antibiotics

Antibiotics are available for the treatment of salmonellosis, and chloramphenicol is often used for this purpose. While antibiotics may be effective in reducing the severity of symptoms or even in curing the disease, no antibiotics are known to eliminate the carrier stage.

Number of Cells Necessary for Infection

In the case of typhoid fever, it is believed that the ingestion of even one living cell of *Salmonella typhi* may cause the disease. As a rule with ordinary salmonellosis, many more than one cell have to be ingested to cause infection in normal adults. Only scanty information is available regarding the number of cells which must be consumed to cause salmonellosis and the number required appears to vary with the particular species of the organism. The information which has been obtained on this subject was determined by feeding volunteers, some of whom may have received immunization treatment years previous to participating in the test and thus these individuals might have required more cells for infection than would be the case with adults who had not been immunized. The number of cells of certain salmonellae serotypes required to cause salmonellosis in normal adults is listed in Table 10–1.

It can be conjectured from the above that foods when first contaminated with these organisms do not contain the required number of these bacteria to cause infection in normal adults. However, if some mishandling of the food occurs under conditions suitable for growth of the original contaminant, the number of organisms can increase to the point that enough are present to cause infection when the food is eaten.

TABLE 10–1

Number of cells of certain salmonellae species which Must be Ingested to Cause Infection

Organism	No. cells
S. meleagridis	7,000,000 – 10,000,000
S. newport	152,000
S. bareilly	125,000
S. anatum	44,500,000 – 67,200,000
S. derby	15,000,000

While comparatively large numbers of *Salmonella* bacteria may be necessary for the infection of normal adults, there are good indications that even a very few of these organisms may cause infection in infants and the very young (Meyer, 1953).[41] The same may be the case with the elderly, although the degree to which this group is subject to infection has not been delineated. The presence of salmonellae in foods in any concentration may, therefore, be considered as undesirable and to constitute a public health hazard.

The Bacteria Causing Salmonellosis

The salmonellae are rod-shaped bacteria which are 1–3 μ in length and 0.5–0.7 μ in width. They are gram-negative and are usually motile possessing peritrichous flagella. Although salmonellae grow readily on or in artificial media or in foods, in nature they are primarily parasites distributed among man and other mammals, birds, and possibly even among some amphibians. The salmonellae grow best in the presence of oxygen but they will grow quite well under conditions in which oxygen is absent. Among the common sugars only glucose is fermented by salmonellae, with the formation of acid and gas. They do not ferment sucrose or lactose. Few *Salmonella* species liquefy gelatin, and they do not produce the enzyme urease. Almost all *Salmonella* species produce sulfide from proteins and all species are able to decarboxylate certain amino acids.

At the present time, more than 1200 different serotypes (as determined by antigen–antibody reactions) of salmonellae have been identified, and it is considered that all species are pathogenic to man.

Salmonellae and their Antigens

Salmonella species and strains are distinguished from one another and classified according to the antigens which they contain and which produce specific antibodies in the blood serum of animals when the inactivated bacterial cells are injected into these animals.

The salmonellae contain both flagella antigens (the symbol "H" is used to indicate flagella antigens) and somatic antigens (the symbol "O" is used to indicate somatic or body antigens). If the antibodies (produced in an animal by the injection of inactivated *Salmonella* cells), corresponding to the antigens contained in a particular species of salmonellae, are added to a suspension of the cells of this particular species, the cells will be caused to clump or agglutinate. This may be observed with the naked eye by suspending the cells in saline on a glass slide and mixing in a drop of the antibody solution. The reaction can also be carried out in a test tube or capillary tube, observation being made by macroscopic examination. Agglutination is best observed with a microscope after mixing the cells and the antibody on a glass slide if the proper magnification is used, about 40×. With this method a suspension of cells without antibody should also be observed as a control.

It has been observed that under conditions in which flagella antigen–antibody reactions take place, the flagella swell and become entangled, forming large, loose masses or clumps of bacteria. When somatic antigen–antibody reactions occur, the cells appear to rush at each other forming fine, tight clumps of the organism.

Since if both flagella and somatic antigens and antibodies are present, the flagella antigen-antibody reaction occurs first, in order to observe somatic reactions of this type it is necessary to first suppress the flagella antigens. This may be done by growing the cells on a culture medium containing phenol or by heating the cells at around 70°C (158°F) in a mixture of ethyl alcohol and water (50%).

Each *Salmonella* serotype contains a number of specific somatic antigens. It has been shown that the different somatic antigens are associated with the cell wall and that they produce different antibodies because of the presence of different oligosaccharides which are a part of the cell wall (Luderitz et al., 1966).[36] A particular somatic antigen can be determined and classified after adsorbing interfering antibodies on cells of one or several related species (not containing the particular antigen being evaluated). This is done by suspending the related cells in a solution of the antibodies produced with the organism being examined, centrifuging out the suspended cells, then using the supernatant antibody for further testing. In this case the *antibody* being evaluated would remain in solution but interfering *antibodies* would be removed with the cells.

The manner in which the antigenic properties of salmonellae (and other bacteria) may be determined and used to identify them as different species may be illustrated as shown in Table 10–2. Suppose that the indicated salmonellae species were related antigenically as listed in Table 10–2. Antiserum No. 1 would agglutinate any of the organisms W, X, or Y. If, therefore, it was desirable to obtain antibody "a" for purposes of testing for the presence of antigen No. 1, the procedure would be to suspend the organisms X and Y in antiserum No. 1, allow the cells to agglutinate, then centrifuge to remove the cells. This would leave an antiserum which contained antibody "a" only, which then could be used to test for antigen No. 1.

In such a manner an antiserum can be prepared which is specific in antibodies for a particular antigen and such a serum can be used to type related bacterial species according to the particular antigens that they contain. This kind of a procedure would have to be used for each specific antigen present in the several bacterial species.

In the salmonellae, it is known that a number of the different antigens present in particular species are due to different, short-chain polysaccharides which are part of the cell wall. If these polysaccharides are known, and in some species many of them have been identified, it is only necessary to add the particular carbohydrate to the serum containing the antibodies to remove the antibody corresponding to this carbohydrate. Thus, the method of reacting antiserum with bacterial cells, to remove some of the antibodies and produce a more specific antiserum, may in some cases be eliminated.

The reaction of the cells of a particular species with the antibodies of a related species is called "cross agglutination." The reaction involving the cells of a particular species with antibodies corresponding to its own antigens is called "homologous agglutination."

Salmonellae are classified as serotypes according to the "O" and "H" antigens that they contain. There are about 600 antigens and since various salmonellae contain a common O antigen they are placed in about 12 groups indicated by capital letters, i.e., A, B, C, etc. Individual "O" antigens are indicated by small arabic numbers. "H" antigens differ according to the state or phase that the organism is in, hence must be listed as phase I antigens, indicated by small arabic numbers or phase II antigens, indicated by small letters or by small letters and small arabic numbers. Classification of the various salmonellae serotypes, therefore, would be carried out according to the following:

Somatic antigens "O"		Flagella antigens "H"	
Group	Specific antigens	Specific antigens	
		Phase I	Phase II
Capital letters	Small arabic numbers	Small arabic numbers	Small arabic numbers and small letters

In order to determine H antigens it is necessary to examine the organism in both phases. Ordinarily most of the organisms of a particular culture will be in either phase I or II (assume that this is Phase I) with some cells of the

TABLE 10–2
Antigen—Antibody Relationships

Organism	Antigens no.	Antibodies	Antiserum no.
W	1 2 3	a b c	1
X	2 3 4	b c d	2
Y	3 4 5	c d e	3

other phase being present (assume that this is Phase II). To obtain cells from the phase represented by only a few cells of the culture (Phase II) the following procedure is used. A ditch is cut in a plate of sterile agar. In this ditch a strip of sterile filter paper is placed which has been saturated with antibodies of the Phase represented by the majority of organisms in the culture (Phase I) and over one end of this strip a smaller sterile filter paper strip (moistened) is placed at right angles and allowed to rest on the agar. The end of the large paper strip away from the smaller paper strip is now inoculated with the culture, and the plate of agar is incubated. The organisms representing the majority (Phase I) will be agglutinated on the large paper strip but the minority cells (Phase II) will move onto the small paper strip which can be removed after incubation and cultured to obtain cells in the minority phase (Phase II) (Collins, 1967; Boyd, 1966).[5,14]

Effect of Temperature on the Growth of Salmonellae

Whereas *Salmonella* species grow best at temperatures near 37°C (98.6°F), these organisms will grow at both higher and lower temperatures than these. Tests have been made in which chicken a là king, custard filling, and ham salad have been inoculated with different species of *Salmonella* and growth determined at various temperatures (Angelotti et al., 1959; Angelotti et al, 1960).[1,2] It was found that while growth rates were slower at the lower temperatures, the organisms grew at 10°C (50°F), and it was necessary to reduce the temperature to below 6.7°C (44°F) before growth ceased entirely. Such information is helpful in evaluating refrigerator temperatures at which foods should be held.

In evaluating the effect of cooking on bacteria in foods it is necessary to know at what temperatures the organisms may be expected to continue to increase and at what temperatures they will start to die off. Using the same *Salmonella* species and the same foods, the authors of the above report found that these organisms might increase in numbers at 45.5°C (114°F) (Angelotti et al., 1959)[2] but definitely decreased in numbers at 46.6°C (116°F). In turkey stuffing marked decreases in the number of salmonellae present were not evident until temperatures in the range of 60°C (140°F) to 65.6°C (150°F) were reached at the slowest heating point.

Foods as Carriers of Salmonella Bacteria

Almost any food may, at times, be contaminated with salmonellae and under conditions of mishandling may become involved in the transmission of salmonellosis. However, it appears that there have been few, if any, outbreaks of this disease caused by the ingestion of fruit. In the past, outbreaks of salmonellosis in which the cause was traced frequently involved eggs and egg products (fresh, frozen, and dried eggs). Since June of 1966 the United States Food and Drug Administration has required that all egg products shipped in interstate commerce be free from salmonellae (It should be realized that "free from salmonellae" does not mean that an occasional *Salmonella* organism would never be present in a sample of food. It does not mean that no salmonellae would be found in 400 g of the food (sixteen 25-g samples).) Pasteurizing by heating at 60–62.2°C (140–144°F) for 3–4 min is presently being used by producers of dried or frozen eggs to eliminate salmonellae. Prior to the requirement for pasteurization of eggs, it was possible to find salmonellae in significant percentages of samples of these products. What the effect of pasteurization has been on salmonellae contamination in egg products is presently not known, but it must have resulted in a lesser frequency of contaminated samples.

It has been shown that, depending upon various conditions, up to more than 30% of processed poultry may contain *Salmonella* bacteria (Bryan et al., 1968; Nickerson et al., 1968; Savage, 1956; Gileroy, 1966).[7,25,53,59] It is evident that the presence of a few infected birds, among the many others handled in a poultry-processing plant, since both equipment and the hands of workers become contaminated, serve to spread the organisms to bird carcasses which originally did not contain them.

In a survey made in Florida (Galton, 1954)[23] it was found that 7.5% of fresh pork sausage manufactured in the plants of nationally known packers and 57.4% of this product manufactured by local organizations contained *Salmonella* bacteria. While it is known that salmonellae may be found not infrequently on beef, the frequency of contamination of this product with salmonellae seems not to have been investigated so extensively as is the case with some other food products.

In the past, shellfish, especially clams and oysters, have had a history of involvement in the transmission of salmonellosis and typhoid fever (Old and Gill, 1940).[55] It is obvious that, since these products are often eaten without cooking, if salmonellae are present, they might cause disease in those partaking of such foods. Clams and oysters have also been connected with outbreaks of infectious hepatitis within the last few years (Mason and McLean, 1962; Morbidity and Mortality Weekly Report, 1964).[40,42]

Fresh crabmeat has in the past been implicated in the transmission of salmonellosis. Since this product is thoroughly cooked before removing from the shell, contamination of the meat with salmonellae could occur only through improper plant sanitation or, more probably, through improper personal sanitation of workers who remove the meat from the shell. Similar instances have occurred with such foods as chicken salad.

In recent years, dried milk powder has been associated with outbreaks of salmonellosis in children and the U.S. Food and Drug Administration has shown that salmonellae were present in these products produced in several different milk-drying (Morbidity and Mortality Weekly Reports, 1966)[44] plants. It is not expected that the level of contamination was high in samples, but such information has not been made available. It must be remembered that, should the reconstituted product be held at room temperature for a few hours, multiplication to larger numbers of these organisms could have occurred. In any case, it is possible that relatively small numbers of salmonellae will cause disease in children. Hence, in foods which may be eaten without cooking, or without further cooking, the presence of even small numbers of salmonellae may constitute a public health hazard.

Not many years ago an outbreak of salmonellosis involving 300 people occurred in three states of this country. The disease was transmitted through the ingestion of a type of smoked fish which is eaten without further cooking (Morbidity and Mortality Weekly Reports, 1966).[45] This product was produced in Canada, and it was found that polluted water was being used for cleanup and other purposes in the plant where the fish was processed.

A chocolate candy product has recently been found to contain moderate numbers of salmonellae and has been withdrawn from the market. Shelled nuts have also, in some instances, been found to be contaminated with salmonellae.

Any food or food product may be contaminated with salmonellae by those who are food handlers and are carriers of these organisms. When such people go to the toilet, they sometimes do not subsequently wash and sanitize their hands to the extent that all pathogenic bacteria of this type are removed. Hence, when they return to their job of handling food, they may contaminate the product with a few or even with many of these organisms. This is probably not often the case with typhoid carriers today since public health officials keep a strict account of people known to have had this desease and prohibit them from working at food-handling jobs.

It has been pointed out that food handlers who work with certain types of food may become carriers of *Salmonella* bacteria without having had the disease. Such people are called "symptomless carriers." This is a distrubing circumstance since there is no good way to identify such individuals, especially with the present-day regulatory requirements for workers in food plants.

The presence of a few salmonellae in foods which are usually cooked may be of greater significance than ordinarily considered to be the case. In the first place it should be remembered that it takes only a moderate amount of mishandling of foods to increase a few salmonellae which were present at the start to many organisms of this kind. It has been shown, for instance, that salmonellae, inoculated into the yolk of an intact egg (a "shell" egg) with a hypodermic needle, will reach very large numbers when held at 20 – 30°C (68 – 86°F) (more than 100,000,000 per gram) (Licciardello et al., 1965).[34] Shell eggs may be used for eggnogs without cooking. Moreover, baked, fried, or scrambled eggs are often not completely cooked to temperatures which would destroy all, or even significant numbers of salmonellae, which might be present. This fact is well established by actual tests (Stafseth et al., 1952; Beloian and Schlosser, 1963)[4,62] and by outbreaks of salmonellosis in hospitals due to *S. derby.*

If a sufficient number of viable salmonellae are present in foods after cooking to cause the disease, or if only a few of these organisms remain viable after cooking and then the food is held for some time at temperatures at which these bacteria will grow [above 6.6°C (44°F)], sufficient numbers may eventually be present to cause the disease even in normal adults. Since hard-boiled eggs are used to make egg salad and since cooking eggs to the hard-boiled stage effectively destroys salmonellae (Licciardello et al., 1965)[34] instances in which this food has caused the disease may have been due to contamination which occurred after the eggs were cooked.

Egg white may be used for such foods as baked Alaska or meringue. Fresh, frozen, or even dried products may be used for these foods. During preparation the egg white on such dishes often received a heat treatment little, if at all, higher than that necessary to bring it to a temperature suitable for fast growth of *Salmonella* bacteria. If foods of this type are not adequately refrigerated after preparation, such products may cause disease in normal adults. There are some cases on record in which egg white, used in this manner, has been responsible for outbreaks of salmonellosis.

In some foods in which dried eggs are used, little heating is required and even in cake mixes the heating received in cooking the product is considered not always to be sufficient to destroy all salmonellae which might be present. In the past a dried baby food, not requiring heating, transmitted salmonellosis to some infants and caused some deaths. Relatively few of these organisms may have been present in the product in this case.

It was previously pointed out that poultry flesh is frequently contaminated with salmonellae. In almost all cases the temperature–time relationships involved in the cooking of poultry should be such that even very large numbers of these organisms would be destroyed if present. There is, however, one type of poultry preparation in which this might not be the case, that is roasted, stuffed turkey or chickens. When the stuffing is added to the body cavity of the bird, bacteria will be wiped off from the interior surface of the bird onto the stuffing. Some of this may end up near the central mass of the dressing or in the slowest heating point in the bird. As the bird is cooked, especially if cold at the start, the temperature of the inner material rises slowly so that slow-heating parts of the stuffing may encounter suitable temperatures for the growth and multiplication of bacteria during a period of several hours. This may result in the presence of large numbers of bacteria. At temperatures around 48.8°C (120°F) and certainly 54.4°C (130°F), vegetative types of bacteria will start to die off. However, the destruction of bacteria by heat is usually not simply a matter of the temperature attained but is a time–temperature phenomenon during which, in the lethal temperature range, a percentage of the bacteria die off in a definite period of time, the bacteria not destroyed remaining viable. If, then, the bacteria which grow are salmonellae and if the time–temperature relationship in the dressing is such that a sufficient number of these organisms survive, even normal adults ingesting this food may be infected with salmonellae. A number of outbreaks of this kind, caused by the eating of dressing from stuffed poultry, especially turkeys, are on record.

Because of the possibilities of transmitting food-borne diseases through improperly cooked poultry stuffing, a metal-stemmed thermometer should be used from time to time to check temperatures in the coldest parts of this material and prior to removal of the bird (or birds) from the oven a temperature of at least 68.3°C (155°F) should have been attained.

When salmonellae are present in significant numbers in the interior portions of beef and pork, the products are usually of the ground type such as fresh pork sausage and hamburger. Fresh pork sausage should always be cooked to the point where all salmonellae would be destroyed not only because of the possibility of transmitting salmonellosis but also because of parasities which might cause trichinosis. With hamburgers, the same statement cannot be made since some people prefer this meat in the rare state. However, cooking times are very short for this food; and, if the product has been adequately refrigerated prior to cooking, as should be the case, there would be no significant multiplication of the organisms and probably little danger of transmitting disease to normal adults. It is doubtful that sufficient numbers of salmonellae would survive in cooked products of this kind to cause disease in normal adults if the product in the fresh state had been handled properly and the cooked product was eaten within a reasonable length of time after preparation. It is quite certain that at times some salmonellae may survive in these foods after cooking. If, therefore, such cooked materials are allowed to stand at room temperature for a number of hours prior to eating, these organisms may multiply to sufficient numbers to cause disease.

Distribution of Salmonellae in Animals

Salmonella bacteria are usually associated with animal products of one kind or another. The question naturally arises as to why this should be the case. The answer lies in the fact that salmonellosis is a disease of man and other warm-blooded animals. Poultry appears to be especially frequently affected by this disease, but hogs and cattle are also subject to infections of this type. Salmonellae may, therefore, be present in rare instances in unbroken shell eggs (the percentage of eggs infected with these organisms is very small) and they also may be present in the intestinal contents of many warm-blooded animals. There may be occasions in which these organisms would be present in the flesh of warm-blooded animals but this would be rare. When present in meat and poultry, most often flesh surfaces become contaminated through direct or indirect contact with intestinal contents.

An examination of hogs and cattle for salmonellae has been made as they proceeded from the farm to market and through slaughter. It was shown that on the farm the feces of a very small percentage contained salmonellae. In the holding pens, in stockyards, the percentage of infected animals increased and by the time that the slaughtering floor was reached a significant portion of the animals, 12 – 18% or more, were infected (Galton et al., 1954).[23]

Since it is known that a small, though definite, percentage of hogs and cattle are infected with salmonellae prior to marketing and poultry flocks are known on occasion to have this disease, the problem appears to originate on the farm. How, then, do poultry, hogs, cattle, and other animals become infected with salmonellae? Certainly salmonellae are widespread in nature but not so widespread as to be found on natural products free from direct or indirect contamination with material derived, in one way or another, from warm-blooded animals. The answer, so far as slaughter animals are concerned, seems to lie in the food or feeds which are supplied to them by man. Essentially all of the edible warm-blooded animals raised by man for food are fed materials which are supplemented with meat meal, bone meal, or fish meal. Bacteriological examinations have shown that percentages, varying from low to high, of these supplements contain *Salmonella* bacteria, a number of different species of these organisms having been isolated therefrom (Leistner, 1961; Watkins et al., 1969).[32,68]

The temperatures used in manufacturing meat, bone, and fish meal and the duration of time that these temperatures are applied are such that all of the *Salmonella* bacteria originally present in the raw materials should have been destroyed during processing. The presence of these organisms in the final products, therefore, can only mean that contamination has occurred after processing. Such materials are sometimes stored in bulk in a warehouse associated with the processing plant, and it may be at this point that contamination occurs. It can be speculated that contamination occurs in this case from droppings of wild birds, rats, or mice; but the actual facts are not known. Recent findings (Morris et al., 1970)[49] indicate that in fish meal plants the product of the first 30–45 min of the day's processing contain salmonellae and that, if this portion is reprocessed, the contamination is reduced to nondetectable levels.

In addition to food supplements for slaughter animals, it has also been shown that many pet foods, of the dried type, may contain salmonellae which cause salmonellosis in pets such as dogs and cats (Mackel et al., 1952).[38] Such infections may be a source of the disease in humans, especially in children, since they handle objects which have been contacted by the mouth of the animal or may otherwise have been contaminated.

In Australia a salmonellosis outbreak in humans was traced to an infection of this kind in pets and subsequent examinations showed that more than 3% of the pets of that particular area were infected with salmonellae (Mackerras and Mackerras, 1949).[39] In another examination of household pets in Florida, it was found that more than 15% of domestic dogs and cats of the particular area were contaminated with salmonellae (Mackel et al., 1952).[38] This source of human infection cannot, therefore, be ignored.

The Control of Salmonellosis

It should be understood that, at present, there is no absolute method of preventing all outbreaks of salmonellosis in humans. At the same time it should be realized that there are many things which can be done to improve sanitation and to prevent the mishandling of foods. If these things were done, as should be the case, this disease, like typhoid, would become one of very low incidence and might eventually be almost eliminated as one of the causes of sickness and sometimes of death.

It has been suggested that salmonellosis might be controlled by (a) the elimination of insects, rats, and mice from areas where they might contact foods, (b) the follow-up testing of humans and animals to determine carriers in cases in which the individuals, flocks, or herds are known to have had salmonellosis, (c) the prevention of the use of animals which have died as foods—apparently this is sometimes done on farms (Meyer, 1953).[41]

These recommendations, although helpful, would seem not to be adequate for the control of salmonellosis. There is no doubt that humans who handle foods and are known to have had intestinal illness should be excluded from the direct handling of foods until such a time that they can be shown not to be a carrier of intestinal disease organisms. However, this would do little to detect symptomless carriers who apparently may be found among both humans and slaughter animals and among food handlers (Cockburn and Simpson, 1961).[13]

Sanitation in food operations is one of the best methods of eliminating salmonellae from the product. Sanitation refers to many things: to food plant construction, to food plant cleanup, to the personal sanitation of those handling foods in manufacturing plants, in food preparation procedures (such as those used by the catering trade, institutions, in restaurants, and even in the home) and to many other factors. The average individual will never be employed in handling foods as a means of livelihood, but we all handle foods in the home. Therefore, we all can do some things to prevent the spread of disease.

The proper cooking and refrigeration of foods is involved in the application of good sanitation, and this is important in the home, as well as in commercial operations. One thing which everyone can, and should do, is to thoroughly wash and cleanse his hands before handling anyone's food or drink. Since salmonellosis is an intestinal disease and the organisms causing the disease are to be found there, clean hands are an absolute requirement for those who contact food which is to be served to others.

The application of adequately low temperatures to foods, at times when food is neither in preparation for eating, about to be eaten, or being eaten, is another method which would often prevent outbreaks of salmonellosis. In normal adults a few salmonellae, if present in foods, probably will not cause disease. What happens in many outbreaks of salmonellosis is that the food is mishandled before it is eaten or is held at temperatures at which salmonellae will grow, and eventually enough of these bacteria are present to cause the disease in adults. When not being processed, cooked, or eaten, all perishable foods should be held at temperatures at or below 4.4°C (40°F). This, of course, does not refer to canned foods which should be commercially sterile or to dried foods in which the moisture content is too low to allow the growth of bacteria.

The proper heating of foods during cooking should not be overlooked as a method of controlling outbreaks of salmonellosis. In cooking there may be some instances, as with the preparing of egg dishes, oyster or clam preparations, where it may not be possible to heat to the extent that it can be said that all salmonellae, if present, have been destroyed. However, this certainly can be done with poultry, and, if cooking of these products in food handling and eating establishments and in the home were always carried out in such a manner as to make certain that adequate temperatures had been reached in all parts, one means of transmitting salmonellosis would have been eliminated.

In 1966, the United Stated Food and Drug Administration of the Department of Health, Education, and Welfare started to enforce a regulation which requires that frozen and dried egg products must be free from salmonellae if these products are to be shipped in interstate commerce. This regulation has resulted in the pasteurization or heating of the liquid material to 60–62.2°C (140–144°F) and holding at this temperature for 3–4 min prior to cooling. Such heat treatment would be sufficient to destroy at least moderate numbers of salmonellae, if present, and while such treatment might not always destroy all of these organisms which might be present and would do nothing to prevent recontamination after pastuerizing and cooling, there is no doubt that such heating will do much to prevent outbreaks of salmonellosis which are transmitted through egg products. Most salmonellae, including *Salmonella typhimurium,* the organisms most often found to be the cause of salmonellosis in this country, are relatively sensitive to heat. A strain of one species, *Salmonella senftenberg* (Osborne et al., 1954),[56] and possibly one other serotype are known to be comparatively heat-resistant. The time–temperature requirements for egg products which would destroy large numbers of the ordinary salmonellae would do little to decrease heat-resistant strains such as *S. senftenberg*. However, the heat-resistant types have been associated with outbreaks of salmonellosis only infrequently, hence pasteurization of egg products will doubtlessly serve as an effective control measure for prevention of this disease.

In the past the egg industry has been hesitant to apply pasteurization treatment for liquid eggs prior to the manufacture of egg products. The reasons for this are: (1) During the heat treatment egg proteins tend to coagulate, causing equipment to become clogged and preventing efficient transfer of heat to the product passing through. (2) The heating tends to affect the functional characteristics of egg proteins, resulting in poorer whipping properties when egg white is used for meringues, baked Alaska, etc., and in lower cake volumes when egg magma or egg white are used in cakes. Recently the U.S. Department of Agriculture, Western Utilization Research Development Laboratories have found that by adding 0.00625% of aluminum sulfate and 0.02% of lactic acid to liquid egg prior to heat pasteurization, coagulation during the temperature–time requirement for pasteurization does not occur and the functional characteristics of egg products, pretreated in this manner, are not affected. At present, it would appear that some manufacturers of egg products are satisfied with pasteurization requirements and some are not.

Cooked crab meat is also now being pasteurized in many instances by heating the product to a minimum temperature of 76.7°C (170°F) in all parts. This is done by immersing the filled metal shipping containers in water well above 76.7°C (170°F) until the required temperature is reached at the center of the product. Although pasteurization has, no doubt, reduced the incidence of food-borne disease as transmitted through this food, there is said to be some loss of quality in crab meat treated in this manner.

While not yet approved by the Food and Drug Administration, another method of destroying salmonellae in egg products or in crab meat is to treat with ionizing radiations (cathode or gamma rays) at relatively low dose levels. It has been shown, for instance, that at a temperature of 54.4°C (130°F) the radiation treatment required to bring about a 7-log cycle reduction of salmonellae in liquid egg is minimal (about 120,000 rad) and coagulation would not be a problem at this temperature. The application of a dose of 200,000–250,000 rad to crab meat would also effectively reduce salmonellae if present and lengthen the refrigerated storage life of the unpasteurized product.

In the long run the best method of eliminating outbreaks of salmonellosis may be to break the chain of infection in slaughter animals. A common source of salmonellae in foods is the animals which constitute the food products involved. As has been pointed out, these animals apparently become infected and indirectly contaminated through the feeds provided by man, especially through the protein supplements and bone meal included in their food. Since survival may occur during processing and recontamination may take place after the heat processing of the products, it is evident that plant processing procedures and plant sanitation procedures and requirements, especially in areas where the dried material is stored and packaged, are not adequate. It is possible that enforcement agencies could do something to cause a change in this situation, although conditions might be difficult to improve because of poor construction and sanitation in many plants producing this type of product. Another way of approaching this problem would be to destroy the salmonellae in fish, bone, and meat meal after the product has been packaged. It has been shown that salmonellae present in materials of this kind may be destroyed by heating but heating in this case has to be carried out with dry air (otherwise the product would get wet and would be subject to microbial decomposition). Heating by dry air, unfortunately, causes losses of some of the very nutrients for which such supplements are added to animal feeds. In the Netherlands it has been shown that salmonellae in fish meal can be effectively destroyed by radiation treatment with gamma rays, without destroying significant amounts of the nutrient materials therein (Thornley, 1963).[65] The situation should be no different for meat and bone meals. Pelleting of dried supplements also destroys Salmonellae.

In Denmark, fish meal coming into the country is required to be essentially free from salmonellae. It would appear that if such requirements were established for animal feeds and their ingredients in the United States, if stockyards were made to clean up to some extent, and if some slaughtering equipment, such as dehairing machines, were made more sanitary, the problem of salmonellosis might eventually be greatly abated. At the present time, United States government regulatory agencies are giving the kind of attention necessary to bring about better sanitation in food-processing establishments in which milk powder and other dried foods are produced. The emphasis presently being placed on salmonellae in foods by enforcement authorities must, in the long run, serve to decrease the incidence of salmonellosis in humans.

Shigellosis

According to public health statistics, there are almost as many yearly cases of human shigellosis, or bacillary dysentery, in the United States as there are cases of salmonellosis. Moreover, the mortality rate is higher with shigellosis than with salmonellosis. For instance, in 1967 there were 13,474 cases of shigellosis and 18,120 cases of salmonellosis reported. The number of deaths attributed to these diseases in 1966 were 86 for shigellosis and 73 for salmonellosis (Morbidity and Mortality Weekly Report, 1968).[47]

It is believed that many more cases of salmonellosis occur than are reported. However, there would appear to be no reason to believe that the same situation does not exist with shigellosis.

While considerable information is available concerning the biochemical characteristics of the *Shigella* bacteria, little information has been published on the possibility of transmitting shigellosis through various foods. Water has often been involved in the transmission of shigellosis. The effect of temperature on the growth of shigellae in foods, temperatures at which these organisms are destroyed in foods, the number of these organisms which must be ingested to cause the disease in normal adults, or other factors pertinent to the control of shigellosis are not known.

Regardless of the apparent relative importance of shigellosis, investigations concerning the growth characteristics of bacteria causing this disease, the foods involved in transmission and the manner in which the particular organisms get into foods, have been comparatively few.

Shigellosis—the Disease

Ordinary symptoms of shigellosis or bacillary dysentery are diarrhea, with bloody stools, abdominal cramps, and sometimes fever. The bloody-stool symptom serves to differentiate shigellosis from salmonellosis. In severe cases it is probable that complications such as septicemia, pneumonia, or other manifestations, such as peritonitis, may cause more drastic affects. Shigellosis, caused by *Shigella dysenteriae,* may produce severe toxemia ending in collapse. There may be such complications as arthritis. Recovery from infection with this organism is slow and relapses are frequent. The incubation period for shigellosis, or the time elapsing after ingesting the bacteria and prior to the onset of symptoms, is said to be as long as 7 days with an average of 4 days.

Persons having had shigellosis may become carriers of the organisms, excreting them in their feces, but the period during which the carrier stage may persist is shorter than with salmonellosis, usually not longer than a few weeks. In rare cases, humans may become persistant carriers after an infection of this kind.

Foods and Shigellosis

Foods suspected of being involved in the transmission of the disease are moist prepared foods, milk, and dairy products. It is also stated that foods of this kind which transmit the disease upon ingestion have been contaminated with human excreta. That being the case, it would seem that almost any food might be involved in transmission and in recent years some foods have been implicated. In 1968, for instance, outbreaks of shigellosis were traced to the ingestion of potato salad and chicken salad (Morbidity and Mortality Weekly Report, 1968).[47]

Since we do not know whether other animals may have this disease, it cannot be said that some foods which have not been contaminated by humans may be implicated in the chain of infection. Also, it seems not to be known whether the *Shigella* bacteria will grow in foods at temperatures much lower than their optimum growth temperature of 37°C (98.6°F). It cannot, therefore, be stated that mishandling of foods is an important factor in causing outbreaks of shigellosis.

If the chief source of contamination of foods which transmit shigellosis is human excreta and the organisms do not grow in foods, we might expect that the ingestion of relatively few bacteria of this kind will cause the disease since, while human excreta may contain large numbers of bacteria per unit quantity, it is difficult to envisage the contamination of foods with anything but trace amounts of materials of this kind even under conditions of gross carelessness. However, it has been stated (Dack, 1956)[16] that the ingestion of between 100×10^6 and 50×10^9 organisms is necessary to cause the disease.

The Bacteria Causing Shigellosis

The *Shigella* bacteria are gram-negative, nonmotile rods $2–3 \times 0.5–0.7\ \mu$ in size. They do not form spores nor do they produce capsules. They ferment a number of sugars but do not ferment lactose. They reduce nitrates to nitrites and ammonia. They do not produce hydrogen sulfide from proteins. Like the salmonellae, they are facultative in their requirements for oxygen, growing both in the presence or absence of oxygen.

Regarding the resistance of shigellae to heat, it may be expected that these organisms are destroyed in a manner similar to the salmonellae as far as the time–temperature relationships are concerned. Also, it might be expected that they would start to die off slowly at temperatures around 46.6°C (116°F).

A comparatively few species of the shigellae are known. Among these are *Shigella dysenteriae* (formerly called *Shigella shigae), Shigella schmitzii, Shigella sonnei, Shigella alkalescens, Shigella flexneri* (formerly *Shigella paradysenteriae), Shigella ambigua,* and *Shigella boydii.* Another species, *Shigella dispar,* is known but it never has been clearly established that this organism is pathogenic. *Shigella flexneri* and *Shigella sonnei* are the strains most commonly involved in causing shigellosis.

Some of the shigellae have a common antigen and some of them do not. Unlike the salmonellae, they cannot, therefore, be identified and classified unambiguously on the basis of the antigens which they produce. Instead, they are identified chiefly by their biochemical characteristics, especially the sugars which they ferment.

The shigellae produce an endotoxin or a toxin which is present within the cell. This is a lipopolysaccharide which is released from the cell when the cell lyses. Cultures of several days of age tend to lyse. In infections caused by *Shigella dysenteriae* the endotoxin affects the intestine causing inflammation and sometimes ulceration. All of the shigellae produce an endotoxin, hence it is probable that in infections these organisms grow in the intestine, eventually lyse, and cause the particular gastroenteritis symptoms which have been described.

Shigella dysenteriae is the species among the shigellae which causes the most severe disease, and it has been shown that when this organism is grown in media it produces an exotoxin (Wilson and Miles, 1964),[69] as well as an endotoxin. The exotoxin is said to be a protein which affects nerve function and is quite toxic when injected into animals.

Control of Shigellosis

Since little is known of the bacteria causing shigellosis, especially those factors concerned with transmission of the disease, it is somewhat difficult to specify control measures. However, certain procedures may be suggested which will help control the disease.

People having had shigellosis may become carriers of the organism causing the disease; it is, therefore, an excellent practice to suspend personnel known to have had an intestinal disturbance from the direct handling of foods. This precaution should especially be practiced in restaurants, food-manufacturing plants, and other institutions preparing foods for public consumption.

It would be in the interests of public health to have the feces of all food handlers known to have had intestinal disturbances checked for the presence of salmonellae and/or shigellae by bacterial and agglutination tests. With conditions as they are in the food industry, it is doubtful that any such steps will ever be taken except with those known to have had typhoid. Any plant foreman, however, can exclude people from the direct handling of foods for some period after known intestinal disturbances.

Since intestinal disturbances are often not reported, and the testing of stools is costly and may involve trouble with employee–employer relationships, one of the best methods of control, therefore, is to insist on rigorous personal sanitation habits as far as the food handler is concerned. This must be enforced at all times, and as will be seen in a later chapter, can be done without duress as far as the worker is concerned. Actually, such sanitary practices often may have beneficial effects for the worker. Good plant, institutional, or community sanitation is also indicated as a method of control, especially the elimination of insects in areas where food is exposed since it has been shown that flies may be carriers of the organisms causing this disease and may transfer these organisms to foods (Tanner and Tanner, 1953).[64]

Since shigellosis is presently considered to be often transmitted by water, adequate sanitary control of municipal water supplies is one of the best methods of control. Also, in those areas where water for drinking and cooking may come from driven or dug wells, the selection of the area where the well is to be located should be made in such a manner as to eliminate possible contamination from nearby sources of pollution. All well water should be tested bacteriologically to assure the potability of the supply, especially at the start and periodically thereafter.

While it has never been shown that the *Shigella* bacteria grow in foods at temperatures below those stated to provide optimum growth rates, it is a good precaution to keep foods at or below 4.4°C (40°F) when not being processed, eaten, or about to be eaten.

There is much needed information regarding shigellosis that could be readily obtained by scientific investigation.

Perfringens Poisoning

Perfringens poisoning has received more attention in recent years as a food-borne disease. This disease is believed to be a food-borne infection due to the growth of *Clostridium perfringens* (formerly called *Clostridium welchii).* Perfringens poisoning has been classified as a food infection on the basis that, whereas sterile filtrates of nutrient materials in which *Clostridium perfringens* has grown will not cause the disease, the disease has been caused in human volunteers fed cultures of these organisms previously isolated from outbreaks of the disease (Dack, 1966).[18] *Clostridium perfringens* is the organism which in the first world war caused so much trouble by infecting wounds and producing gas gangrene. Whereas, a few years ago it was not recognized that *Clostridium perfringens,* if present in foods in sufficient numbers, would cause food-borne infections, it is said that today as much as 10% of the food-borne diseases in England may be of this type and that in the United States the number of known cases of this kind increased from 25 in 1952 to 95 in 1957; and in 1969, 18,527 cases of perfringens poisoning were reported. *Clostridium perfringens* is an organism normally found in the human lower intestine, yet if the right types and numbers of this organism are present in foods it apparently can grow in the upper intestine and cause gastric disturbances.

Recent experiments have demonstrated that cell free extracts and culture fluids of *Cl. Perfringens* may cause fluid accumulation and diarrhea in ligated intestinal loops in rabbits, thus indicating that perfringens food poisoning may be classified as food intoxication rather than a food infection.

Perfringens Poisoning—the Disease

The symptoms of this disease are acute abdominal pain, diarrhea, and nausea, but rarely vomiting. The onset of symptoms occurs 8–22 hr after the food has been eaten (Hall et al., 1963).[26] The disease is not of long duration and usually has no lasting effect, although one death due to this type of poisoning was reported in Britain in 1965.

Clostridium perfringens—the Organism

Clostridium perfringens is a spore-forming, gram-positive, rod-shaped organism. It is a strict anaerobe, is normally found in soil, in water, and even

TABLE 10–3

Toxins and Related Agents Produced by Strains of *Clostridium Perfringens*

β, γ, and η toxins; have not been characterized
α toxin; a lecithinase
ϵ and c toxins; lethal necrotizing substances
κ toxin; a collagenase
λ toxin; decomposes gelatin
μ toxin; an hyaluronidase
ν toxin; is a deoxyribonuclease
θ toxin; an oxygen-labile hemolysin

in the intestinal contents of man and other animals, as well as in flies (Hobbs, 1969).[27] In size *Clostridium perfringens* is 4–8 × 0.8–1.0 μ. Whereas, it is usually rodlike in form, it sometimes produces branch-like filaments (involution forms). The rods may occur singly or side by side. The organism is nonmotile (one of the few nonmotile clostridia), it produces a β type of hemolysis on blood agar (complete hemolysis of the blood cells surrounding a particular colony) and it liquefies gelatin. In liquid media containing other nutrients and glucose, maltose, lactose, or sucrose, *Clostridium perfringens* produces acid and gas. Nitrites and nitrates are reduced by this organism and some ammonia may be formed. Hydrogen sulfide is not formed by the growth of this organism in the presence of proteins or sulfur-containing amino acids but hydrogen is formed from carbohydrates. As with other clostridia, *Clostridium perfringens* is unable to produce the enzyme catalase. The organism will grow at pH values as low as 5.7 but the optimum pH for growth is 7.2. There is growth in the presence of concentrations of 5% of sodium chloride but not in 10%. There are six known types of *Clostridium perfringens* (A, B, C, D, E, F) some of which are said to grow at low temperatures. These organisms are differentiated according to the different toxins, or so-called toxins, which they produce (Hobbs, 1969).[27]

The various types of *Clostridium perfringens* produce a number of toxins (Wilson and Miles, 1964).[69] These are listed in Table 10–3.

Effect of Temperatures on *Clostridium perfringens*

Very little seems to be known about the extremes of temperatures at which *Clostridium perfringens* of the food-borne disease type, will grow. The optimum temperature for growth is 43.3–47.2°C (110–117°F) (Hobbs et al., 1953).[28]

It has been suggested (Nygren, 1962)[54] that *Clostridium perfringens* produces phospholipase c, which decomposes lecithin forming phosphoryl choline, which is the cause of intestinal disturbance. It has been shown that phosphoryl choline probably is not to be the cause of perfringens poisoning.

The number of spores which must be ingested to cause perfringens poisoning is not known but it has been stated (Dack, 1966)[18] that the disease can be caused when the organism is present in foods in concentrations of many millions per gram.

It is known that the spores of *Clostridium perfringens* are fairly heat resistant although exact determinations with specific numbers of spores at definite temperatures seem not to have been made. It is known that the spores may be boiled for several hours and survive or remain viable. Considering this, it would appear that some spores of *Clostridium perfringens* are at least as heat resistant as those of the most heat-resistant type of *Clostridium botulinum,* type A.

Foods Usually Involved in Perfringens Poisoning Outbreaks

Perfringens poisoning has been caused by the ingestion of prepared meats or meat products (usually beef or veal) and poultry. It has been shown that raw, unprocessed beef and veal usually contain the organism. *Clostridium perfringens* is also usually present on fresh meats. What has happened in past perfringens poisoning outbreaks has usually been that the beef, veal, or poultry is cooked and then is left without refrigeration for comparatively long periods of time prior to eating or prior to serving. What probably happens in such cases is that the spores of the organism survive the heating applied during cooking, then when the food is left at temperatures suitable for growth, the spores germinate, the cells multiply and grow to numbers which are infective or produce toxin. Most other organisms which might grow and interfere with the growth of *Clostridium perfringens* would have been killed off during cooking.

It is said that it has not been possible to produce perfringens poisoning in humans by feeding sterile filtrates of beef infusion or chicken broth previously cultured with this organism. The disease can be caused by feeding cultures of *Clostridium perfringens* although infectivity may be lost during the holding of cultures (Hobbs et al., 1953).[28] An outbreak of this disease in 1965 involving 366 students at a university cafeteria was traced to gravy rather than meat (Morbidity and Mortality Weekly Reports, 1965).[43]

Control of Perfringens Poisoning

The control of perfringens poisoning would seem to be relatively simple, namely, the prompt application of temperatures below 4.4°C (40°F) or above 65.5°C (150°F) to foods which after cooking are not to be eaten immediately, and the prompt application of temperatures below 4.4°C (40°F) to foods left over from a meal. Incubation of foods on the back portion of a stove, as is sometimes done in the home, may be considered to be very bad practice.

Since institutional-type steam tables are often used in such a manner as to be, in effect, good bacterial incubators, great care should be exercised to see (a) that only small portions of meat or gravy are held in steam table containers at any one time; (b) that all of the product in a container is served before a new container is added (this is because of the fact that some of the product from the old container may be added to that in the new container and end up with long residence on the steam table); (c) that products are held in such a manner on steam tables that all parts of the food are maintained at 65.5°C (150°F); and (d) that all steam table containers are thoroughly washed and dried prior to reusing for the purpose of holding food.

Vibriosis

In comparatively recent years a food-borne disease caused by a vibrio has become noted as a cause of outbreaks of gastroenteritis in Japan. It is said that outbreaks of this disease involving as many as 20,000 people have occurred (Kazuyoshi and Matsuno, 1961).[30]

It is not known that outbreaks of vibriosis have occurred in the United States. The organism causing vibriosis is apparently of marine origin and during the last few years it has been recovered from marine specimens taken off the coast of Florida and Washington and from Chesapeake Bay (Ward, 1968; Liston and Matches, 1968).[35,67] It has also been shown to be present in coastal waters in some concentrations and in concentrations of 10^3–10^4 per gram in some samples of commercially prepared crab meat. It is reasonable to speculate, therefore, that this organism may have been responsible for some food-borne disease outbreaks, the cause of which was not determined.

The Disease

The symptoms of vibriosis are epigastric pain, nausea, vomiting, and diarrhea with occasional blood and mucus in the stools. A fever of 1–2 degrees F is experienced in 60–70% of the cases. The incubation period is 15–17 hr and the symptoms last for 1–2 days. The organism, *Vibrio parahaemolyticus,* has been isolated from foods, some of which were known to have been eaten prior to the development of symptoms and from the stools of patients having the disease. Also, the disease has been caused by feeding the viable vibrio.

The Organism—*Vibrio parahaemolyticus*

The organism is a short, curved rod 1–5 × 0.3–0.6 μ in size. These vibrio are motile with a single polar flagellum and they are facultative anaerobes which are gram-negative. They require 2–4% of sodium chloride but grow in media containing as much as 11% of salt. These organisms grow best at temperatures between 30 and 40°C (86–104°F). Strains of the organism may reduce nitrate and liquefy gelatin. Certain sugars may be fermented without acid or gas formation. The cells are not particularly heat resistant and they do not form spores (Sakazaki et al., 1963).[58]

Foods Involved in Outbreaks of Vitriosis

Raw fish and molluscs (cuttlefish, squid, and octopus) have been the foods most often involved in causing vibriosis. Macaroni and vegetable salads have also been implicated. It is believed that in cases in which the eating of nonmarine foods cause the disease, contamination from human carriers or with sea water may have been involved. This organism will grow in white-fleshed fish, reaching concentrations of 10^7–10^8 per gram in 7 hr when inoculated at levels of 10^3 per gram. It does not grow in tuna or mackerel apparently because the pH of these fish is 6.0 or lower.

Control of Vibriosis

Since cooking temperatures and procedures are sufficiently high and severe to destroy large numbers of vibrios, a control measure would be the elimination of raw fish consumption, especially during those months in which inshore waters are warm and comparatively large numbers of this organism may be present.

Vibrio parahaemolyticus is a mesophile, hence probably will not grow at 10°C (50°F) or below (minimum growth temperature not known). Refrigerated handling of raw materials and especially of such products as cooked lobster, crab meat, and shrimp is, therefore, a method of controlling the growth of these organisms.

Sanitary procedures in plants producing cooked crustacean products would eliminate these organisms from such foods. This includes the sanitation of personnel and also suitable plant cleaning and sanitation procedures. In connection with the latter, sea water should not be used to clean plant equipment since the available water of this kind is often polluted.

Cholera

Cholera is rarely encountered in the United States; and, as a matter of fact, public health records which go back through 1953 show no reported incidence of the disease. Actually it never has been of public health significance in the United States. In some parts of the world, however, especially in India, China, and some parts of South East Asia, epidemics of this disease occur frequently.

Cholera—the Disease

The incubation period is 1–3 days, sometimes as long as 5 days. The disease is characterized by diarrhea with copious watery stools, vomiting, and prostration. Mucus is present in the stools, but little, if any, blood and no pus. Eventually the patient becomes dehydrated, since he is unable to retain water taken by mouth, the pulse becomes weak, and there may be muscular cramps. Secondary infections may occur but are not common.

It is considered that in epidemics of this disease, cases occur in which the only symptoms are malaise and simple diarrhea.

Cholera is caused by the ingestion of contaminated food or water, usually the latter, or by the ingestion of food or water contaminated directly by feces from a person or persons who have had the disease. It is believed that insects such as flies may spread the disease (from contaminated materials to foods) but the importance of this means of transmission has not been established.

In epidemics of cholera the death rate may be as high as 60%, especially in areas where patients do not receive adequate treatment. However, if patients receive suitable medical attention, the case fatality is usually below 5% even in epidemics.

The Organism Causing the Disease

The organism causing cholera belongs to the groups of bacteria known as *Vibrio* and has been given the specific name of *Vibrio comma.* It is a short curved, rod-like organism with a single terminal flagellum. It is aerobic, motile, will grow on simple or ordinary bacteriological media and is gram negative. It produces both body (somatic) and flagella antigens and can, therefore, be identified by agglutination tests made with the antibodies of these antigens.

The Control of Cholera

In the United States diseases spread in the same manner as cholera have been controlled by community control of drinking water supplies. This has involved much investigation and research. For many years sanitary surveys of drinking water supplies have been made to determine that there is no possibility of contamination from polluted rivers, streams, or other waters draining into the reservoirs where water supplies are stored. Surveys are also made to determine that no sewerage system or drainage from single outdoor toilets has access to drinking water supplies.

Since it would be infrequent that actual disease bacteria would be identified in water with bacteriological tests, even if the water were polluted, tests to determine other bacteria more frequently associated with human intestinal contents and with pollution in general have been developed and applied. These are the tests for the coliform group of bacteria which included *Escherichia coli, Aerobacter aerogenes,* and related organisms.

Some waters are bacteriologically pure enough to be used for drinking without treatment, while others must be filtered and/or treated with chemicals to precipitate a sludge and then filtered. Generally, filtered waters are then treated with chlorine or ozone to destroy any pathogenic organisms which might have survived and passed the filtration procedure.

Trichinosis

Trichinosis is neither a bacterial nor a viral disease, yet it is entirely food-borne so far as it is known, hence some description will be given here. The disease is caused by the roundworm, *Trichinella spiralis,* and in humans, is largely transmitted through pork. In 1967, in the United States, there were 66 cases of trichinosis reported, and in 1966 there were 115 cases and 3 deaths reported. It is possible that many more cases than this occur yearly without being called to the attention of public health authorities. In European countries fewer cases of trichinosis per unit of population are experienced than in the United States. The reason for this is that the incidence of contamination of hogs with the roundworm is lower and methods of detecting the organism in slaughtered hogs are applied in Europe.

Trichinosis—the Disease

The symptoms of trichinosis are said to be highly variable and to vary with the number of larvae cysts ingested. If large numbers of cysts are eaten, the patient will develop symptoms of nausea, vomiting, and diarrhea (1–4 days after ingestion). If few cysts are ingested, these symptoms may be absent. On the seventh day after eating, the larvae migrate from the intestines to the muscles; and this usually produces muscular stiffness and pain accompanied by remittent fever which may reach temperatures as high as 40°C (104°F). Occasionally, transient skin rashes occur. The fever may persist for some weeks. The muscles affected by pain depend upon whatever muscles are involved and pains in the back, in the muscles involved in chewing, swallowing, or breathing, or even the muscles regulating eye movement are common.

Edema is said to be the next most common symptom to muscular pains. The eyes and the upper eyelids are the most frequent sites of edema although it may occur in other areas. After encystment of the larvae in the muscles, the symptoms subside and consist mainly of vague muscle pains. Autopsies have indicated that in the United States about 16% of humans have had mild or severe cases of trichinosis (Cecil and Loeb, 1959).[8]

The Organism Causing Trichinosis

The live larvae of *Trichinella spirilis* are present in meat as a non-calcified cyst when ingested by the human. In the small intestine of the human the cysts are dissolved by the fluids and digestive enzymes present. The larvae then anchor themselves to the mucosa of the small intestine from which they derive the necessities of active life, oxygen and food. Within 2 days, they develop into sexually mature male and female adults. The male is about 1.5, and the female 3–4 mm in length. The male dies after copulation with the female and the eggs develop and hatch in the larval uterus (Cecil and Loeb, 1959).[8] About 1500 larvae from each female are then discharged into the lymph system of the host (the human) at the rate of about one per half hour. Total deposition of larvae required about 6 weeks after which the female roundworm dies.

Once the larvae enter the general circulation of the human, they are distributed to all tissues and eventually break out of the capillaries between muscle fibers. Inflammation induced by the larvae in the musculature causes the body to produce a cyst around the coiled larvae which the body eventually calcifies in a period of 6–18 months. The larvae may live and retain viability in this state for periods up to 10 years.

Trichinae can infect a wide variety of animals both carnivorous and herbivorous. As far as the human is concerned, however, pork is the common carrier of the disease, although other meat, such as bear meat, has been implicated. The incidence of infection in hogs is said to be 0.5% for grain-fed animals and 5% for those fed on uncooked garbage. Presumably the uncooked garbage is infectious to hogs because it contains uncooked or improperly cooked pork.

It has been found that as much as 10% of the fresh sausage in large city markets may be infected with trichinae.

Control of Trichinosis

Probably the best and most effective control of trichinosis in humans is the adequate cooking of all pork meat and all products containing pork. Trichinae are destroyed at a temperature of 58.3°C (137°F); and all pork products cooked in the home, in restaurants, or in institutions should be brought to a temperature of 60°C (140°F) or higher. All people should refuse to eat any fresh pork product which shows pinkness or shows signs of fresh (uncooked) blood after cooking.

In the United States, one of the methods which has been instrumental in controlling trichinosis has been the regulations of the Meat Inspection Division of the Department of Agriculture. This government agency regulates the slaughtering and processing of all meat and meat products shipped interstate in this country and eliminates diseased animals, and controls plant sanitation and many other factors associated with the slaughtering and processing of meats.

Specific regulations of the Meat Inspection Division for controlling trichinae in products containing pork that may be considered ready to eat are: (1) The pork shall have been held at −15°C (5°F) for 20–30 days; at −23.3°C (−10°F) for 10–20 days or at −28.8°C (−20°F) for 6–12 days, the time depending on the size of the product, or (2) that the pork product shall have been mixed with curing agents and held for at least 40 days at a temperature not lower than 7.2°C (45°F), or (3) that ready-to-eat products shall be heated to 58.3°C (137°F).

It has been estimated that about 70% of the pork raised in this country is processed in plants which come under the inspection of the Meat Inspection Division. Actually, in plants producing ready-to-eat products not inspected by MID, it is probable that, excepting bacon, all parts of such items should have been heated to a minimum of 58.3°C (137°F). The problem, therefore, is concerned mainly with raw pork products and the safety of such foods is almost entirely the responsibility of whoever cooks the product, and that of the person who is eating it. The warning "to eat no red or pink fresh pork products" cannot be overemphasized.

In some countries microscopic examination of a certain excised section of the muscle tissue of all hogs is made before the carcass is released for sale as fresh pork. However, in the United States such a method is not considered to be economical or practicable. The cysts of trichinae do not calcify in pork and are almost invisible to the naked eye so that even though all carcasses shipped interstate are inspected by veterinarians for disease, this type of infection would not always be determined by this kind of inspection.

It is considered that the cooking of garbage which is to be fed to hogs is a method of controlling trichinosis. There is no doubt that such procedures contribute to the control of trichinosis and a number of states have laws which require that such procedure be used. However, since pigs may eat rats and other animals which may be infected with trichinae, the cooking of garbage, prior to feeding to hogs, is not an absolute method of controlling trichinosis in man.

Amebiasis

As in the case of trichinosis, amebiasis is not a disease caused by bacteria. It is not the case that amebic dysentery or amebiasis is a disease of the tropics. The disease is world-wide in distribution and may be encountered as far north as the Arctic circle. In 1967, there were 3157 cases of this disease reported with 54 deaths.

Whereas amebiasis may be transmitted through drinking water, it may also be transmitted through food, hence must be considered as food-borne.

Amebiasis—the Disease

Amebiasis is caused by a single-celled animal, the protozoan *Entamoeba histolytica.* The symptoms of the disease vary greatly from patient to patient and even in severity in the same patient at different periods. Some patients have no particular complaints but are carriers and may have such anatomical conditions as liver abscesses.

Diarrhea is a common symptom of amebiasis but about one-third of patients are said not to have this symptom. The diarrhea may be severe and persistant or mild and occasional. Watery stools may be common. Abdominal distress, pain, and distention are often encountered. Fatigue, some fever, and backache frequently are complained of along with gastrointestinal symptoms.

Amebiasis may cause ulceration of the colon but this varies greatly in extent and intensity. In patients who die from the disease, ulcers are often present throughout the colon and extend to other parts of the intestine. There also may be lesions or abscesses in the liver and lesions of the lungs or brain.

Amebiasis is treated by administering certain chemicals, for instance arsenicals or iodine-containing compounds, or through the administration of certain antibiotics, especially the tetracyclines (Cecil and Loeb, 1959).[8]

The Control of Amebiasis

The control of amebiasis is essentially a matter of the application of good sanitary procedures to water supplies and to foods.

Adequate sanitary surveys, frequent testing for organisms indicative of pollution, and filtration and treatment with chlorine or ozone, when necessary, are imperative for all community water supplies. Where well water is utilized, the individuals using such a source should make sure that wells are dug or driven in areas away from drainage from private septic tanks or cesspools. Also, any individuals utilizing well water should have samples tested by the state authorities at intervals, to determine that it is not polluted or does not show indications of pollution. The water from streams or lakes, the sanitation of which is unknown to the user, should not be utilized for drinking, or for purposes of cooking, or for irrigation.

Disposal of human wastes so that flies cannot contact such material and then contact foods is also a measure contributory to the control of amebiasis. Also sewage should not be used as fertilizer.

While it is said that chlorination will not kill the cysts of *Entamoeba histolytica,* if water is tested for indicator organisms and found to be potable, it is extremely doubtful that it would be involved in the transmission of amebiasis.

People known to have or to have had amebic dysentery should be rigidly excluded from food-processing operations, from work in restaurants, institutions, or even in households if the work is of such a nature that it might result in the direct or indirect contamination of foods, until such a time as it can be shown that the stools of such individuals no longer contain the cysts of the organism causing the disease. This is a very important feature in the control of amebiasis.

Brucellosis, Undulant Fever, or Malta Fever

At one time brucellosis was mostly food-borne (through the drinking of raw or improperly pasteurized milk) and there were several thousand cases of this disease reported yearly in the United States. Today this disease is chiefly transmitted through contact with animals (slaughterhouse workers, farmers, etc.) and the incidence of brucellosis has greatly diminished. In the United States, in 1967, 265 cases of brucellosis were reported and in 1966 there were 262 cases and 3 deaths due to this disease (Morbidity and Mortality Weekly Report, Annual Supplement, 1968).

Brucellosis—the Disease

The incubation period for brucellosis is usually 5–21 days, and it is said that in some cases development of symptoms after infection may require as long as 6–9 months.

The onset of brucellosis may be sudden with chills, fever, and sweating, or the patient may have an ill-defined illness for weeks. Nocturnal sweating, abdominal pains, headaches, and muscular pain (also pains in the joints) are frequent symptoms. Intermittent fever, up to 40°C (104°F), and sometimes sustained fever, are frequently encountered. Appetite is often lacking. The disease may also be the cause of abortion in pregnant women and animals. Serious complications sometimes are caused by brucellosis and under such circumstances death may ensue. The organisms causing brucellosis get into the lymph system and invade the organs. Recovery from the disease usually takes place within 3 months to 1 year but symptoms may last for many years. Treatment with antibiotic and sulfa drugs is usually specified; but, due to the intracellular location of the organism, such treatment is not very effective.

The Bacteria Causing Brucellosis

The brucellae are small (about 0.5–0.7 $\times$ 1–1.2 μ), gram-negative, nonmotile, nonspore-forming rods. Some of these organisms grow best, when isolated, at reduced oxygen tension or in the presence of 10–25% of carbon dioxide, but other isolates grow well aerobically. They do not ferment sugars nor do they liquefy gelatin. They produce hydrogen sulfide from proteins.

For purposes of identifying these organisms, the cells are isolated from the blood of infected animals or humans on culture media after which agglutination tests are made. There are three species: *Brucella abortus* which causes disease in cattle, *Brucella melitensis* which causes disease in goats, and *Brucella suis* which causes disease in hogs. All of these are pathogenic to man.

Control of Brucellosis

Important in the control of brucellosis as a food-borne disease is the pasteurization of milk and cream at suitable temperatures and times. Since there is evidence that this organism will survive at least 6 months in cheddar cheese, milk used for the manufacture of such products should be pasteurized or the cheese stored for adequate periods prior to distribution. Also important in the control of this disease is the elimination of infected animals from herds or flocks. Since the disease can be detected in animals, infected specimens can be eliminated. Also, since vaccination with attenuated organisms may be used to provide an immunity, prevention of the disease in animals is possible.

Since these organisms grow in intercellular areas, they must be present in the flesh and especially in the organs of infected animals. It would seem, therefore, that the cooking of the flesh of warm-blooded animals to the well-done stage would be another method of control, although transmission of the disease through the consumption of animal flesh or organs would appear to be rare.

Streptococcal Poisoning

At various times it has been reported that in food which appeared to be the cause of gastroenteritis in humans, the only organism present in large numbers were the fecal streptococci. These organisms belong to Lancefield's serological O group and there are two species, *Streptococcus faecalis,* varieties *liquefaciens* and *zymogenes,* and *Streptococcus faecium,* variety *durans.* The former species is normal to the human intestinal tract, the latter species is found in the intestinal tract of the hog.

While certain authors (Dauer, 1961)[19] have found that gastroenteritis in humans similar to that caused by *Clostridium perfringens* may be caused by the enterococci, more recent work (Shattock, 1962)[60] indicates that cultures of these types when fed to humans have no physiological effect indicative of disease.

Owing to the doubtful nature of this disease, no further discussion of the subject will be attempted.

Erysipeloid

Erysipeloid is a food infection not involving the gastrointestinal tract. The disease causes infection of the hands of food handlers, especially those who contact fish, meat, or poultry. The incidence of this disease is seasonal, almost all cases occurring during the warmer months.

The Disease

The skin lesion first appears as a pink area which causes an itching or burning sensation. This soon changes in appearance to reddish-blue which develops in size and become painful. There is no purulence. There may be more than one area affected and the lesion persists for about 7 days or more after which it is disappears. Most infections of this type are said to respond to treatment with pencillin (Proctor and Richardson, 1954).[57]

The Organism Causing Erysipeloid

The organism causing Erysipeloid have been classified as three species, *Erysipelothrix rhusiopathiea, Erysipelothrix muriseptica,* and *Erysipelothrix erysipeloidis.* All of these are the same species or strains of the same species. The organisms are pleomorphic ranging from coccoid rods to filamentous forms. They grow best at pH values between 7.3 and 9.0 and are inhibited below pH 6.4. These organisms have some tolerance to sodium chloride and produce acid in media containing sugars. The name *Erysipelothrix insidiosa* has been suggested as the correct nomenclature (Langford and Hansen, 1953).[31] These organisms also cause mouse septicemia and swine erysipelas.

Food Causing the Disease

Workers handling fish, meats, or poultry ordinarily are subject to erysipeloid disease, the infection occurring after some cut or abrasion of the skin has been sustained. Some cases in humans handling vegetables have also been cited. It is apparent, therefore, that the organism must be present, at least in low concentration, on foods of certain types. The fact that the incidence of the disease is greater in warmer months would suggest that there may be some build-up or increase of these organisms on foods or on equipment contacting foods, providing contamination of wounds with greater concentration of the organism hence increasing the probability of infection.

Control of Erysipeloid

Since it is probable that there are increased concentrations of the organism causing erysipeloid of foods or equipment during summer months, the application of suitably low temperature refrigeration to raw materials is indicated as a method of control. Also, the application of frequent cleaning and sanitation of equipment contacting foods should decrease the concentrations of the causative organisms, hence tend to decrease the incidence of the disease.

It is highly probable that if food handlers were required to wash their hands and dip them in a suitable disinfectant at frequent intervals, and especially after suffering minor wounds or abrasions of the hands, the incidence of erysipeloid would be greatly decreased.

The Sanitary Control of Shellfish

(National Shellfish Sanitation Program Manual of Operations, 1965, Parts I, II, III) [50-52]

One of the methods of controlling food-borne infections involves circumstances usually not concerning the ordinary food handler, but is concerned with surveys and tests made by the U.S. Public Health Service and authorities appointed by various states in coastal areas. This system of disease control refers to the growing and handling of shellfish and to the waters in which shellfish are grown. This subject is of sufficient importance to public health to deserve some consideration in a discussion of food microbiology.

The United States' catch of oysters amounts to more than 60,000,000 lb yearly and, in addition, more than 50,000,000 lb of clams are harvested yearly. These figures refer to oyster meats and clam meats without the shell. It should be realized though that the figures for clams refer to several distinctly different species.

In the past, shellfish, especially molluscs such as oysters and clams, have been associated with many outbreaks of food-borne disease. In Europe and Scandinavia, numerous oubtreaks of typhoid, salmonellosis, and even cholera have been traced to the consumption of oysters or mussels. Oysters are usually eaten raw on the half shell, without cooking, and hard-shelled clams are also frequently eaten raw on the half shell. When oysters are cooked they are usually prepared as an oyster stew, a procedure of cooking in which the heating is not nearly adequate to destroy many of the bacteria which might be present therein. Even when the soft-shelled clam is cooked by steaming, the heating may not be sufficient to destroy all the pathogenic bacteria which might be present, if the shellfish were taken from polluted areas, and certainly may not be sufficient to inactivate all viruses.

The types of species of shellfish, then, that in the United States are of particular concern to public health authorities from the standpoint of food-borne disease are the oysters; *Crassostrea virginica, Crassostrea gigas,* and *Ostrea lurida;* the hard-shelled clams, *Venus mercenaria* and *Protothaca staminae,* and to a lesser extent *Mya arenaria,* the soft-shelled clam.

It should be understood that oysters and clams may contain pathogenic bacteria because of the fact that, in order to grow them, harvest them, and protect them from natural predators, they must be raised in areas near the shore in comparatively shallow waters. These waters must, of necessity, considering the density of coastal populations, be near settled communities, and the difficulty of keeping these areas free from contamination with sewage from such communities is considerable.

It is only due to the vigilance of the United States Public Health Service and to the efforts of state shellfish authorities, over which the United States Public Health Service has a kind of supervision, that outbreaks of food-borne disease, due to the ingestion of certain molluscs, have occurred only infrequently.

Regarding shellfish sanitation regulations, in order to understand what is done and why, a brief description of the growing of oysters will be included.

Oysters and clams are bivalves, the two parts of the shell (upper and lower) are hinged and joined at a point, in the middle of one side, by a ligament which tends to force the shell open against the pull of the adductor muscle which is used to close it.

The visceral mass, which includes the intestinal tract, liver or digestive organs, heart, and reproductive organs, lies in the area from about the center of the shell to near the hinge and is enclosed by the mantles, which are ribbon-like tissues extending from the visceral mass along the margin of the shell to again join the visceral mass at the other end. The mantles are attached to the visceral mass at both ends and are lightly attached to the shell in areas away from the visceral mass. At the posterior end, the mantle is more or less fused to form a projection called the neck. This projection is quite large in some species, such as the soft-shelled clam and barely distinguishable in hard-shelled clams and oysters. In the neck, there are two tubes, the syphons. Through these syphons water is brought into the shell and visceral mass area (the inlet syphon) and after circulation over the visceral mass is again forced to the outside (the outlet syphon). By this means both food and oxygen are provided for sustenance.

Attached to the visceral mass is the gill structure by which oxygen for respiration is removed from the water. The gills are composed of small tubes lined with cilia and lie within the mantles extending along the posterior and ventral margins of the visceral mass. The syphons also contain cilia, and the movement of cilia in the inlet syphon and gill tubes causes water to enter the shell cavity between the mantle and the viscera and the movement of the cilia in the outlet syphon forces the water to the outside.

The gills remove oxygen from the water and pass it into the blood and they also strain out the microscopic plant life in the water which serves as food for these molluscs. An opposing set of cilia in the gills force the strained-out food against the water current within the shell, then pass it along to the palps (flesh-like folds over the top anterior visceral mass, also having cilia) which pass it along to the mouth. In the palps, sand and extraneous material are rejected in such a manner that they fall into the shell cavity. Feces from the intestinal tract are also emptied into the shell cavity and will be eventually forced to the outside through the outlet syphon.

Water temperatures are important to the activity of oysters and other bivalves. At 4.4°C (40°F) an oyster takes in almost no water, at 20°C (68°F) it will syphon in and out about 2 liters per hour, and at 30°C (86°F) it will syphon about 2-1/2 liters of water per hour. Above this temperature, the rate of syphoning is slowed down and at 40°C (104°F) it ceases to function.

Oysters and clams can be kept alive for periods of at least 1–2 weeks under conditions in which they are protected against drying out and the temperature is held below 4.4°C (40°F), but above freezing, because they are relatively inactive biological specimens. Bivalves can live for comparatively long periods of time without food. Furthermore, they are able to obtain enough oxygen for survival respiration from oxygen dissolved in the water present in the gills. Oysters and clams, therefore, are often shipped and handled in the shell in the live state, and one should not eat any bivalve specimen which died in the shell before cooking.

When oysters or clams are fed a suspension of bacteria 50–80% of these bacteria pass through the gills. The action of the cilia in the gills tend to concentrate or remove bacteria from the water which is syphoned into and out of the shell cavity. For this reason, the bacterial counts made on bivalve tissues usually indicate higher numbers than do those made on the water in which these specimens are grown or from which they are harvested.

When bacteria are ingested by bivalves coliform types and the disease-causing salmonellae are not destroyed but pass out into the central body cavity with the feces. The shell liquor of bivalves, therefore, contains many of the various types of bacteria present in the water from which they were harvested. It has been shown, for instance, that certain species of *Salmonella* bacteria will survive in oysters for as long as 60 days.

Oysters, hard-shelled and soft-shelled clams are often handled in the shell in the live state. In such circumstances in the plant these molluscs may be washed with potable water, culled for size, and placed in burlap bags or barrels for purposes of shipment. Since these molluscs soon die at higher temperatures, they must be kept at temperatures below 4.4°C (40°F) and preferably near 1.7°C (35°F) until prepared for eating.

Oysters are shucked, or removed from the shell, in some cases, for eating in a form other than on the half-shell, and hard-shelled clams are also shucked, especially for use in Manhattan-style chowders. Soft-shelled clams are frequently shucked for preparing as fried clams. After shucking, the meats may be washed and aerated or the washing water may be agitated by passing air bubbles through the water.

United States Federal and State Control of Oyster and Clam Harvest Areas

Since clams and oysters, of the type indicated, grow near the shore in comparatively shallow waters, and since such waters are generally near populated areas, they are apt to become polluted. In order to prevent the transmission of disease from the ingestion of such specimens, the United States federal and state authorities maintain strict control over the sanitary quality of waters where these shellfish are harvested.

The United States Food and Drug Administration exercises supervision over the sanitary quality of shellfish shipped in interstate commerce. This service evaluates whatever standards and requirements are set up by the state authorities where the shellfish are grown and makes recommendations to state authorities for improvements, changes in procedures, increased supervision, or whatever may be necessary to attain the objective of bringing shellfish of suitable sanitary quality to the consumer.

The state makes sanitary surveys of the areas where the shellfish are grown, inspects shellfish packing houses periodically, and makes bacteriological examinations of the waters over shellfish growing areas and of the shellfish product itself. The shucked shellfish freshly removed from the shell or taken from products present in packing house shipping containers are also examined bacteriologically.

The state sets up regulations and requirements for the enforcement of the state shellfish control measures and for the certification of shellfish for shipment in interstate commerce. The U.S. Food and Drug Administration also keeps a list of certified states and of certified producers and packers within the state. Failure of any state to comply with the minimum requirements results in all shippers in the state and the state itself being removed from the certification list. An embargo is then placed on all interstate shipment of products of this kind from this state or produced in the state. Since the U.S. Food and Drug Administration is an enforcement agency, shellfish products, shipped interstate under such conditions, would be seized and destroyed. Shippers or growers within a state who do not comply with state regulations are also removed from the certification list and prevented from shipping products interstate. Also under such circumstances the state authorities would, no doubt, prevent intrastate shipment of these molluscs or mollusc products.

Any area which is to be used for growing oysters or clams or from which these shellfish are to be harvested must be subjected to frequent sanitary surveys by qualified state personnel. In this survey such things are considered as (1) the extent of pollution of waters tributary to the growing area, (2) the direction and speed of water currents and the possibilities of natural or self-purification between the point of pollution and shellfish growing areas, (3) the degree of dilution of polluted waters over the growing area and the evidence of residual pollution in the growing area. The latter is determined by the results of bacteriological tests on the water to detect the number of coliform bacteria per 100 ml of water or of shellfish liquor or as the number of coliform bacteria per 100 g of shellfish meats.

On the basis of the results of sanitary surveys of the growing area and surroundings and of bacteriological tests on the waters of the growing area, such areas are classified as either (1) approved, (2) restricted or moderately polluted, or (3) closed or grossly polluted.

An approved area is one in which the sanitary survey discloses no likelihood that human fecal discharges will reach the area in dangerous concentrations before sufficient time has elapsed to render such discharges innocuous. Also, the waters of such an area according to standard tests must show a mean coliform density not in excess of 70/100 ml. The samples on which such bacteriological tests are made must be collected over the various stages of the tide from those parts of the shellfish growing area most likely to be exposed to pollution with fecal material. Also, during any one or more active shellfish seasons the conditions disclosed by the sanitary survey must not have worsened.

A grossly polluted area is one in which the sanitary survey shows obvious gross pollution by direct discharges of sewage, or is one in which there is exposure to even slight pollution more or less continuously from nearby sources, or is one in which there is exposure to occasional direct contamination with human fecal discharges, or is one in which the standard bacteriological tests indicate a median coliform density in excess of 700/100 ml. Any of these conditions is sufficient to have a shellfish growing area classified as grossly polluted.

A restricted or moderately polluted area is one in which the sanitary survey indicates that it is questionable as to whether there is ever any pollution with material from human fecal discharges or in which the standard bacteriological tests on the water indicate a mean coliform density greater than 70/100 ml and not more than 700/100 ml.

Grossly polluted areas are closed to the harvesting of oysters or clams. These products may not be taken from such areas for consumption by humans excepting under special conditions. These special conditions are: when the water reaches a temperature of 5°C (41°F), or less, provided that permission is received from the state and provided that the mean coliform density in the shellfish meats, as indicated by standard bacteriological tests, is not greater than 20/100 g. It should be pointed out that this is tantamount to saying that the removal of oysters or clams from grossly polluted growing areas for purposes of consumption by humans can never be done. For instance, even in shellfish removed from approved areas the mean coliform density is rarely, if ever, as low as 20/100 g. Oysters cease to syphon water at about 5°C (41°F); but the probability that the coliform bacteria, which they certainly would contain if the shellfish were held in polluted waters, would die off even in a considerable period of time, is negligible.

Oysters or clams may be taken from a polluted area and relaid in an approved area provided that (1) permission is received from the state, (2) state officials supervise the removal and relaying, and (3) provided that the removed shellfish will remain in the approved area for not less than 14 days at water temperatures greater than 10°C (50°F) before they are removed to be sold for human consumption. It should be noted that under these conditions (14 days in approved waters at 10°C (50°F) or higher) the shellfish would have purified themselves or would have syphoned out any pathogenic types of bacteria. Pathogenic bacteria would not be present in the approved waters.

The boats harvesting oysters or hard-shelled clams also come under state and federal regulations. Such boats must be so constructed and maintained as to prevent contamination of the shellfish in the boat with enteric pathogenic bacteria. The shellfish must be held on perforated platforms on the boat in such a manner that this product is not contaminated with bilge water (the drainage which collects in the deepest section of a boat) or with water from unapproved areas from over the side. Also, all water used for cleaning the boat must not be polluted.

Shellfish harvesting boats must carry chemical toilets in order that the waters of approved growing areas will not be polluted with the fecal discharges of those working on the boat.

Previously, coliform counts were required which provided a MPN not greater than 230/100 g for oyster meats or oysters from areas not considered to be polluted. Such counts were considered to indicate that the product would cause no public health hazard. In recent years, it has been found that products which are not polluted do not conform to such standards. Other standards based on fecal coliforms *(Escherichia coli)* have, therefore, been adopted. These standards are as follows (National Shellfish Sanitation Program Manual of Operations, Part I, 1965)[50]:

Satisfactory. "Fecal coliform density of not more than 230 MNP per 100 grams and 35°C plate count of not more than 500,000 per gram will be acceptable without question."

Conditional. "Fecal coliform density of more than 230 MPN per 100 grams and/or 35°C plate count of more than 500,000 per gram will constitute a conditional sample and may be subject to rejection by the State regulatory authority. If these concentrations are found in two successive samples from the same shipper, the State regulatory authority at the source will be requested to supply information to the receiving state concerning the status of operation of this shipper. Further shipments to receiving markets by the shipper concerned will depend upon satisfactory operational reports by the shellfish regulatory authorities at the point of origin."

Note. Escherichia coli in the above standards is defined as coliforms from positive presumptive broth tubes which will produce gas from E.C. liquid media (E.C. media contains tryptone, lactose, bile salts, K_2HPO_4, KH_2PO_4, and NaCl) within 24 hr when incubated at 44.5 ± 0.2°C (112.1°F ± 0.36°F) in a water bath or in an air incubator at 45.5 ± 0.2°C (113.9 ± 0.36°F). Total counts are determined by standard plate count procedures for shellfish.

Oyster- and clam-shucking plants also come under state and federal regulation. To begin with, the sanitary condition of the plant is regulated even before operations are allowed to start. These regulations refer to such things as the construction and drainage of floors, screening, toilet facilities, hand-washing facilities, heating, ventilation, lighting, equipment, benches, and uniforms for personnel. No person known to have or to be a carrier of a communicable disease or no person known to have infected hands or to have hand lesions may work in an oyster- or clam-shucking plant. All equipment and utensile must be cleaned and sanitized within 3 hr after termination of the day's work. The specifications of sanitizing are the same as those indicated for dishes and utensils in the Chapter on food preparing and serving establishments (See Chapter 12).

The state issues permits to shellfish producers and packers who comply with their regulations and the packer's permit number must appear on each package of shellfish which is shipped interstate. This is a U.S. Public Health Service regulation but the states usually require the same thing for intrastate shipment of clams and oysters. The shellfish dealer must also keep a completed record of all of his activities in handling these shellfish. He must record (1) the date of sale of any product of this kind, (2) the date of purchase and the person from whom the shellfish were purchased, (3) the person to whom the shellfish were sold. The shucking-plant operation must also have a permit and is subject to the same regulations regarding shipment of the shucked product as are those who handle oysters or clams in the shell. Reshippers of shellfish or of shellfish products must comply with the same regulations.

As has been previously indicated, oysters and clams, placed in approved waters, will purify themselves. Under good drinking (syphoning) conditions, the bacteria ingested by molluscs and surviving passage through the intestines will be discharged into the shell cavity in about 5 hr and eventually syphoned to the outside. Shellfish of this kind may, therefore, be purified by floating or holding in tanks of water which contains no pathogenic microorganisms or in which the pathogenic microorganisms that might have been present have been killed. The tanks used for floating shellfish must be so constructed that they may be properly drained. The water used to float shellfish must not be more than moderately polluted and this water must be kept chlorinated so that the residual of available chlorine never falls below 0.05 ppm. The oxygen content of the water must also be kept at 30% of saturation or higher in order that the shellfish be able to obtain enough oxygen to remain active. The reason, then, that moderately polluted water may be used for purposes of purification is that it has to be chlorinated beyond the break point to start with and this is sufficient to destroy enteric pathogens which might be present.

The procedure of floating is carried out by placing the oysters or clams in a tank and filling it with seawater which should be above 15.5°C (60°F) or allowed to come to this temperature. The water is then chlorinated with 2–5 ppm of available chlorine (this may be done during addition of the water). This water is allowed to remain for a period of 24 hr. Shellfish of this kind will not syphon water when the chlorine content is greater than 0.2 ppm but after the addition of water containing 2–5 ppm the chlorine soon (in several hours) reacts with organic matter in the water and on the shells of the molluscs to the point where the shellfish start to syphon. After holding in this manner for 24 hr, the process may be repeated and the shellfish held for a second 24 hr under these conditions.

By an alternate method the seawater used for purification may be purified in a reservoir and treated with an excess of chlorine. After a period of 30 min to 1 hr the chlorine in the water is dissipated either by aerating or by adding a solution of sodium thiosulfate which neutralizes the chlorine. The water is then run through the purification tank containing the shellfish for a period of 24 hr or more, in such a manner as to keep the molluscs well covered with water and at such a rate as to supply sufficient oxygen to these specimens. (In some areas water purified by ultraviolet light treatment is being used.)

It is the syphoning of nonpolluted water, which purifies the shellfish, not the chlorine. The chlorine is used only to destroy those enteric pathogens which might have been present in the water, used for purification, when it was removed from the ocean.

The purification of oysters and clams must be carried out under state supervision and daily bacteriological tests must be made on the water used for purification and the treated shellfish. The meat of treated specimens must meet the standards previously indicated as satisfactory.

Viruses

For many years, the food industry, together with various government agencies, has been working to produce food products of better quality. Much of the increased quality or acceptability of foods has been achieved by minimizing, to the smallest possible extent, the deleterious effects of heat, freezing, drying, and storage on the organoleptic and nutritional quality of foods. Although methods have been developed and employed to improve the sanitary aspects of food quality during processing, paradoxically, some of the processes necessary for the production of better quality foods tend to optimize the survival of organisms of public health significance. It is interesting that some workers in the field of public health believe that the effective measures taken to increase the sanitary quality of foods, water, and sewage disposal was one of the main reasons that poliovirus emerged as an important pathogen in recent years. The over-all result of increased sanitation was to lower the incidence of immunity caused by early subclinical infections and thereby greater numbers of individuals were exposed, for the first time, to a subclinical infection which resulted in paralytic poliomyelitis in some individuals having no immunity.

Because of the rapid increase in commercially available frozen foods and other convenience comestibles, the question arises as to whether such products may be involved or potentially involved in the dissemination of viral disease. Although viruses are known not to multiply in foods, they may be introduced in a number of possible steps in the food chain from harvesting to ingestion. For instance, arthropods, humans, or even animals may carry or harbor viruses and introduce them into foods. From an epidemiological point of view, there are known outbreaks of food-borne diseases in which a bacteriological agent has not been established as the infective agent; in such instances viruses have not been ruled out as the causative agent (Dack, 1963).[18] Further epidemiological studies have indicated that the Picornaviruses which includes ECHO, polio, and the animal foot-and-mouth disease viruses may be transmitted by foods (Cliver, 1965).[12] Therefore, the environmental factors affecting the survival of viruses and their transmission in food systems that are not sterilized, but processed in such a manner as to allow the survival of microorganisms, should be investigated. This would include foods preserved by freezing, freeze-drying, drying, or radiopasteurization.

Currently some investigators in the field of food-borne viral diseases have been concerned mainly with the transmission of infectious hepatitis by foods (Cliver, 1966).[11] Recently Lynt (1966)[37] has investigated the survival of enteroviruses in various commercially frozen foods stored at 10°C (50°F) or at − 20°C 4°F). Also, Cliver (1965),[9] Cliver and Yeatman (1965),[12] and Sinskey et al. (1971)[61] have been conducting research concerned mainly with the development of techniques to determine the recovery of small numbers of viruses such as enteroviruses and reoviruses. These techniques may be applicable for the detection of viruses in foods. Methods to increase the recoverability of small numbers of viruses from foods by employing such procedures as membrane filtration and ultracentrifugation were investigated. However, to date no indices have been established for foods that would indicate the number of viruses that might constitute a potential public health hazard if ingested. Research with animal viruses has demonstrated the survival of foot-and-mouth disease virus in cured and uncured meats (Cottral et al., 1960).[15] Outbreaks of ornithosis (psittacosis) have been attributed to poultry plant workers handling virus-infected turkeys (Delaplane, 1958).[20]

Infectious hepatitis virus predominates as the primary infective agent associated with outbreaks of food-associated viral disease. Food types most commonly implicated with outbreaks of infectious hepatitis include shellfish, milk, and dairy products. Frequently food-handlers may become carriers of the virus. Many times epidemiological evidence has indicated that food-handlers have worked while ill and probably contaminated various food samples.

Presently, the fact that no suitable laboratory tissue culture system is available for the cultivation of the hepatitis virus *in vitro,* and the restricted host range of the virus has hampered research work. Furthermore, the long incubation period, an average of 30 days with a range of 15–50 days, has hampered epidemiological studies.

Probably the second most important group of viruses that may be transmitted by foods are the enteroviruses. The enteroviruses are considered to be a subgroup of the picornaviruses. All of these viruses have the common properties of a diameter of approximately 38 mμ cubic symmetry, an RNA core, and lack of structural lipid. Other important groups include reoviruses, arborviruses, myxoviruses, adenoviruses, and papovaviruses.

Many of the animal viruses play an important role in the economic sense. Most noted are foot-and-mouth disease virus, rinderpest, picornaviruses of cattle and swine, and Newcastle disease virus.

Since viruses are not extremely heat resistant, the types of foods that they may be associated with are probably those foods that are fresh or are sublethally processed, i.e., frozen, dehydrated, and pasteurized.

Presently, there is limited knowledge in the general area of viruses in foods. Most research has been directed toward the development of isolation techniques for the recovery of viruses from foods. Information concerning this phase of the problem is needed since viruses are probably present in low numbers and, furthermore, do not multiply in foods. Present research has concentrated on this problem by attempting to develop classification methods for the removal of food debris, followed by concentration methods such as centrifugation and membrane or ultrafiltration. To date, however, very little information is available concerning the survival of viruses in foods during processing. Research in this area is needed for the development of processing techniques that will control viral transmission by foods.

Viruses have variable resistance to heat and freezing. For instance, the enteroviruses are resistant to heating at 50°C (122°F) for 1 hr in the presence of 1 *M* $MgCl_2$ (Wallis and Melnick, 1962),[66] while the influenza virus is very sensitive to heating (Francis and Maasab, 1965).[22] Coxsackie virus is quite resistant to heating as compared to the other viruses and will survive 70°C (158°F) and 80°C (176°F) for 30 min if the virus is suspended in milk, cream, or ice cream (Kaplan and Melnick, 1951).[29] Also, coxsackie viruses have been shown to maintain infectivity if stored in 50% glycerol or horse serum at room temperature for 70 days or if stored in a refrigerator for longer than 1 year. The mouse polyoma virus will withstand freezing, thawing, and lyophilization, without undergoing an appreciable loss in infectivity (Brodsky, 1959).[6]

Recently Sullivan (1970)[63] has developed a method for isolating viruses from beef that involve formation of a slurry, followed by filtration. The method was employed to examine market-purchased ground beef for viruses. Numbers of 1–195 viral PFU/g were isolated from 3–12 loaves of meat. Poliovirus and echovirus were the types isolated. The significance of these in foods remains to be determined.

References

1. R. Angelotti, M. J. Foter, and K. H. Lewis, Time–temperature effects on salmonellae and staphylococci, *Foods II. Behavior at Warm Holding Temperatures, Thermal Death Time Studies,* U.S. Dept. of Health, Education, and Welfare S.E.C. Tech. Rep. F 60–5, 1960.
2. R. Angelotti, E. Wilson, M. J. Foter, and K. H. Lewis, Time–temperature effects on salmonellae and staphylococci, *Foods I. Behavior in Broth Cultures and Refrigerated Foods,* U.S. Dept. of Health, Education and Welfare Sanitary Engineering Center, Tech. Rep. F 59–2, 1959.
3. Annual Supplement Summary 1967, *Morbidity and Mortality Weekly Report,* U.S. Dept. of Health, Education and Welfare Public Health Service, **16**, No. 53, 1968.
4. A. Beloian and G. C. Schlosser, Adequacy of cooking procedures for the destruction of salmonellae, *Amer. J. Pub. Health.* **53** (1963) 782.
5. W. C. Boyd, "Fundamentals of Immunology," 4th ed. Interscience, New York, 1966.
6. I. Brodesky,P. R. Wallace, J. W. Hartly, and W. T. Lane, Studies of mouse polyoma virus infection, *J. Exp. Med.* **109** (1957), 439–447.
7. F. L. Bryan, J. C. Ayres and A. A. Kraft, Salmonellae associated with further-processed turkey products, *Appl. Microbiol.* 14, (1968).
8. R. L. Cecil and R. F. Loeb, "A Textbook of Medicine," 10th ed., Saunders, Philadelphia, 1959.
9. D. O. Cliver, Factors in the membrane filtration of enterovirus, *Appl. Microbiol.* **13** (1965), 417–425.
10. D. O. Cliver, *"Radiation Preservation of Foods,"* Nat. Res. Council, Publ. No. 1273, 1965.
11. D. O. Cliver, *Pub. Health Rep.,* **81** (1966), 159–165.
12. D. O. Cliver and J. Yeatman, Ultracentrifugation in the concentration and detection of enteroviruses, *Appl. Microbiol.* **13** (1965), 387–392.

13. W. C. Cockburn and E. Simpson, Food poisoning in England and Wales, *Min. of Health and Publ. Health Lab. Serv. Monthly Bull.* **10** (1950), 223.
14. C. H. Collins, "Microbiological Methods," Plenum, New York, 1967.
15. G. E. Cottral, B. F. Cox, and D. E. Baldwin, The survival of foot-and-mouth disease virus in cured and uncured meat. *Amer. J. Vet. Res.* **21** (1960), 288.
16. G. M. Dack, "Food Poisoning," Univ. of Chicago Press, Chicago, 1956.
17. G. M. Dack, Problems in food-borne diseases, *in* "Microbiological Quality of Foods," pp. 41–49. Academic Press, New York, 1963.
18. G. M. Dack, Importance of food-borne disease outbreaks of unrecognized causes, *Food Technol.* **22** (1966), 1279.
19. C. C. Dauer, 1960 Summary of disease outbreaks and a year resume, *Pub. Health Rep.,* **76** (1961), 915.
20. J. P. Delaplane, Ornithosis in domestic food: Newer findings in turkeys, *Ann. N. Y. Acad. Sci.* **70** (1958), 495–500.
21. R. H. Diebel and J. H. Silliker, Food poisoning potential of the enterococci, *J. Bacteriol.* **85** (1963), 827.
22. T. Francis and H. F. Maasab, "Viral and Rickettsial Infections of Man" (F. L. Horsfall, and I. Tamm, Eds.), 4th ed. Lippincott, Philadelphia, 1965.
23. M. M. Galton, W. D. Lowery, and A. V. Hardy, Salmonella in fresh and smoker pork sausage, *J. Infec. Dis.* **95** (1954), 232.
24. M. M. Galton, W. V. Smith, H. B. McElreath, and A. B. Hardy, *Salmonella* in swine, cattle and the environment of abattoirs, *J. Infec. Dis.* **95** (1954), 236.
25. W. F. Gileroy, M. P. Hines, M. Kerbaugh, M. E. Green, and J. Koomen, *Salmonella* in two poultry processing plants, *J. Amer. Vet. Med. Ass.* **148** (1966), 550.
26. H. E. Hall, R. Angelotti, K. H Lewis, and M. J. Foter, Characteristics of *Clostridium perfringens* strains associated with food and food-borne disease, *J. Bacteriol.* **85** (1963), 1094.
27. B. C. Hobbs, *in Food-borne infections and intoxications* (Hans Riemann, ed.), pp. 131–171. Academic Press, New York, 1969.
28. B. C. Hobbs, M. E. Smith, C. L. Oakley, H. G. Warrack, and J. C. Cruickshank, *Clostridium welchii* food poisoning, *J. Hyg.* **51** (1953), 74.
29. A. S. Kaplan, J. L. Melnick, Effect of milk and other dairy products on the thermal inactivation of coxsackie viruses, *Amer. J. Pub. Health* **44** (1951), 1174–1184.
30. A. Kazuyoshi and M. Matsuno, The outbreaks of enteritis-type food poisoning due to fish in Japan and its causative bacteria, *Jap. J. Microbiol.* **5** (1961), 337.
31. G. C. Langford, Jr., and P. A. Hansen, The species of *Erysipelothrix, Antonie van Leewenhoek* **20** (1954), 87–92.

32. L. Leistner, The occurrence and significance of *Salmonella* in meat animals and animal by-product feeds, *Proc. 13th Res. Conf. Amer. Meat Inst. Found.*, p. 9. 1961.
33. K. H. Lewis, Nature and scope of the food-borne disease problem, *Current Concepts in Food Protection,* U.S. Dept. of Health, Education and Welfare Publ. Hlth. Service, Cincinnati, Ohio, 1964.
34. J. J. Licciardello, J. T. R. Nickerson, and S. A. Goldblith, Destruction of salmonellae in hard-boiled eggs, *Amer. J. Pub. Health* **55** (1965), 1622.
35. J. Liston and J. R. Matches, Study of the basic microbiological and biochemical factors involved in the irradiation of marine products, *8th Annual A.E.C. Food Irradiation Contractors Meeting,* U.S. Atomic Energy Commission, Division of Isotopes Development and Biology and Medicine, 1968, p. 77.
36. O. Luderitz, A. M. Staub and O. Westphal, Immunochemistry of O and R antigens of *Salmonella* and related *Enterobacteriae, Bacteriol. Rev.* **30** (19–6), 192.
37. J. R. Lynt, Survival and recovery of enterovirus from foods, *Appl. Microbiol.* **14** (1966), 218–222.
38. D. C. Mackel, M. M. Galton, H. Gray, and A. V. Hardy, Salmonellosis in dogs, *J. Infec. Dis.* **90** (1952), 12.
39. M. J. MacKerras and I. M. MacKerras, Salmonella infection in Queensland, *Aust. J. Exp. Biol. Med. Sci.* **27** (1949), 163.
40. J. O. Mason and W. R. McLean, Infectious hepatitis traced to consumption of raw oysters, *Amer. J. Hyg.* **75** (1962), 90.
41. K. F. Meyer, Food poisoning, *N. Eng. J. Med.* **249** (1953), 804–812, 843–852.
42. Morbidity and Mortality Weekly Report, U.S. Dept. of Health, Education, and Welfare, Public Health Service, 13(7) (1964) 53.
43. Morbidity and Mortality Weekly Reports, U.S. Dept. of Health, Education and Welfare, Public Health Service, **14** (1965), 385.
44. Morbidity and Mortality Weekly Reports, U.S. Dept. of Health, Education and Welfare, Public Health Service, **15** (1966), 185.
45. Morbidity and Mortality Weekly Reports, U.S. Dept. of Health, Education and Welfare, Public Health Service, **15** (1966), 385.
46. Morbidity and Mortality Weekly Report, Annual Supplement, Summary 1967, U.S. Dept. of Health, Education and Welfare, Public Health Service, **16** (1968), 7.
47. Morbidity and Mortality Weekly Report, U.S. Dept. of Health, Education and Welfare, Public Health Service, **17** (1968), 418.
48. Morbidity and Mortality Weekly Report, U.S. Dept. of Health, Education and Welfare, Public Health Service, **19** (1970), 151.
49. G. K. Morris, W. T. Martin, W. H. Shelton, G. W. Joy, and P. S. Brach-

man, Salmonellae in fish meal plants: Relative amounts of contamination at various stages of processing and a method of control, *Appl. Microbiol.* **19** (1970), 401.

50. National Shellfish Sanitation Program, Manual of Operations, Part I, Sanitation of Shellfish Growing Areas, U.S. Dept. of Health, Education and Welfare, Public Health Service, 1965.
51. National Shellfish Sanitation Program Manual of Operations, Part II, Sanitation of the Harvesting and Processing of Shellfish, U.S. Dept. of Health, Education and Welfare, Public Health Service, 1965.
52. National Shellfish Sanitation Program, Manual of Operations, Part III, Public Health Service Appraisal of State Shellfish Sanitation Programs, U.S. Dept. of Health, Education and Welfare Public Health Service, 1965.
53. J. T. R. Nickerson, J. J. Licciardello and S. A. Goldblith, Treatment of Poultry with Ionizing Radiations to Destroy Salmonellae, Final Report for the U.S. Atomic Energy Commission, Division of Biol. and Med., Contract No. AT(30–1)–3727, 1968.
54. B. Nygren, Phospholipase C-producing bacteria and food poisoning: An experimental study on *Clostridium perfringens* and *Bacillus cereus, Acta Pathol. Microbiol. Scand. Suppl.* **160** (1962), 162.
55. H. N. Old and S. L. Gill, Typhoid fever epidemic caused by carrier bootlegging oysters, *Amer. J. Pub. Health.* **30** (1940), 623.
56. W. W. Osborne, R. P. Straka, and H. Lineweaver, Heat resistance of strains of salmonella in liquid whole egg, egg yolk and egg white, *Food Res.* 19 (1954) 451.
57. D. M. Proctor and I. M. Richardson, A report on 235 cases of erysipeloid in Aberdeen, *Brit. J. Ind. Med.* **11** (1954), 175.
58. R. Sakazaki, S. Iwanami, and H. Fukumi, Studies on the enteropathogenic, facultatively halophilic bacteria, *Vibrio parahaemolyticus* 1. Morophological, cultural and biochemical properties and its taxonomic position, *Jap. J. Med. Sci. Biol.* **16** (1963), 161.
59. W. Savage, Problems of salmonellae food-poisoning, *Brit. Med. J.* **2** (1956), 317.
60. P. M. Shattock, "Chemical and Biological Hazards in Foods," (J. C. Ayres, H. E. Snyder, and H. W. Walker, Eds.), pp. 303–319. Iowa State Univ. Press, Ames, Iowa, 1962.
61. A. J. Sinskey, G. C. Y. Chu, and D. I. C. Wang, Concentration and purification of viruses by ultrafiltration, *Chem. Eng. Series* **67**, No 168 (1971) 75.
62. H. J. Stafseth, M. M. Cooper, and A. M. Wallbank, Survival of *Salmonella pullorum* on the skin of human beings and in eggs during storage and various methods of cooking, *J. Milk Food Technol.* **15** (1952), 70.
63. R. Sullivan, A. C. Fassolitis, and R. B. Read, Method for isolating viruses from ground beef, *J. Food Sci.* **35** (1970), 624–626.

64. F. W. Tanner and L. P. Tanner, "Food-Borne Infections and Intoxications," Gerrard Press, Champaign, Illinois, 1953.
65. M. J. Thornley, Microbiological Aspects of the Use of Radiation for the Elimination of Salmonellae from Foods and Feeding Staffs, Int. Atomic Energy Agency, Tech. Rep. Ser., No. 22 (1963), 81.
66. C. Wallis and J. L. Melnick, Cationic stabilization—a new property of enteroviruses, *Virology* **16** (1962), 504–506.
67. Q. B. Ward, Isolation of organisms related to *Vibrio parahaemolyticus* from American estruarine sediments, *Appl. Microbiol.* **16** (1968), 543.
68. R. J. Watkins, A. I. Flowers, and L. C. Grumbles, Salmonellae organisms in animal products used in poultry feeds, *Avian Dis.* **3** (1969), 290.
69. G. S. Wilson and A. A. Miles; "Principles of Bacteriology and Immunity," 5th ed., Williams & Wilkins, Baltimore, Md., 1964.

CHAPTER 11

Food Intoxications

Staphyoococcal Poisoning

Staphylococcal poisoning is the disease most often wrongly referred to as ptomaine poisoning. This is a food intoxication in which the bacteria grow in the food and form a toxin which is excreted into the food itself.

One of the complicating features associated with staphylococcal poisoning is that when the organisms grow in foods they produce no pronounced odor or taste; and a food having hundreds of millions of staphylococci per gram may taste, smell, and appear to be little different from that in which none of these organisms have grown.

Staphylococcal poisoning is probably the most common type of food poisoning reported in the United States, although it would seem that perfringens poisioning may be catching up with this disease in frequency. During the first 6 months of 1968 there were 2391 cases of staphylococcal poisoning reported (Anonymous, 1969)[9] (confirmed and not confirmed) which was about 25% of all food-borne illnesses. This disease is ordinarily not fatal, and it is considered that death from this disease occurs only when the patient is already in a weakened or sickly condition when the poisoning occurs (it is probably that there were many more cases than 2391 since this is not a notifiable disease).

Staphylococcal Poisoning—the Disease

The symptoms of staphylococcal poisoning are nausea, vomiting, abdominal cramps, prostration, and diarrhea (Cecil and Loeb, 1959).[18] While the symptoms last, suffering may be acute; but this usually involves a period of only a few hours and in rare cases several days. Generally, patients recover without complications. The incubation period after ingesting toxin is 1–7 hr, usually 3–6.

Bacteria Causing the Disease

The bacteria causing staphylococcal food poisoning are the same as those causing suppurating infections, boils, and carbuncles. All of these organisms are classified under the genetic and specific names of *Staphylococcus aureus,* although it is evident that there must be a number of different strains of this organism.

The organism known as *Staphylococcus aureus* is spherical or ovoid in shape, nonmotile, and in liquid culture media or suspended on a slide arranges itself in grape-like clusters, in small groups, or sometimes in pairs. These organisms are also said to sometimes form short chains. The individual cells are 0.8–1.0 μ in diameter. As is the case with the streptococci, the staphylococci are gram-positive.

The staphylococci grow best in the presence of oxygen, but they will also grow in the absence of oxygen and are, therefore, facultative aerobes. The pH range over which they will grow is relatively narrow, 4.8–7.6; therefore, they grow best in foods which are neutral or nearly so and will not grow in acid foods. Like some of the streptococci, *Staphylococcus aureus* grows over a rather wide temperature range in foods as will be seen later. These bacteria will also grow in media containing concentrations of sodium chloride as high as 10% or even higher.

Staphylococci of the *aureus* type liquefy gelatin and reduce nitrates to nitrite and ammonia. When grown on solid culture media, they produce golden, yellow, and sometimes white colonies.

Staphylococcus aureus produces three types of hemolysins all of which may be fatal when injected into laboratory animals; they also produce a leukocidin.

When grown in foods or in liquid media, *Staphylococcus aureus* produces toxins which can be filtered away from the bacterial growth.

The toxins can be extracted from foods with water or saline solution and filtered away from the food particles or bacteria. Further purification of such extracts have shown that these toxins are neither hemolysins nor leuckocidins. *Staphylococcus aureus* also produces an enzyme, coagulase, which will cause clotting of blood serum.

Foods Most Frequently Causing Outbreaks of *Staphylococcus* Poisoning

Almost any moist food which will support the growth of *Staphylococcus aureus* may cause staphylococcal poisoning; and, as will be seen later, even dried foods may be involved in the transmission of the disease. The types of foods most frequently cited as having caused this disease upon ingestion are: ham and ham products; bakery goods, such as eclairs, which have custard fillings; chicken products, especially chicken salad, potato salad, and cheddar cheese. Potted meats, meat pies, gravy, prepared meat specialties, spray-dried eggs, milk, and even ice cream have sometimes been implicated in the transmission of staphylococcal poisoning.

It is interesting to speculate as to why some of these foods should so often be associated with the transmission of staphylococcal poisoning. Two factors have tended to cause ham and ham products to be frequently involved in the transmission of staphylococcal poisoning. Ham usually contains 2–3% of salt which tends to eliminate the growth of many saprophytic types of bacteria which might otherwise grow and cause decomposition of the product but would have no otherwise harmful effect. This provides a substrate, the ham, in which the *Staphylococcus* will grow well and can do so without interference from the growth of other bacteria which might produce antibacterial substances inhibitive to staphylococci. Also, in handling ham and ham products people are likely to consider that, since the main ingredient of these foods has been cured and smoked, it is quite stable and does not need to be refrigerated. This is far from the case.

Custard fillings would make a good substrate for the growth of many bacteria, both saprophytic and pathogenic. However, when this material is prepared it is heated to some extent. This heating is enough to kill off the ordinary rod-like, gram-negative forms of bacteria, both those causing decomposition, the ordinary saprophytic forms, and those causing such diseases as salmonellosis. Again, therefore, the staphylococci have a chance to grow without interference from competing bacteria. Custard or cream-filled pastry have caused public health problems, especially in situations in which bakery personnel and retailers have mishandled these products by holding them without refrigeration. This particular type of mishandling of foods has decreased somewhat in recent years due to warnings by public health officials, and some of the larger bakeries have discontinued the production of such products during summer months. However, it is still not uncommon to see custard and cream-filled pastry exhibited in an unrefrigerated store window, such foods eventually being sold to some unwary customer. This practice should be discontinued.

The reason why cheese has been sometimes involved in causing staphylococcal poisoning seems to be that milk, in bulk, is sometimes allowed to stand around, especially in tank trucks, for long periods before it is used. Under such circumstances the temperatures of the raw material rises allowing the bacteria in these products to multiply to the exent that the milk contains large numbers of microorganisms.

The toxins produced by staphylococci are somewhat resistant to heat. When present in foods (Read and Bradshaw, 1966),[30] the bacteria themselves are comparatively heat resistant for nonsporeforming types but are not so heat resistant as the toxin (it has been found that the purified toxin may be quite heat labile (Bergdoll, 1968)[14]). Considering the heat resistance of the toxins in foods, it is possible to have a staphylococcal poisoning outbreak from foods in which the toxin is present, having resisted processing procedures, but in which the bacteria themselves have been killed during the heating involved in processing. There is a case on record in which dried milk, shipped to Puerto Rico from the United States, caused an outbreak of staphylococcal poisoning involving 1500 people. No viable (living) staphylococci could be found in the product, but the toxin was there (Armijo, 1957).[12] The bacteria, in this instance, had grown in the milk and produced the toxin prior to product drying. When the milk was dried, the bacteria were killed by the heating involved, but the heating was not of sufficient severity in time and temperature effect to destroy the toxin.

Many people seem to be of the opinion that it is the mayonnaise which supports the growth of staphylococci in foods, such as chicken salad and potato salad, both of which may be a good growth medium for staphylococci. This is far from the truth. The pH of mayonnaise, commercial or homemade, is never above 4.0; and at this pH staphylococci just do not grow. Tests with the strains of staphylococci producing type-B toxin have shown that they will not grow in mayonnaise (Martins, 1964).[25] However, when mayonnaise is added to other materials, such as potatoes, chicken, or even ham, the pH of the mixture is much higher than that of the mayonnaise itself; and the mixture becomes a suitable medium for the growth of *Staphylococcus aureus.*

Source of Organism

The human is considered to be the most important source of *Staphylococcus aureus* in foods. It has been found that approximately 40% of normal human adults harbor these organisms in the nose and throat, hence the finger tips of humans are often contaminated with these bacteria (Williams, 1963).[36] This may be one reason why certain salads are likely to be involved in transmission of the disease. Chicken for salad, for instance, must be removed from the bone by hand, affording excellent opportunity for contamination of the meat with *Staphylococcus aureus.* Moreover, many times personnel will use their hands to mix the ingredients of salads, again providing opportunity for inoculation with *Staphylococcus aureus* and other human pathogens.

Humans with minor abrasions on their hands, which may become somewhat suppurative, are sometimes allowed to handle foods. The contact of foods with such minor sores may literally add millions of *Staphylococcus aureus* cells to foods.

At least in a few instances the source of the organism has been the cow, the organism coming directly from the udder, particularly if the animal has mastitis.

Properties of Staphylococcal Enterotoxin

As previously pointed out, staphylococcal poisoning is due to toxins secreted in foods when the causative organism grows therein. The toxins have been shown to be proteins of low molecular weight, 30,000–35,000, consisting only of amino acids. The amino acid composition and the terminal amino acids of the toxins are known; but the particular sequence of amino acids in the molecules is not known (Bergdoll, 1968).[14]

It has been found (Casman and Bennett, 1965)[17] that large numbers of *Staphylococcus aureus* must be present in foods to cause poisoning. It was estimated that 1–4 μg of type-A staphylococcal toxin would be required to cause symptoms. Foods which caused staphylococcal poisoning were found to have a staphylococcal count of 50×10^6–200×10^6 per gram, a very high count indeed. However, it must be remembered that *Staphylococcus aureus* may grow to these large numbers in foods without causing changes in the odor, taste, or physical appearance, thereby providing no warning signal to the consumer of such foods. Other findings (Bergdoll, 1968)[14] would indicate that less than 1 μg of A enterotoxin may be required to cause sickness in humans, the latter being more sensitive to the toxin than monkeys. A recent outbreak caused by whipped butter (Anonymous, Morbidity and Mortality Weekly Report, 1970)[11] would indicate that very little type-A toxin may be necessary to cause sickness.

Effect of Temperature on Staphylococcal Toxin

Compared to heat-labile bacterial toxins, such as those produced by the various strains of *Clostridium botulinum,* staphylococcal toxins are heat stable. These toxins also have been reported to have the property of being inactivated faster in some ranges of lower temperature than in some ranges of higher temperature. The inactivation times at different temperatures are listed in Table 11–1 for enterotoxin B (Satterlee and Kraft, 1969).[32]

Effect of Temperature on Staphylococci

It has been shown (Angelotti et al., 1959)[2] that in three foods, chicken à la king, custard filling, and ham salad, *Staphylococcus aureus* grew best at

TABLE 11–1

Staphylococcal Enterotoxin B in Phosphate–Saline Buffer (pH 7.4)

Temperature	Time to inactivate 50% of Toxin (min)	
	Crude toxin	Partially purified toxin
60°C (140°F)	204	204
80°C (176°F)	14	12
100°C (212°F)	37	24
110°C (230°F)	17	16

35°C (95°F) in custard filling, reaching somewhat higher cell concentrations. At 6.7°C (44°F) there was somewhat better growth in chicken à la king than in the other foods.

At the higher temperatures, such as might be encountered in cooking or in the early stages of cooking, *Staphylococcus aureus* showed only slight increases in ham salad at 44.4°C (112°F), but in chicken à la king reached comparatively high numbers and even higher cell concentrations in custard filling at this temperature. In chicken à la king *Staphylococcus aureus* decreased in numbers at 45.5°C (114°F) but in custard filling increased to comparatively high numbers at this temperature. In custard filling at 46.6°C (116°F) the cell count of *Staphylococcus aureus* remained constant for some time, then decreased slowly. This temperature, therefore, must be considered as the lowest at which staphylococci will be destroyed in foods and probably it would be safer to consider the limiting temperature to be higher than this.

During the roasting of medium-sized and very large turkeys, it was shown that there might be no or very slight increases in the numbers of *Staphylococcus aureus* cells inoculated into the stuffing, until the bird had been cooked for some time; but decreases in the numbers of cells did not start until stuffing temperatures had reached 48.9°C (120°F) in some cases and 57.2°C (135°F) in others.

Immunological Tests to Determine the Presence of Staphylococcal Enterotoxin (Casman and Bennet, 1965)[17]

Bacteriological tests to determine enterotoxic staphylococci do not provide absolute indication of their presence in foods, hence certain immunological tests for the presence of toxin have been developed.

At the present time the purified toxins (the antigen) of types A, B, C, and D are available and also the purified antitoxins (antibodies) corresponding to these toxins.

In order to make the test, the suspected food must be extracted to obtain the toxin and the extract further purified and concentrated. This may be

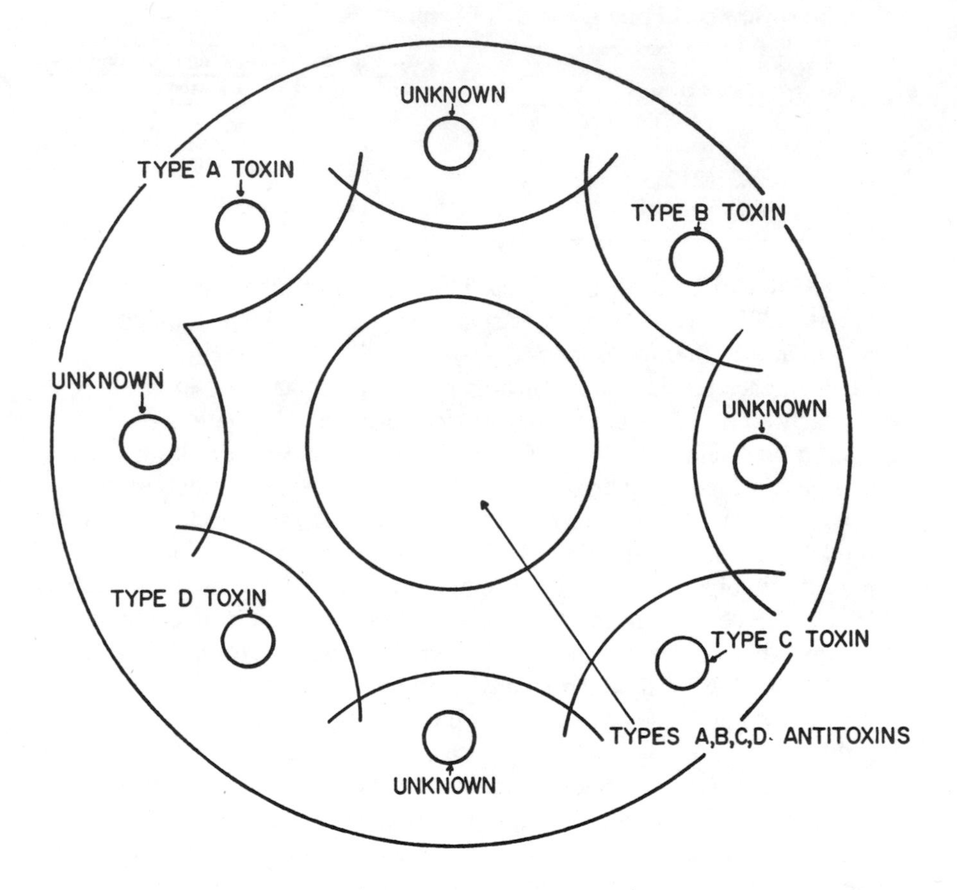

FIG. 11–1. Modified Ouchterlony techniques for identification of staphylococcal enterotoxin.

done, for instance, by extracting the food with 0.2 *M* sodium chloride, adjusting the pH to 7.4 to 7.5 centrifuging and dialyzing against polyethylene glycol to concentrate the extract. The extract is then adsorbed on a column of Sephadex G-100 and eluted with 0.2 *M* NaCl and the eluate concentrated by dialyzing against polyethylene glycol.

A modified Ouchterlony technique (Bergdoll et al., 1965)[15] may now be used to identify the toxin. A center well is prepared in an agar plate in which a mixture of the various antitoxins is added. Peripheral wells containing a known toxin, the unknown extract, a known toxin of another type, and so on are prepared. If after incubation a confluent line of precipitation is obtained with a known antigen or toxin and the extract, then the extract is identified as containing that particular toxin (See Fig. 11–1).

The test for staphylococcal toxin must be considered as a good method of determining that foods had caused a staphylococcal poisoning outbreak or might be the cause of a food poisoning outbreak of this kind. However, there are several strains of types of *S. aureus* which will produce enterotoxin. It is already known that four different types of enterotoxin, A, B, C, and D can be produced by different strains of *Staphylococcus aureus*. Recent examination of cultures of coagulase-positive staphylococci isolated from foods would appear to indicate that there are, at least, several other types of enterotoxin which may be produced by staphylococci. Such facts serve to limit the effectiveness of toxin–antitoxin reactions as a test for wholesomeness of food. Since type-A staphylococcal toxin is that most frequently encountered as the cause of staphylococcal poisoning outbreaks, a test of this kind would usually provide significant results (this does not exclude the fact that a test of this kind might fail to identify the cause of a particular food poisoning).

Control of Staphylococcal Poisoning

It is usually the case that mishandling of food, at one point or another, has taken place when outbreaks of staphylococcal poisoning occur (Anonymous, 1969).[10] This means that the foods, at some point, have been held for at least several hours at temperatures well above 6.6°C (44°F). This disease, therefore, could usually be controlled by applying suitable refrigeration temperatures to foods at all times when not in the process of manufacture, not being prepared for eating, or about to be eaten.

Since the organisms causing staphylococcal poisoning are comparatively heat resistant for vegetative types, it is probably not possible to be certain that all staphylococci which might be present in foods will be destroyed when some methods of cooking are employed. It is usually possible, however, to make certain that temperatures will be reached during cooking at which most of these organisms will have died. The attainment of temperatures of 68.3°C (155°F) in the coldest part of the stuffing during the roasting of stuffed birds would accomplish this.

It is also possible to make certain that during the cooking or holding of foods the product is not held for long periods of time at temperatures at which staphylococci will grow. It must be remembered that once staphylococci have produced toxin in foods it will not be possible to destroy the toxin by ordinary cooking, even though the organisms themselves may be killed by the heating applied. Also, type -A enterotoxic staphylococci produce toxin during the exponential growth phase, hence may produce toxin at relatively low cell concentrations. Toxin production by type B is said to occur in the stationary phase of growth, hence at relatively high cell concentrations.

One point, in the chain of preparing foods for serving, where foods are frequently mishandled, is when held on steam tables. Steam tables should be operated in such a manner that the temperature in all parts of the foods held therein never falls below 60°C (140°F). This means that large masses of food should never be held on a steam table in a single container unless the material is added to the container at a temperature of 60°C (140°F) or higher and also unless the container is almost completely immersed in the heating medium (steam or water) which is at temperatures sufficiently high to maintain this temperature in the produce 60°C(140°F).

Foods also should not be held on steam tables for long periods of time both because of the possibility of cooling off to temperatures suitable for the growth of food-borne disease organisms and since foods held in this manner tend to quickly deteriorate in quality.

The personal sanitation of food handlers in all food operations is extremely important in the prevention of staphylococcal poisoning since the human is the chief source of contamination of foods with the organisms causing this disease. Food handlers once having left the food operation, be it food manufacturing or the cooking or dispensing of foods, for purposes of personal necessities or a rest or coffee break, should be forced to thoroughly wash and disinfect their hands before returning to their particular duties. It is the responsibility of supervisory personnel to see that this is done. People whose hands come in contact with foods in any type of food operation should be discouraged by supervisory personnel from handling or picking their nose when actually working on foods. This is obviously a requisite to good food sanitation since so many humans harbor *Staphylococcus aureus* in their nose or throat.

Of great importance in the prevention of outbreaks of staphylococcal poisoning due to the ingestion of foods is the elimination of personnel from direct contact with foods when infections are present on their hands. This would include people with even small abrasions or wounds which are purulent or pussy and certainly would include any person who has boils.

To illustrate the importance of keeping staphylococci out of edible foods or in as low concentrations as is possible, let us assume that one food had 100 cells of *Staphylococcus aureus* per gram while another had 1,000,000 per gram. Also, assume that 100,000,000 cells per gram of this organism will cause food poisoning. Under conditions of mishandling of these foods during which they would multiply or double every 40 min, we would have the following situations: It will have taken the high-count food only 4 2/3 hr under conditions of mishandling to have reached the point when it would cause a food poisoning outbreak, whereas under the same conditions the low-count food would require 13 1/3 hr to reach the same condition. Moreover, there are greater safety factors than this in foods containing only very small numbers of pathogens which might cause food poisoning outbreaks. Oftentimes when only few of pathogenic bacteria of this kind are present in foods they will be overgrown by competitive bacteria, those types which ordinarily spoil foods and do not cause disease. If pathogens are present in foods in large numbers at the start of a situation in which foods might be mishandled, they will often predominate, overgrowing the other or ordinary spoilage types of bacteria.

It is obvious that better sanitation and food handling procedures may be necessary in some plants producing cheese and dried milk in this country. It would seem to be the responsibility of public health officials and food enforcement agencies, both State and Federal, to determine that the equipment in such plants is of sanitary design and construction, that such plants and equipment are kept clean and frequently sanitized, and, especially, that the milk used for such products (used for drying or for making cheese) is processed before it has been held for significant lengths of time at temperatures above 6.7°C (44°F).

Botulism Poisoning

Not many cases of botulism are encountered yearly in any country. Before much was known about its cause and its prevention there were years in the United States of America in which more than 50 cases of botulism were reported (Anonymous, 1968)[8]; in recent years the average number of known cases per year has been much less than this. It appears that every few years there is an increased number of cases of botulism, such as in 1963 when 46 known cases and 14 deaths were reported in the United States (Angelotti et al., 1960).[1]

What makes botulism so extremely horrifying is not the frequency of outbreaks but the high mortality rate which over the years has been about 57% of all cases in the United States.

Botulism is considered to be a food intoxication, the bacteria growing in foods and producing a toxin therein which is extremely poisonous to man and many other animals. While this is the case, at least with some types of the organism causing the disease, it is known that no toxin is released into the surroundings until the cells grow and lyse. This does not mean that botulism is not a food intoxication or that botulism is a food infection, since if the cell has grown in the food the toxin will doubtlessly be present, some of the cells having lysed.

Botulism—the Disease

The symptoms of botulism are vomiting (said to be caused by end products other than toxin present in the ingested food), constipation, ocular paresis (difficult eye movements), diplopia or double vision, pharyngeal paralysis, abdominal distention, a red raw sore throat, secretion of a thick viscid fluid, and sometimes dysphonia or difficulty of speaking (Rogers et al., 1962).[31] When severe, as is often the case, the breathing mechanism is affected. Attempts have been made to get around this by tracheotomy and the use of an iron lung, but the heart eventually becomes affected and ceases to function.

Botulism toxin affects the peripheral nerves of the automatic system at the point of the myoneural junctions or synapses. It does not affect the central nervous system. Nerve function is stimulated chemically by acetyl choline formed in the nerve sheath. What appears to happen is that the toxin prevents the formation of a compound (sympathin) which is required for the synthesis of acetyl choline, hence no acetyl choline is available to stimulate nerve function to activate automatic muscle systems. The lungs, heart, blood vessels (dilation and contraction), and other organs, the function of which is regulated by the automatic muscles, therefore, cease to act.

As has been pointed out, botulism toxin is extremely poisonous, as a matter of fact, one of the most poisonous substances known to man. Based on calculations in which the poisonous dose for a mouse is interpreted in terms of the total dose required to kill a number of mice weighing as much as a man, less than 0.2 μg of the toxin will kill a 200-lb man.

The Bacteria Causing Botulism (Wilson and Miles, 1964)[37]

The various types of *Clostridium botulinum* are gram-positive rods which form spores. Since they are spore formers, they are heat resistant. They are motile having a number of peritrichous flagella. These organisms are strictly anaerobic, growing only in the absence of oxygen or under conditions in which oxygen in the food or growth medium is tied up by compounds present in the medium. All of the *Clostridium botulinum* types ferment some sugars but some types are proteolytic and are not saccharolytic, while others are saccharolytic and not proteolytic. These organisms liquefy gelatin but do not reduce nitrates nor do they form indole from tryptophan.

All types of *Clostridium botulinum* are considered to be soil organisms, this being their natural habitat. Type E *Clostridium botulinum* and some other types are found in fresh water lakes and streams, in fish and in the ocean and the mud of the ocean floor and of lakes (in oceans usually fairly close to land but sometimes far out). It is considered that these organisms get into such areas through run off of water and soil which flows into oceans and lakes (Johannsen, 1963).[22]

Clostridium botulinum type A is proteolytic and not saccharolytic. This organism forms spores which are the most heat resistant of any of the spores of the various types. Also, the toxins produced by this organism together with that of type E are the most toxic to man of any of the various botulinum toxins, hence poisoning by these types of *Clostridium botulinum* have most frequently been fatal.

Clostridium botulinum type B is proteolytic and not saccharolytic. The spores of this particular organism are not as heat resistant as those of type A nor is the toxin produced by it nearly so lethal to man as that of type A. While over the years there have been numerous cases in man of type B botulism, the mortality rate of this type of poisoning is much lower than is the case with types A or E botulism (Anonymous, 1968).[8] It is probably because type B botulism appears to predominate among outbreaks in some countries that the mortality rate of botulism is much lower in these areas than in the United States where, when cases of this disease occur, they are frequently due to types A or E.

Clostridium botulinum type C is nonproteolytic and the spores are not especially heat resistant. This organism has caused botulism in man but it usually causes the disease in animals or especially in birds. In birds type C causes a disease known as limberneck and often death. Apparently, this organism grows in the muddy bottom of some ponds and birds foraging in such mud become poisoned.

There is a C_α and a C_β type which are immunologically different. Man is somewhat resistant to the toxin of *Cl. botulinum* type C, but he is less resistant to C_α toxin than to that of C_β.

Clostridium botulinum type D is not proteolytic nor are the spores especially heat resistant. Only one outbreak in man has been reported as due to this organism. This type of *Cl. botulinum* has chiefly caused disease in animals. At one time it was considered that type D was confined to Africa but in recent years it has been isolated in Australia and also from material obtained off of the coast of Mississippi (Ward et al., 1967).[35] Man is considered to be fairly resistant to the toxin of *Cl. botulinum* type D.

Clostridium botulinum type E is not proteolytic nor are the spores very heat resistant. It is saccharolytic fermenting several sugars including mannose. This organism and its toxin was discovered in 1937. Type E has been associated with fish or fishery products and is known to be present in some lakes, the water and mud of some sea areas as far out as 40 miles from the coast, and in the fish and shellfish in these areas. Man is quite sensitive to the toxin of *Clostridium botulinum* type E which is probably as toxic to man as that of any other botulinum type.

Clostridium botulinum type F is proteolytic (some strains are not proteolytic) and the spores are fairly heat resistant. This organism was isolated in 1960 and only two outbreaks due to this organism have been reported (Dolman and Murakami, 1961; Anonymous, 1966).[7,19] Apparently, man is more resistant to the toxin produced by this type of *Clostridium botulinum* than to the toxin of certain other types.

Clostridium botulinum Toxin

It is known that, like that of *Staphylococcus aureus,* the toxin of *Clostridium botulinum* is a protein consisting only of amino acids. Also, as in the case of *Staphylococcus aureus* toxin, the sequence of amino acids in the molecule is not known (Schantz, 1964).[33] The toxin of *Clostridium botulinum* type A has been reported to have a molecular weight of about 900,000. Types B, C, and D toxin are considered to be of about the same molecular weight. However, the true molecular weights are not established and remain a controversial problem. The toxin of type E is said to have a molecular weight of about 19,000.

Considering the fact that in any case the molecule is quite large and should not pass through either the stomach or the upper part of the intestine, it has always been somewhat of a mystery as to how it gets into the system to poison peripheral nerves. It is speculated that it does so through small abrasions in the gastrointestinal tract which allow more or less free access to the blood and especially to the lymph system.

Generally, it is not believed that the organisms *(Clostridium botulinum)* themselves invade the tissues or organs, grow there, and produce toxin. This would seem to be the case since we must all, in eating raw vegetables of one kind or another, have ingested some spores of *Clostridium botulinum,* yet we do not get botulism from eating these foods. A recent report (Ingram and Roberts, 1966)[29] stated that these organisms may invade the spleen, liver, etc. and that what happens is that the toxin, ingested along with the spores, causes the defense mechanism of the white and red blood cells to be inactivated, whereupon the bacteria invade the tissues. It is also known that spores of *Clostridium botulinum,* fed orally to animals in very large numbers, may eventually be isolated from such organs as the liver and spleen. However, the toxin itself will cause death, when administered in extremely small doses since this has been shown to be the case with animals fed or injected with crystalline toxin.

Clostridium botulinum type-E toxin differs from that produced by other types in that a precursor molecule is produced by the organism which is activated when treated with the proteolytic enzyme trypsin (Duff et al., 1956),[20] this enzyme breaking the precursor molecule down into smaller units of chains of amino acids. For some years it was known that extracts of

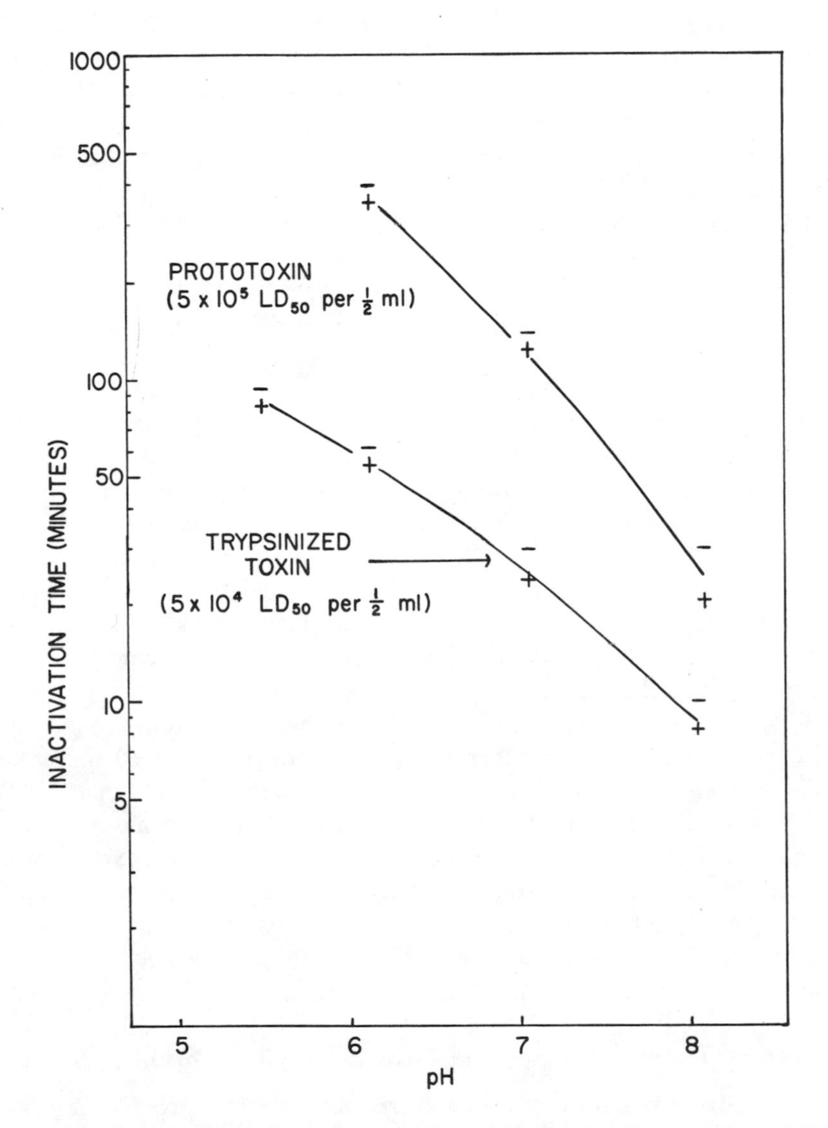

FIG. 11–2. Effect of pH on thermal inactivation of Type-E botulinum toxin at 150°F in TPG–haddock broth.

this toxin from foods having caused outbreaks of botulism were less toxic to animals than should be the case according to the severity of the disease in humans. It was then discovered that extracts could be made about as toxic to animals as was indicated by outbreaks in humans by treatment with trypsin. It is now considered that what happens when a human ingests food in which *Clostridium botulinum* type E has grown is that the precursor compound is broken down to the more active toxin by proteolytic enzymes in the small intestine, causing it to be more poisonous. The other strains may secrete proteolytic enzymes which carry out this reaction. The toxin produced by *Clostridium botulinum* is quite heat labile, but the amount of heating required for destruction seems to vary with the type (A through F) producing the toxin.

The time required to inactivate specific amounts of A and E types of *Clostridium botulinum* toxin at various temperatures are known (Schantz, 1964; Licciardello et al., 1967; Licciardello et al., 1967).[23,24,33] It is fortuitous that these toxins are heat labile; the heating of home-canned vegetables prior to ingestion may have prevented cases of botulism and saved lives. There is one case on record in which home-canned green beans were tasted, then heated, and served to a family. The taster came down with botulism and eventually died. Others eating the beans were not affected.

The medium in which *Clostridium botulinum* type-E toxin is heated is important since it is more heat stable at pH 6.0 than at pH 7.0 and less stable at pH 8.0 than at pH 7.0. This work was done in fish press juice; (see Fig. 11-2). This finding has some significance since in flesh-type foods, by the time that toxin has been produced in inoculated packs, there has been considerable growth of bacteria other than clostridia which tend to shift the pH to the alkaline side.

Whereas tests for toxicity are usually made by intraperitoneal injection of mice, it is claimed by some that injections of this kind are 10 times more toxic than when the toxin is administered orally and by others that this material is 1000 times more toxic when injected intraperitoneally than when administered orally. The results of recent tests with type-E toxin, in which some mice were injected intraperitoneally and others were administered the toxin by intubation through the mouth into the stomach, would indicate that the toxin is 50–100 times as potent injected intraperitoneally as when administered orally (Licciardello et al., 1967).[23]

Effect of Temperature on the Growth of *Clostridium botulinum*

In the 1930's it was believed that *Clostridium botulinum* would not grow at temperatures below 10°C (50°F) since tests made in attempts to grow types A and B in defrosted frozen foods showed that this was the case (Prescott and Geer, 1936).[28] Under some conditions these types may grow at temperatures as high as 50°C (122°F).

In recent years it has been found that *Clostridium botulinum* types E and F will grow and produce toxin in comparatively long periods at temperatures as low as 3.3°C (38°F) (Schmidt et al., 1961; Eklund et al., 1967).[21,34]

Regarding temperatures at which the cells of *Cl. botulinum* would be killed off, it must be remembered that this is a spore-forming organism. With bacteria which form spores, cells in vegetative form are destroyed by heat at much lower temperatures than are the spores. It would be surprising if vegetative forms of the low-growth temperature types did not start to die off at 50°C (122°F). However, since some spores are invariably found with vegetative forms of the same organism, heat resistance must be based on the effect of different temperatures on spores and not on the heat liability of vegetative forms.

The spores of *Clostridium botulinum* type A have the highest heat resistance. When extremely large numbers of spores of this type are present (60 billion), all of them will not be killed until all parts of the food have been heated for about 5 hr at 100°C (212°F) or all parts have been heated at 121°C (250°F) for about 2.5 min. It is the comparatively high heat resistance of *Clostridium botulinum* type A spores which causes canned foods processed in the home, without proper consideration of the times and temperature used, to be so hazardous from the standpoint of public health. Spores of types B and especially *Clostridium botulinum* type E are not so heat resistant as those of type A.

The spores of *Clostridium botulinum* type F are said to be fairly heat resistant and may be comparable with those of type A. However, since types A, B, and E are most frequently found in foods, waters, and soil deposits, etc., in the United States it may be considered that these are the types of greatest public health significance and methods of controlling botulism may be based on the characteristics of these types of the organism.

Treatment of Botulism

For many years it was considered that antitoxins were not helpful in treating those who had contacted botulism. This is now being proved to be probably not the case. Certainly it is now known that even when the patient shows definite symptoms with type-E botulism he can be saved by injection of type-E antitoxin. The evidence is not so clear in cases of type A botulism and it is considered that the toxin may be soon fixed after ingestion, so that treatment with antitoxin does little good. Ordinarily, however, when symptoms are recognized, treatment with polyvalent antitoxin is used.

Foods Involved in the Transmission of Botulism

From 1899 through 1967 there were 1669 cases of botulism in the United States. Among these 69.8% of the foods involved were not identified. The eating of vegetables was the cause of 17.8% of the cases. Especially neutral vegetables, such as green beans, corn, beets, spinach, asparagus, and mushrooms have been involved and usually these materials have been home canned or preserved as is the case with fruits. Fruits have accounted for 4.1% of the cases, fish for 3.6%, condiments for 2.2%, beef products for 0.8%, milk products for 0.5%, and poultry products for 0.1% of the cases of botulism in the United States.

It should be stated that, whereas years ago commercially produced canned foods had sometimes caused botulism, no case of this disease was caused by the consumption of commercially canned foods in the United States for more than 30 years. However, in 1963 there were three cases and two deaths due to commercially canned tuna fish. The fact that *Clostridium botulism* had grown in this particular product and produced toxin was not due to underprocessing of the canned tuna but was due to the use of defective cans. Can leakage after processing was apparently the cause of the presence of viable clostridia in the tuna after processing (Anonymous, 1963).[5]

An outbreak of botulism due to canned liver paste, also occurring in 1963, was caused by underprocessing; but this product was manufactured in Canada and imported into this country (Anonymous, 1963).[6]

In 1965 there were two cases of botulism caused by the consumption of commercially processed ham packed in another country (Anonymous, 1968).[8]

Generally, it can be said that commercially canned foods are as safe as any foods and present no public health hazard from the standpoint of botulism. On the other hand, home canned foods are responsible for most of the cases of botulism occurring in this country. This is mainly due to the fact that cold pack or other unsuitable methods are often employed for heat processing under circumstances in which the temperature and time of heating are not sufficient to destroy all spores of *Cl. botulinum* which might be present.

It may also be noted that in those instances in which canned foods have caused botulism, the product is usually the type which is neutral or low in acid. The organism *Clostridium botulinum* does not grow in acid foods (pH of 4.5 or below) and it has been stated (Baird-Parker and Freame,

1967)[13] that it will not grow at pH values below 5.0. While this is the case, there have been several instances in which acid foods, below pH 4.5, have caused botulism (apricots, figs, pears, and tomatoes) (Meyer, 1953; Anonymous, 1969)[9,27] In such instances, it is believed that some other organism (mold, yeast, or bacteria) first grew in the product and caused the pH of the food to shift toward the alkaline side, thus allowing *Clostridium botulinum* to grow, or possibly that other organisms first grew, producing compounds which favored the growth of *Clostridium botulinum* at a lower pH than 4.5.

Two cases of botulism from ingestion of frozen chicken pies have been reported (Anonymous, 1960).[4] These are the only cases of this disease which have been associated with frozen food, and they could just as well as have been caused by the ingestion of fresh chicken pies. In this case a number of frozen pies were heated in an oven for an hour or more, all but two were removed and eaten, the remaining two were left in the oven over night. The oven had a pilot light and thus the oven served as an incubator to keep the temperature at a point suitable for the growth of *Clostridium botulinum* over a long period of time. The next day the pies were taken on a picnic, heated slightly, and tasted by two people. They were not eaten since they now did not taste right. However, both of the tasters ingested enough toxin from the small amount consumed to come down with botulism.

It should be emphasized that almost no food product can be mishandled to the extent that were the chicken pies, just cited, and not become a public health hazard. If the consumer persists in holding perishable foods at temperatures suitable for the growth of bacteria, for more than very short periods of time after they are cooked, he can expect, sooner or later, to become the victim of one or more types of food-borne disease organisms. No manufacturer is able to keep all food-borne disease organisms out of all types of foods; and, even if this were possible, some of these organisms might be added to the product during preparation for eating. It is, therefore, the responsibility of the housewife, restaurateur, or caterer to see that foods are handled in such a manner that they are exposed to temperatures suitable for the growth of bacteria for periods of 1–2 hr at most.

Fish and marine products have fairly frequently caused botulism. Part of this is due to the fact that Indians, and especially Eskimos, frequently hold fresh salmon eggs, seal meat, whale meat, or seal flippers at rather high ambient temperatures for long periods of time and in some cases even allow them to putrefy, before they are eaten. Since *Clostridium botulinum* types E or F, which grow at comparatively low temperatures, may be present in such products, it is to be expected that occasional cases of botulism will occur among these groups as long as such practices are used.

Another class of products which have caused outbreaks of botulism in the civilian population of the United States has been smoked white fish and smoked white fish chubs. There are several reasons for this. To begin with the fish themselves may come from areas of the Great Lakes known to be comparatively highly contaminated with type E *Clostridium botulinum* (Bott et al., 1964).[16] Also, in the past, the plants in which these products are produced have not been as clean and as sanitary as should have been the case. This may now be changed since after the 1963 outbreak the Food and Drug Administration and other authorities started a campaign to clean up these processing plants. Another factor which may be contributory to public health hazards with these products is that the smoking of foods tends to mask off-odors which would oftentimes be caused by the growth of other bacteria by the time that *Clostridium botulinum* had grown to a sufficient extent to cause botulism in man. Hence, people may eat spoiled smoked products and not realize that the product is spoiled. Finally, in essentially all of the outbreaks of botulism caused by smoked fish the products causing the trouble have been held at ambient temperatures for long periods of time (several days either during transportation, in a warehouse, or by the consumer). Perishables of this type cannot be held out of refrigeration without becoming hazards to public health.

Control of Botulism

Since more than 90% of all cases of botulism today are caused by the consumption of home-canned foods, it would appear that some attempt should be made to educate the housewife to the hazards of improperly processed home-canned foods. The housewife should realize that the heating of canned or glass-packed foods in boiling water, especially nonacid foods (foods other than tomatoes or fruit), is not adequate to destroy the organisms causing botulism. Pressure cooking or cooking at 10–15 lb steam pressure must be employed for neutral types of foods such as green beans, corn, mushrooms, and many other vegetables. Moreover, the time required at a particular steam pressure will depend upon the size of the containers heated in the pressure cooker. The housewife has no way to determine, off hand, what times at a particular steam pressure should be employed in the heat processing of containers of foods, especially of the neutral type, of different sizes. The Bureau of Agriculture publishes several pamphlets on the home canning of foods, which any housewife may obtain at the cost of a few cents (Anonymous, 1936).[3] Such bulletins may also be obtained from Superintendent of Documents, Washington, D.C. Unless the housewife wishes to take a chance on her own demise, or that of some members of her family, she would be unwise to proceed with the home canning of foods without obtaining these pamphlets of instruction from the Bureau of Agriculture.

Whereas there have only been two or three small outbreaks of botulism due to the consumption of commercially canned foods during the past period of approximately 40 years, it would seem that all states should have laws which would require all canners to show that they had processed all batches of canned foods sufficiently to obtain what is known as a "minimum botulinum kill." The state of California has such a law and there is no doubt that since the 1920's this law has served to promote the public health of people in this country in general (Meyer, 1931).[26]

As has been previously indicated, the toxin produced by the various types of *Clostridium botulinum* is quite heat labile. How many cases of botulism have been prevented by the fact that preserved vegetables are usually heated to the boiling point of water before they are served, will never be known. Heating neutral products such as canned vegetables, mushrooms, etc., to the boiling point of water before serving is a method of controlling botulism. It is very unlikely that commercially canned products would cause any trouble in this respect, but this is especially important in home-canned foods and certainly would do no harm in the case of commercially canned foods of neutral type.

Since some types of *botulinum* (E and strains of type F) will grow at temperatures as low as 3.3°C (38°F), the application of suitably refrigerator temperatures (3.3°C or below) to foods at all times when not being prepared for eating or being eaten, is a method of control. This is especially important with flesh-type foods, notably fish and marine products. In the ordinary circumstance marine products would become so spoiled that the usual person would refuse to eat them by the time that toxin was developed, but it cannot be assumed that all people would react in this manner and this is especially the case with lightly smoked and lightly salted products in which spoilage off-odors tend to be masked by the smoky odor and taste.

Fresh meat products have no record of having caused botulism. Whether this is because they are usually cooked prior to eating is not known. Prepared meat products, however, such as sausage, potted meats, ham, and patés have been involved in the transmission of the disease. It is, therefore, indicated that such food items, if not heat processed in sealed containers, should be held at temperatures below 3.3°C (38°F) until shortly prior to preparing for eating. Even with fresh meat and poultry longer storage life and better edible quality will be obtained by holding at the temperatures indicated; and, since this would be an added safety factor, it must be considered to be good practice.

It has been found that at the time of the botulism outbreak caused by the smoked whitefish chubs some of the plants producing these products were extremely unsanitary and might even serve as a source of contamination with *Clostridium botulinum* type E. Good sanitation practices in food plants may, therefore, be considered to contribute significantly to the control of botulism.

References

1. R. Angelotti, M. J. Foter, and K. H. Lewis, Time-Temperature Effects on Salmonellae and Staphylococci in Foods. II. Behavior at Warm Holding Temperatures. Thermal Death Time Studies. U.S. Dept. of Health, Education and Welfare, Sanitary Engineering Center Technol. Rpt. F60–5, 1960.
2. R. Angelotti, E. Wilson, M. J. Foter, and K. H. Lewis, Time–Temperature Effects on Salmonellae and Staphylococci in Foods. I. Behavior in Broth Cultures and Refrigerated Foods, U.S. Dept. of Health, Education and Welfare, Sanitary Engineering Center Technol. Rep. F59–2., 1959.
3. Anonymous, Home Canning of Fruits, Vegetables and Meat, U.S. Dept. of Agric. Farmers Bulletin 1762, 1936.
4. Anonymous, Morbidity and Mortality Weekly Reports **9**, No. 27, U.S. Dept. of Health, Education and Welfare, Public Health Service, 1960.
5. Anonymous, Morbidity and Mortality Weekly Reports **12**, 126, U.S. Dept. of Health, Education and Welfare, Public Health Service, 1963.
6. Anonymous, Morbidity and Mortality Weekly Reports **12**, 386, U.S. Dept. of Health, Education and Welfare, Public Health Service, 1963.
7. Anonymous, Morbidity and Mortality Weekly Report, **15**, 359, U.S. Dept. of Health, Education and Welfare, Public Health Service, 1966.
8. Anonymous, *Botulism in the United States,* Review of cases, 1899–1967 and *Handbook for Epidermiologists, Clinicians and Laboratory Workers,* U.S. Dept. of Health, Education and Welfare, Public Health Service, 1968.
9. Anonymous, Morbidity and Mortality Weekly Reports, **18**, 105, U.S. Dept. of Health, Education and Welfare, Public Health Service, 1969.
10. Anonymous, Morbidity and Mortality Weekly Report, **18**, 97, U.S. Dept. of Health, Education and Welfare, Public Health Service, 1969.
11. Anonymous, Morbidity and Mortality Weekly Report, **19**, 271, U.S. Dept. of Health, Education and Welfare, Public Health Service, 1970.
12. R. Armijo, Food poisoning outbreak associated with spray-dried milk, an epidemiologic study *Amer. J. Pub. Health.* **47** (1957), 1093.
13. A. C. Baird-Parker and B. Freame, Combined effect of water activity and temperature on the growth of *Clostridium botulinum* from spore and vegetable cell inocula, *J. Appl. Bacteriol.* **30** (1967), 420.
14. M. S. Bergdoll, Bacterial toxins in foods, presented at the I. F. T. Annual Meeting in Philadelphia, 1968.
15. M. S. Bergdoll, R. B. Concordia, and M. A. Remedois, Identification of a new enterotoxin as enterotoxin C, *J. Bacteriol.* **90** (1965), 148.

16. T. L. Bott, J. S. Deffner, E. M. Foster and E. McCoy, Ecology of *Cl. botulinum* in the Great Lakes, *in* "Botulism," (K. H. Lewis and K. Cassel, Jr., Eds.), U.S. Dept. of Health, Education, and Welfare Public Health Service, 1964.
17. E. P. Casman and R. W. Bennett, Detection of staphylococcal enterotoxin in food, *Appl. Microbiol* **13** (1965), 181.
18. R. L. Cecil and R. F. Loeb, "A Textbook on Medicine," 10th ed., Saunders, Philadelphia, 1959.
19. C. E. Dolman and L. Murakami, *Clostridium botulinum* type F, with recent observation on other types, *J. Infec. Dis.* **100** (1961), 107.
20. J. T. Duff, G. Wright, and A. Yorinsky, Activation of *Clostridium botulinum* type E toxin by trypsin, *J. Bacteriol.* **72** (1956), 455.
21. M. W. Eklund, F. T. Poysky, and D. I. Wieler, Characteristics of *Clostridium botulinum* type F isolated from the Pacific Coast of the United States, *Appl. Microbiol.* **13** (1967), 1316.
22. A. Johannsen, *Clostridium botulinum* in Sweden and the adjacent waters, *J. Appl. Bacteriol.* **26** (1963), 43.
23. J. J. Licciardello, C. A. Ribich, J. T. R. Nickerson, and S. A. Goldblith, Kinetics of the thermal inactivation of type E *Clostridium botulinum* toxin, *Appl. Microbiol.* **15** (1967), 344.
24. J. J. Licciardello, J. T. R. Nickerson, C. A. Ribich and S. A. Goldblith, Thermal inactivation of type E *Cl. botulinum* toxin, *Appl. Microbiol.* **15** (1967), 249.
25. S. I. Martins, A study of staphylococcal enterotoxin, Sc.D. Thesis, Massachusetts Institute of Technology, 1964.
26. K. F. Meyer, The protective measures of the State of California against botulism, *J. Prev. Med.* **5** (1931), 261.
27. K. F. Meyer, Food poisoning, *N. Engl. J. Med.* **249** (1953), 765, 804, 843.
28. S. C. Prescott and L. P. Geer, Observations on food poisoning organisms under refrigeration conditions, *Refrig. Eng.* **32** (1936), 211.
29. "Proceedings of the Fifth International Symposium on Food Microbiology, Moscow," (M. Ingram and T. A. Roberts, Eds.), Chapman Hall, London, 1966.
30. R. B. Read, Jr., and J. G. Brandshaw, Thermal inactivation of staphylococcal enterotoxin B in veronal buffer, *Appl. Microbiol.* **14** (1965), 130.
31. D. E. Rogers, M. G. Koenig, and A. Spickard, Chemical and laboratory manifestations of type E botulism in man, "Botulism" (K. H. Lewis and K. Cassel, Jr., Eds.), U.S. Dept. of Health, Education and Welfare Public Health Service, 1964.
32. L. D. Satterlee and A. A. Kraft, Effect of meat and isolated meat proteins on the thermal inactivation of staphylococcal enterotoxin B. *Appl. Microbiol.* **17** (1969), 906.
33. E. J. Schantz, Purification and characterization of *Cl. botulism* toxins, "Botulism" (K. H. Lewis and K. Cassel, Jr., Eds.), p. 91, U.S. Dept. of Health, Education and Welfare, 1964.

34. C. F. Schmidt, R. V. Lechowich, and J. F. Falinazzo, Growth and toxin production by type E *Clostridium botulinum* below 40°F. *J. Food Sci.* **26** (1961), 1.
35. B. Q. Ward, B. J. Carroll, E. S. Garrett, and G. B. Reese, Survey of the U.S. Gulf Coast for the presence of *Clostridium botulinum, Appl. Microbiol.* **15** (1967), 629.
36. R. E. O. Williams, Healthy carriage of *Staphylococcus aureus, Bacteriol. Rev.* **27** (1963), 56.
37. G. S. Wilson and A. A. Miles, "Principles of Bacteriology and Immunity," 5th ed. Vols. 1 and 11, Williams & Wilkins, Baltimore, Md., 1964.

CHAPTER 12

Sanitation in Food-Manufacturing Plants and in Food-Preparation and Serving Establishments

It has been indicated (Anonymous, 1968)[9] that more than one-half of food-borne disease outbreaks in the United States of America are caused by foods eaten in restaurants, schools, or some institution other than the home. This implies that mishandling of foods is most often the reason why such outbreaks occur. Failure to apply adequate sanitary methods in the production and preparation of foods may be considered to be mishandling. From this standpoint, therefore, the application of good sanitary procedures in food-manufacturing and in food-serving establishments is an extremely important feature of the whole process. The application of good sanitation procedures to foods, in the manufacturing and preparation for serving, also has much to do with what might be called the edible quality of the ultimate product; and this fact is another important reason why sanitation in food handling is such an important factor.

While there are differences in the manner in which sanitary procedures may be conducted in food-manufacturing plants, as compared to food-preparation and serving establishments, in general, that which is useful in the one situation will provide satisfactory results in the other. For this reason the two applications will be jointly considered.

Supplies Available in the Areas of Manufacturing Plants or Eating Establishments

It is evident that anyone setting up a restaurant, food-manufacturing plant, or institution where food might be served would be sure to check the availability of certain sources of energy in the area. It would be unwise to establish a food plant, or institution where people are to be fed, without determining that electrical power and fuel for heating and cooling were available at prices which were economically suitable, although for restaurants there might be other factors which would tend to offset the cost of these utilities.

Water supplies are important features to the operation of any food-manufacturing plant or food-serving establishment. It is generally the case that sources of energy and water supplies are available in areas where restaurants or institutions are started. The food-manufacturing plant, however, could be set up in a more or less isolated area; hence, water supplies, especially potable water supplies, may not be available.

The water (which will be used for cleaning of equipment, utensils, etc. or for adding to foods), in food plants or food-serving establishments must be potable and must not contain pathogenic microorganisms. If the results of periodic tests to determine the quality of the water available in food-establishment areas are not attainable from local health authorities, then these tests must be carried out at the instigation and request of management involved in setting up the business or institution. Essentially, potable water for foods should not contain more than one coliform organism per 100 ml of water (Anonymous, 1960).[4]

Water which is to be used for boiler supplies, and especially those used for cooling (refrigeration, air conditioning, etc.), need not be potable; but, when the plant or eating establishment is built, it must be established that there is no possibility of accidental piping connection between potable and nonpotable water supplies and that the two systems are clearly identified as to which system is potable and which is not.

Waste disposal is important both to the manufacturing of food and to its serving. It must, therefore, be determined that sewage-disposal systems of suitable types are available for human wastes. Much of the organic wastes from foods, packaging materials, etc., rapidly accumulate during the process of food serving or of food manufacturing. Wastes must not be allowed to accumulate in the food-operation area. Therefore, it must be made certain that waste disposal systems are available in the area where the food operation is to be established or the operator or manufacturer must set up his own waste-disposal system.

Surroundings of the Building

The landscaping surrounding food plants, institutions, or restaurants should be made as attractive as is possible. For restaurants and institutions it is obvious that pleasant surroundings will serve to attract customers or to satisfy inmates. When the surrounding of the food-manufacturing building is pleasant, the psychological effect on the employees is such that they will now exert some effort to be neat and clean in the plant. If the surroundings are not well kept, the psychological effect is to tend to cause the worker to be slack and untidy as to his own cleanliness, as well as to that of the plant itself.

Surrounding areas of buildings in which foods are processed, prepared, or served should not consist of plain soil. In such cases the outside area becomes a source of dust within the plant which may, and probably will, contaminate the foods processed or prepared within. Also, dust from plain soil will cause the interior of the building to need cleaning and dusting more frequently than might otherwise be the case. Plain soil is also rarely entirely level, and eventually potholes fill up with water and may become breeding places for insects, hence a source of insects within the plant. Such conditions are, therefore, to be avoided.

Blacktopping of the area surrounding the building may be done and, if done properly so that drainage is adequate, is preferable to plain soil. However, blacktopping will not have the best psychological effect on personnel; and in hot weather may become a cause of high temperatures and uncomfortable conditions within the plant.

Under ideal conditions the grounds surrounding the building will consist of well-mown lawns with whatever shrubbery may be included for embellishment. This will have the most satisfactory effect on workers and on conditions within the plant.

The surrounding areas of the food plant or food-serving building should not be used as a storage place for boxes or crates used to hold raw materials, nor should the area be used as a space to store equipment or machinery which is not being utilized. Materials of this kind when stored in this manner become a shelter for rodents, hence may be a contributory cause of such animals within the plant. Such conditions are, therefore, not to be tolerated.

Building Construction
(Assoc. Food Industry Sanitation, 1952)[10]

Whether it is to be used for food preparation and serving or for food-product manufacturing, the building is preferably constructed of brick, concrete, or a combination of the two. Wooden buildings are more difficult to make rodent proof and more difficult to maintain in a condition of good repair. If the building is of wooden construction, it should have stone or concrete foundations elevated at least 18 in. above the ground and with retaining walls extending 2 ft under ground. If the building is connected with a basement, the walls of such should also be constructed of cement. Openings between studs and floor joints just above the sill should be filled with cement to a height of 4 in. above the floor.

Especially in areas where food is prepared or processed or where equipment or utensils are washed, floors should preferably be constructed of acid-resistant unglazed ceramic tile set in an acid-proof binder. Cement is probably the second best material for floor construction in such processing areas, but cement is subject to cavitation and eventually shallow pits and holes are formed. These holes accumulate water which provides an environment in which bacteria can grow and create odor nuisances and the possibility that food products handled in the area will be contaminated. Floors covered with epoxy or polyester resins are satisfactory if used over a suitable base.

Wooden floors may be suitable for areas where some types of materials used in connection with the processing of foods may be stored. Even in such areas, wooden floors are less desirable than cement since they are difficult to clean, may harbour insects, and are more subject to deterioration in general.

It is a fact that in the food-serving areas of restaurants, hardwood flooring may be used with good results although there would be no reason why cement floors or other impervious material covered with a suitable carpeting would not serve as well.

Processing areas and even areas where materials are stored should be so constructed that the inside junction of the wall and floor is watertight and formed in such a manner as to have a coived (curved) base for at least 6 in. above the floor. This facilitates cleaning.

The inside wall in food processing, preparing, and utensil or equipment-cleaning areas should be lined with a material which is easy to clean, preferably smooth glazed tile, for a distance above the floor at least equivalent to splash height.

The junction of the roof and wall should be insect, rodent, dust, and water-proof.

In food-processing, preparing, and utensil-washing areas the floors should be sloped toward the drains to prevent the accumulation of water. If the slope to the floor drains is too steep, some operations, such as moving trucks, become difficult to perform. For this reason detailed instructions have been worked out for the degree of incline which should be used (Association of Food Industry Sanitations, Inc., 1952).[10] Drains, the purpose of which is to eliminate water in processing areas, must be separate from toilet sewer lines to a point outside the building and should be so joined as to prevent the possibility of backup of sewage into the building. It is considered preferable to have closed under-floor sewer drains having a diameter of at least 4 in. Drains should be of sufficient carrying capacity as to prevent the accumulation of water on the floors of the working or walking area.

There should either be drains in walk-in refrigerators, or the floor should be sloped to the door in such a manner that washwater can be disposed of readily. This facilitates the periodic washing of walls and floors. It is unacceptable to have the drains in such areas connected with the floor-drainage system. They should instead empty into a sink or container which empties into the floor-drainage system and be connected in such a manner that there is no possibility of back-up into the refrigerator itself.

Wall openings, such as doors and windows, should be covered with tight-fitting screens (16-mesh) to prevent the entrance of insects and other pests. Air curtains which operate when doors are opened are helpful in eliminating insects.

In areas where foods are processed or prepared or where finished or unfinished materials are stored, window ledges should be sloped to prevent their use by personnel for storing materials or articles.

The processing area of food plants or of food-serving establishments should be well ventilated to prevent the condensation on ceilings and pipes of moisture which may drip onto foods being processed. Also, ceilings in such areas should be high enough to facilitate ventilation. Hoods, installed to remove condensates, should have forced-air movement of sufficient volume to prevent condensation. Where high ceilings and overhead openings are used for ventilation, such openings should be constructed and screened in such a manner as to keep out rain and prevent the entrance of birds.

The degree of lighting, in foot candles, has been specified for particular areas (Association of Food Industry Sanitation, Inc., 1952).[10] Both glare and deep shadows should be avoided. However, comparatively poor lighting may be used for such areas as storage warehouses and comparatively good lighting will be required for areas where foods are inspected. Stairways should be sufficiently lighted to make stairs or steps clearly visible, in order to prevent accidents. Cleanup crews or personnel should be provided with portable floodlights since it is important that they determine that areas shaded from direct illumination have been adequately cleaned and sanitized.

Toilet facilities must be provided in different locations for males and females and must be adequate in capacity to satisfy the needs of all personnel. The number of toilets, urinals, and lavatories which should be provided according to the number of personnel has been specified in standard texts (Anonymous, 1935; Parker and Litchfield, 1962).[1,13] The same facilities, adequate in the number of units, must be provided in food-serving areas. Toilet and lavatory facilities must be constructed with consideration for sanitary design.

Rooms enclosing toilet and lavatory facilities must have self-closing tight-fitting doors. Such areas must be kept clean and in good repair. Covered receptacles for waste materials, such as used paper towels, must be provided in toilet areas. Soap, towels, preferably paper, and toilet tissues must be provided in toilet areas; and hot and cold water should be available for wash bowls.

It is considered that hand-washing facilities for all personnel contacting foods should be present in areas where foods are processed or prepared in order that supervisors may determine that such personnel wash their hands upon returning to the working area, once having left.

In all food-manufacturing plants and in all food-preparing operations there must be raw materials available. In those operations involving the use of foods of high enough water activity (moisture content) to be subject to deterioration through the growth of microorganisms, sufficient refrigerated storage must be available to store all perishable products. Moreover, if products are to be stored in refrigerators in bulk, some method of precooling the product must be available since large masses of foods cool slowly when stored under refrigeration. Refrigerated storage rooms should be held at temperatures as near to the freezing point of the product as is possible without actually freezing the product. Sufficient refrigeration capacity must be available to precool to and maintain temperatures of 4.4°C (40°F) or below for all perishable products stored.

Where dry materials, such as cereals, breading materials, etc. are stored, the area should be refrigerated to maintain temperatures of approximately 12.8°C (55°F) or below to prevent the growth of insects from eggs which may be present in these products. Such areas also should be rooms within the plant which are so constructed and operated as to prevent the entrance of both insects and rodents.

Locker rooms should be supplied for personnel for purposes of storing clothes not used during working hours. If this is not done, clothes will be strewn around working or resting areas and create a nuisance. Male personnel locker rooms should be separate from those for females. Personnel should not be allowed to eat in locker room areas; nevertheless, receptacles for wastes should be provided in such places. Locker rooms should be kept neat and clean at all times.

If personnel are to eat meals which they have themselves brought to the food plant or are to eat materials purchased from machines of the sandwich-machine type, rooms, entirely separated from food processing, preparing, or storage areas and apart from utensil-washing areas or locker rooms, should be available for this purpose. Such rooms must be provided with receptacles for wastes and must be kept scrupulously clean and sanitary. Failure to keep such areas clean will result in the accumulation of scraps which will serve as sustenance for rodents and insects and also cause personnel to become careless with regard to plant cleanliness and sanitation.

Pipelines in food plants and food preparing areas should not pass over locations where foods are prepared for processing or cooked in open containers since there may be leakage of the material carried in the pipes into the product. Foods processed or prepared in areas where there are overhead pipelines may also be contaminated with peeling paint or corroded metal. Piping carrying water should be large enough to provide adequate supplies to all parts of the plant in which it is required for processing, cleaning, or other purposes. Potable water supplies should be in no way connected with nonpotable supplies, such as those which might be used in condensers for cooling. Also, potable water supplies should be installed in such a manner as to prevent their contamination through back siphonage. This may be prevented by the installation of equipment which is designed to prevent such possibilities or by the installation of siphon breakers in water lines for such equipment as flumes, gutters, sinks, tanks, toilets, etc.

Pipelines carrying food materials, such as milk, syrups, etc. should be composed of noncorrodable metal such as stainless steel or glass. The joints and unions of such piping should be constructed in such a manner that the piping can be removed for cleaning, especially if in-place cleaning is not possible. Also, there should be no exposed threads on the inside of the piping, where food materials may become lodged and be difficult to remove. Pumps forcing food materials through pipelines should be composed of noncorrodible metal and should be constructed in such manner that they may be easily taken apart and cleaned.

There should be no dead ends in pipelines carrying foods, where materials may enter and become trapped for some period of time. Foods trapped in such a manner are subject to bacterial growth and decomposition and may contaminate materials passing through the piping system by back siphonage.

Food-Plant and Food-Preparing Equipment

The equipment and machinery used in food processing and in preparing foods for eating should be composed of suitable materials and be of such design as to be easily cleaned and sanitized. This is especially true of equipment which contacts foods.

In almost no case should equipment be composed of wood since wood is subject to contamination and decay. Even cutting boards are preferably composed of hard plastic which is impervious to water.

Black iron and cast iron may be used for certain types of machinery which do not contact foods directly such as can sealers, retorts, etc. Such material is subject to corrosion and has rough surfaces, hence cannot be used in areas where food materials accumulate because of the difficulty encountered in cleaning.

Plain steel is also a poor material for equipment contacting foods since it is subject to corrosion.

Galvanized iron should not be used for the construction of food-processing or food-preparing equipment or for covering tanks or tables, since the zinc soon wears off exposing the iron which becomes rusty. The zinc itself may also cause discoloration of certain foods. Zinc is subject to dissolution when alkaline detergents are used for cleaning purposes.

Copper is used for certain types of manufacturing equipment. It has the desirable characteristic of being a good conductor of heat. However, it must be kept scrupulously clean, otherwise oxides will accumulate which are undesirable in foods; and the metal itself, as well as the oxides, may cause a destruction of vitamin C in foods and the oxidation of fats. Alkaline materials used on copper equipment may cause the formation of a dark-colored, complex (copper–ammonium ion), hence cause discoloration of foods. Considering all things, including the cost of cleaning, it probably is advisable to forego the use of copper in food plant and food preparing areas.

Monel metal, an alloy consisting mainly of nickel and copper, has been used in some food-manufacturing operations. This material is resistant to corrosion and is not affected by brine solutions. Since copper is present in monel metal, vitamin C and fats may be affected. It is also the case that the nickel content of monel metal has caused it to become very costly.

Probably the best all-around metal for the construction of equipment which contacts foods is stainless steel. This material can be finished to a high polish, hence is easily cleaned and sanitized. Stainless steel is also not corroded by most food materials. It has the disadvantage that it is not the best conductor of heat; and, where chlorides (brines) may come in contact with it, a special-formula stainless steel must be used since ordinary stainless is corroded by chlorides. Steam kettles used for food handling are preferably constructed of stainless steel, although there may be instances in which aluminum can be used for construction of steam kettles.

Aluminum conducts heat well but is subject to corrosion by alkalis and fruit acids. Also, aluminum is not so easily cleaned as is stainless steel since it cannot be polished to the same degree as stainless. Some anodized aluminum is now produced which is corrosion resistant.

Glass may be used for piping, especially where it can be cleaned in place. Neither glass nor enamel is suitable for lining metals since both tend to chip off, exposing the metal to corrosion.

Rubber is sometimes used for belting on which foods are transported from one area to another. Rubber is not cleaned so easily as stainless steel. However, it may be necessary to use it in some instances. Stainless-steel mesh belting should be used where possible. Teflon-coated belts are now available and are easily cleaned and sanitized.

Food-plant and food-preparing equipment should be constructed in such a manner that it is easy to clean and does not provide areas which cannot be cleaned or which may provide a source of contamination to foods or food products.

Piping should be of sanitary design. There should be no joints with threads exposed on the inside where food materials can accumulate. Instead, at the joint, one pipe end should abut the other, being held together by a clamp. Piping should also be installed so that there are no dead ends. Food materials undergo microbial decomposition in such dead ends and when there is a surge on the line portions of the decomposed material slope over into the uncontaminated product passing along in the main stream of the pipe line. Pipes should not be joined to tanks holding liquid materials in such a manner that the pipe end extends into the tank itself. When used in this manner, some of the liquid material remains in the pipe end, when the liquid level falls below that of the pipe line, and is subject to microbial decomposition, again providing the possibility of recontamination of fresh material entering the tank. Pipe lines entering tanks should be flush with the inner surface of the tank.

Pumps should be so constructed that they can be taken apart for purposes of cleaning. Tanks, hoppers, flumes, thermometer wells, pots and pans, and other equipment should not be constructed with angled corners. Hoppers should have the general shape of an inverse cone rather than that of an inverse pyramid, so that materials will not be caught in corners. Actually hoppers should not be used for wet materials, if avoidable, since such usage provides the possibility of the holding of perishable materials for long periods and the possibility of mixing poor-quality materials with good-quality products. If hoppers are used and have angled corners, such corners will serve to catch and hold foods and will make cleaning and sanitizing difficult. Tanks, flumes, and pots and pans should not have angled corners at the point where the sides meet the bottom. The sides and bottoms of such equipment should be curved at the junction point, thus facilitating cleaning and sanitizing and affording no apertures to catch and hold food materials which might decompose.

Sanitary Operation of Food Plants and Food-Preparing Installations

Raw Materials

All foods and food products used in the preparation of processed foods or as foods for serving should be of good quality, clean, wholesome, unspoiled microbiologically and safe for human consumption. Foods and food ingredients should be obtained from sources considered to be safe by health authorities (Anonymous, 1943; Anonymous, 1953; Anonymous, 1966).[2,3,8]

Milk and milk products used in the preparation of foods or for serving should meet the standards of quality specified by state and local laws and ordinances. Only pasteurized fluid milk and milk products (prepasteurized) should be used or served. Fluid milk or milk products should be served in the original container or taken from an approved bulk container (Anonymous, 1953).[3]

Bivavle shellfish which is to be used as a raw material or served with or without cooking must have been taken from sources approved by the state shellfish authority. It is important that it be determined that such products have come from or through shippers who are certified by the U.S. Public Health Service (Anonymous, 1965; Anonymous, 1965).[5,6]

All meat, meat products, poultry, and poultry products used as components of manufactured foods, and especially those prepared by caterers or served in restaurants, should be obtained from commercial processing plants which are under the supervision of the Meat Inspection Division or the Poultry Inspection Service of the U.S. Dept. of Agriculture.

Bakery products containing custard or cream fillings to be used in restaurants or in other serving areas should be obtained only from producers known to refrigerate such products at 4.4°C (40°F) or below after they have been prepared.

If nonacid canned foods are used as food ingredients in the production of food products or for serving in eating areas, it is the responsibility of the food plant or food-serving establishment supervisory personnel to determine that a minimum botulinum kill (see Chapter 2) has been used during the thermal processing of such foods.

Holding of Raw Materials to be Used as Food Ingredients or to be Served as Foods

It is imperative that all raw food materials which are perishable and which are to be held for future use, be stored at temperatures of 4.4°C (40°F) or below. This includes all flesh-type foods (meats, fish, and poultry), eggs, many dairy products, fruits and vegetables. Also, such foods should only be held for periods of time which would preclude significant deterioration of quality prior to utilization.

Refrigerators used in food preparation and in food-serving establishments may be of two types, the home-type and the walk-in refrigerator. In either case raw foods or left-over cooked foods tend to become lost. Since both kinds of foods are subject to microbial and other kinds of deterioration during long holding in the refrigerated state, it is necessary to keep a strict and accurate inventory of whatever was put into the refrigerator and when it was put there.

Cooked products can be contaminated by contact with raw foods, hence it is good practice to store the two types of foods in separate refrigerators. Whenever this is not possible, it is desirable to have some method of separating the two kinds of products from one another, within the refrigerator, by some type of impervious barrier.

A considerable proportion of food-borne diseases can be traced to the handling of foods at improper temperatures. Since pathogenic bacteria are not known to grow at temperatures below 3.3°C (38°F), refrigeration at suitable temperatures becomes an important safety factor. Large batches of warm or hot food will not cool sufficiently fast, even in refrigerators of large capacity, to prevent the rapid growth of bacteria. It is, therefore, necessary to portion or containerize such bulky material so that small quantities are present in any one container. The refrigerator used should have the capacity to cool the total mass of material placed within it to a mass mean temperature of 4.4°C (40°F) within 2 or 3 hr.

All cabinets used to display perishable foods such as salads and certain types of pastry should be refrigerated and have the capacity to maintain temperatures of 4.4°C (40°F) or below.

When hot foods are held on a steam table large masses of meat, fish, or poultry should not be placed in any particular container. The steam table set-up should be such that it provides for the maintenance of temperatures of 60°C (140°F) or higher in all meats or gravies held in this manner.

Precautions to be Observed During the Preparation of Foods

Whether foods are being prepared for freezing preservation, for catering services, or to be subsequently served in eating establishments, certain precautions should be observed.

Poultry often contain *Salmonella* bacteria; and, since in the process of preparing stuffing and filling it into a bird this material may become contaminated with either salmonellae or staphylococci, it is necessary that those in charge of the cooking of such products take precautions to make certain that during the process the temperature of all parts of the stuffing reaches a temperature of 65.5–68.3°C (150–155°F). Also, no part of the stuffing should remain at temperatures between 15.5 and 43.3°C (60 and 110°F) for long periods of time during cooking. This means that frozen, stuffed birds require extra precautions during cooking. A metal-stemmed, dial-type thermometer should be used to probe various parts of the stuffing during the heating process involved in cooking.

Pork sometimes contains trichinae *(Trichinella spriallis);* and such meat must be cooked in all parts to temperatures which will destroy this organism, thus all parts of the cooked pork should have reached 65.5°C (150°F) or higher when the cooking is finished. Also, since hamburger is sometimes eaten in the rare or even uncooked conditions, in grinding meats different grinder heads should be used for pork and for beef. Use of the same grinding head for both products is not recommended since, while thorough cleaning and sanitizing after grinding pork would eliminate the presence of trichinae, certification of the thoroughness of cleaning is difficult or impossible.

Fresh eggs in the shell, on rare occasions, contain salmonellae; also, when the egg is cracked and emptied, it may be contaminated with salmonellae from the outer shell. In cooking eggs as by scrambling or by soft boiling or frying the temperature in all parts of the product is not raised to the point where all salmonellae would be destroyed, especially if present in considerable numbers (Stafseth et al., 1952).[15] Such foods, therefore, should either be served promptly after cooking or held at temperatures near 60°C (140°F) until eaten or held at 3.3°C (38°F) until rewarmed.

It has been shown (Licciardello et al., 1965)[12] that in cooking eggs to the hard-boiled state, the temperature at all parts is raised to the point that all salmonellae will be destroyed, even though these organisms are present in large numbers at the start. In the preparation of egg salad, which requires much handling, contamination with salmonellae or with staphylococci may take place. The strictest care should, therefore, be used to make sure that, after such preparation, products, which provide a good growth medium for these organisms, are placed at ambient temperatures of 3.3°C (40°F) or below and held at these temperatures until served for eating.

Chicken salad, chicken à la king, and similar preparations require much handling since the chicken must be cooked and boned prior to the preparation of these products. Again the chicken may be contaminated with salmonellae or staphylococci during preparation hence the necessity for not allowing boned meat or the prepared food to stand at room temperature for periods longer than 1 hr prior to placing at refrigerator temperatures of 3.3°C (40°F) or below.

Due to the salt content of ham, which tends to eliminate the growth of competing bacteria, staphylococci find this material a suitable substrate for growth. Sliced ham or ham products should be placed at 3.3°C (40°F) or below immediately after packaging or after preparation.

All food and drink served in eating establishments should be handled in such a manner as to provide a minimum opportunity for contamination with microorganisms. Butter should be served as packaged patties or handled with tongs if unpackaged. Sugar should be served as package portions or in covered pour containers in which spoons cannot be inserted. Bread or rolls should be handled with tongs, or tissue paper, preferably on individual plates, since serving in a common container to particular tables may result in the handling of individual portions within the container by those at a table who do not eat it.

Cleaning and Sanitizing

In food plants, all equipment contacting foods (tables, pans, knives, belts, conveyers, hoppers, fillers, kettles, and other equipment) should be thoroughly cleaned and sanitized at least twice for each 8-hr shift. The floors and walls of food-processing areas should also be cleaned at least once daily. Some types of piping may be cleaned and sanitized in place. Equipment, such as pans or pails, which will hold liquids should be inverted after cleaning and sanitizing and should not be nested.

In large plants there should be a cleanup crew with a foreman to oversee the cleaning operation. Personnel in charge of cleanup should inspect the equipment and processing areas daily to determine that an adequate cleaning and sanitizing job has been done. Inspection is preferably done at the start of the day's operation; and, if the cleanup has not been done satisfactorily, processing should be held up until the situation has been remedied. Some supervision should also be given to the cleaning operation at the time that it is being done. In small plants or in food-serving or catering establishments such supervision may have to be done by the plant foreman.

Inspection of storage areas for such raw materials as sugar, starch, salt, spices, etc. should be frequently made to determine that they are not being infested with insects or rodents and that they are not being contaminated with dirt or dust from the outside or from some operation within the plant.

Reports should be made to management of undesirable situations in plant processing, storage, or personnel facility areas. It is the responsibility of management to determine that undesirable conditions are speedily corrected.

When possible, laboratory tests for pathogenic microorganisms should be made frequently on raw materials to determine that they are in compliance with Federal, State, or local ordinances or laws and to determine that the use of these products involves no public health hazard.

The water used for cleaning should have a temperature of about 54.4–60°C (130–140°F). Cleaning agents should be carefully selected on the basis of the type of soil or dirt encountered in processing, the kind of metal used for equipment construction, whether burned-on material or calcified deposits will be encountered and so on. The cleaning agents available are in general: strong alkalis (NaOH), mild alkalis (Na_3PO_4, Na_2CO_3); strong acids (H_2SO_4, HCl), mild acids (gluconic), and surface-active agents.

In sanitizing, if chemicals are to be used, chlorine or iodine and their derivatives are ordinarily employed since these are the compounds allowed by the U.S. Food and Drug Administration. Chlorine may be used as a solution of gaseous or liquid chlorine, as a solution of sodium or calcium hypochlorite, or as one of the chloramines. Iodine is used as an iodophor.

The U.S. Public Health Service definitions of sanitizing, which is to be done after thorough cleaning, are as follows (Anonymous, 1965).[6]

1. Immersion for at least 1/2 min in clean hot water at 76.7°C (170°F) or higher.

2. Immersion for a period of at least 1 min in a solution containing 50 ppm of available chlorine at a temperature of at least 23.9°C (75°F).

3. Immersion for a period of at least 1 min in a solution of an iodophor containing at least 12.5 ppm of available iodine, at a pH of 5.0 or less and at a temperature not below 23.9°C (75°F).

4. Heating in a cabinet in hot air at a temperature not below 82.2°C (180°F) for a period not less than 20 min.

5. Heating in a steam cabinet at a temperature not less than 76.7°C (170°F) for a period of at least 15 min or for 5 min at 93.3°C (200°F).

If chemical disinfecting agents are to be applied by spraying or swabbing, as for equipment which cannot be immersed, they must be applied at twice the strength stated above and under the same conditions of temperature and pH.

Rinsing in boiling water is another acceptable method of sanitizing equipment.

Waste Disposal

All garbage and waste materials containing foods should be kept in leakproof metal or plastic containers with tight-fitting covers while they are held on the premises of the food plant or food-serving establishment. Rubbish should be held in a similar manner.

Each garbage or rubbish container should be thoroughly cleaned inside and outside after emptying, and the water used for such cleaning should be disposed of as sewage. An area separate from that used for washing dishes, utensile, or equipment should be provided for the washing of garbage or rubbish containers; and the proper facilities for the washing of these containers should be made available.

Containers of garbage and rubbish which accumulate between periods of removal from the premises, should be stored in vermin-proof rooms well separated from food-preparation or food-serving areas. The floors of such waste storage areas should be impervious to water as should the walls for distances up to splash height. Periodic cleaning should be carried out in storage areas of this kind.

Adequate care should be taken to prevent the entrance of rodents and insects into waste storage areas and to prevent vermin from breeding. In instances in which such pests become established in waste storage areas an extermination service must be employed to eliminate them.

Food Plant Personnel and Others Who Handle Foods

The health of those who handle foods is an important factor in the prevention of food-borne disease. At the present time routine testing of personnel handling foods for communicable diseases is not considered to be practicable. However, no person known to have a communicable disease or afflicted with sores, infected wounds, or abrasions, or having a respiratory infection should be allowed to handle foods in a food-processing or food-preparing establishment. Food handlers known to have had intestinal disturbances should be tested, to determine that they are not carriers of disease organisms, before they are allowed to return to work.

The outer garments of all personnel engaged in handling of foods, or of food ingredients should be kept reasonably clean. It is preferable to supply clean outer garments daily to such personnel. This serves not only to prevent indirect contamination of foods but also has a good psychological effect on the worker with regard to his or her attitude towards cleanliness. There may be operations in which gloves are worn by workers, in such cases they should be of the disposable type or of the type which may be readily cleaned and sanitized. In any case, if gloves are used, they should be changed for clean gloves at least once daily.

The hair of all workers handling foods should be covered by hair-nets, bands, or caps. These coverings should be provided for the workers and it is the responsibility of supervisors to make sure that they are worn.

Personnel should not be allowed to eat or smoke in food-processing, food-preparing, utensil- or equipment-washing areas or in rooms where food ingredients or raw food materials are stored.

The personal cleanliness of food handlers is a factor closely associated with good sanitation. In that connection the subject of hand cleaning and disinfection is interesting. It appears that on the hands of humans there is a transient bacterial flora, picked up with dirt, by handling objects, or contact with other humans, namely, soil of various types. There is also a resident flora existing in sebaceous glands, hair follicles, ridges, and furrows, etc. This resident flora may be staphylococci of the food poisoning types, *Es-*

cherichia coli, or other species of bacteria (Crisley and Foter, 1965).[11] Since *Escherichia coli* may be among the resident flora, it can be assumed that, at least on some occasions, the resident flora may consist of salmonellae.

The transient flora is best removed by washing with soap or detergents and can be readily removed in this manner. This is not the situation as far as the resident flora is concerned. Actually the resident flora is not easily removed and is most readily removed by friction as by rubbing the hands together under running water or by scrubbing the hands with a stiff brush under running water.

There are a number of chemicals which may be used to disinfect hands. One group of such compounds are known as bisphenols. These are diphenols with one of the hydroxyl groups neutralized by the alkalai of the soaps or detergents in which they are incorporated. These compounds have a desirable disinfecting property in that there is a residual bactericidal effect which may operate for some time on skin surfaces after they have been applied (Shermano and Nickerson, 1954).[14] They have, however, some undesirable properties in that their activities are reduced or neutralized by food materials and by nonionic detergents and emulsifying agents. Also, the bisphenols are primarily effective against gram-positive bacteria and not very destructive as far as gram-negative bacteria are concerned. This means that they would destroy the type of bacteria which usually cause suppurating infections and which cause staphylococcus poisoning but would not destroy salmonellae, shigellae, vibrio, and some other bacteria which cause food-borne infections.

In solutions, the bisphenols are ordinarily used in concentrations of 1–3%. Hexachlorophene and bithionol are examples of this type of disinfectant. Hexachlorophene (2,2′-dihydroxy-3,3′ 5,5′, 6,6′ hexachlorodiphenylmethane) has the following formula:

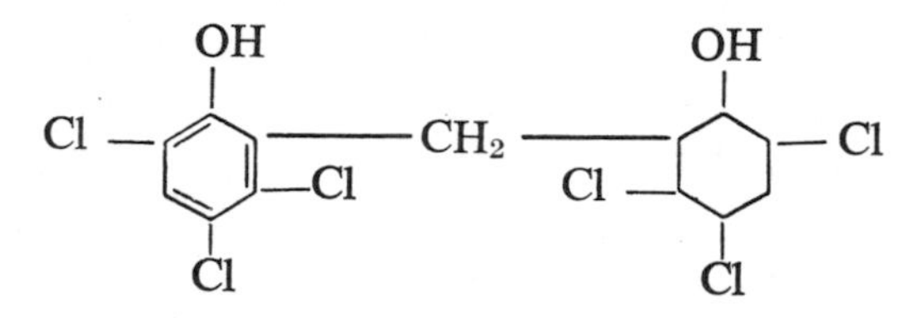

These compounds are nonirritating when used in soaps.

The iodophors are chemical complexes of diatomic iodine and solubilizing agents, or carriers, which are usually nonionic surfactants. Apparently the element iodine is added directly to the surfactant. Under such conditions a part of the iodine becomes firmly bound in the complex and is unavailable as a germicidal agent, but part of the iodine is available and active in destroying microorganisms. The iodophors are active against both gram-positive and gram-negative bacteria, fungi, and viruses. While much more active on the acid side of neutrality, these compounds have microbicidal activity at neutrality or on the alkaline side of neutrality. There is little or slight residual germicidal effect when these compounds are applied. Solutions of the iodophors have little odor and do not stain the skin when applied.

The chlorocarbonilides comprise another class of chemicals which may be used as disinfectants. The compound 3,4,4 trichlorocarbonilide, C1— —NHCONH— C1, and 3 triflor, 4,4 dichlorocarbonilide are examples of this type of chemical disinfectant. These compounds are good germicidal agents for some types of organisms and are not affected by soap. However, they are active against gram-positive bacteria but are ineffective against gram-negative bacteria.

Another group of chemical compounds which are used as disinfectants are the quarternary ammonium compounds. These are amine compounds in which the nitrogen in the molecule has a covalence of 5. Benzalkonium chloride,

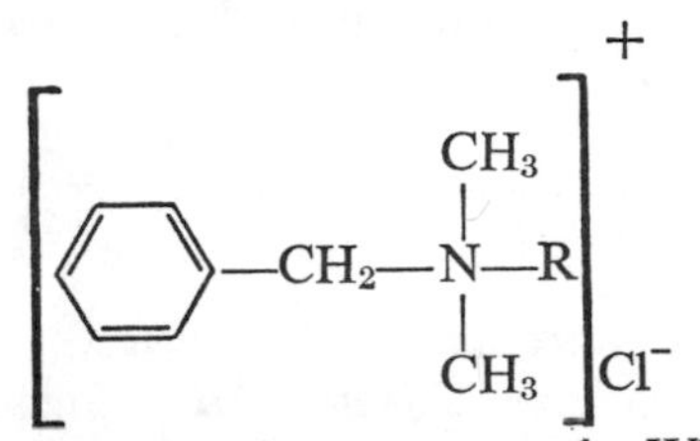

is an example of the quarternary ammonium compounds. While these compounds are active against both gram-positive and gram-negative bacteria, their effectiveness is decreased or neutralized by soap, organic matter and hard water (Crisley and Foter, 1965).[11]

Chlorine may be used as a disinfectant as the liquid element or gas in solution, as a solution of calcium hypochlorite, $Ca(OC1)_2$, sodium hypochlorite, NaOC1, or one of the chloramine compounds chloramine T;

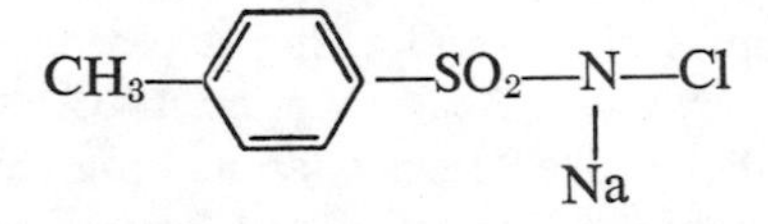

Chlorine in this form is effective against both gram-positive and gram-negative bacteria, but its germicidal effect is neutralized by organic matter.

Considering the foregoing and the fact that the U.S. Food and Drug Administration has approved only chlorine and the chlorine-liberating materials and the iodophors for the sanitizing of equipment, it would seem that, because of its superior qualities, the latter compound would be the first choice for sanitizing and the chlorine compounds the second choice.

For the sanitizing of the hands of food handlers the iodophors would also be first choice and the chlorine compounds second, especially since the former compounds are less irritating to the skin. It may also be the case that when clean hands, free of soap, are to be sanitized that the quarternary ammonium compounds might be acceptable if solutions were made up in soft water and if the hands were thoroughly dried after dipping in the disinfectant. While the quarternary ammonium compounds are not allowed in foods, under such conditions of use negligible amounts would be transferred to whatever food was being handled.

Finally, it should again be emphasized that in areas where foods are being handled as a part of a food-processing or food-preparation operation, wash basins with hot and cold water and soap, a container of disinfecting solution, paper towels, and containers to hold used towels, should be provided at stations near the food-handling operation or operations. The purpose of set-up's of this kind is to allow supervisory personnel to make sure that each person washes and sanitizes his or her hands when returning to work once having left the food-processing area.

References

1. Anonymous, American Standard Safety Code for Industrial Sanitation in Manufacturing Establishments, Amer. Standards Ass., **24** (1935), 1.
2. Anonymous, Ordinance and Code Regulating Eating and Drinking Establishments, U.S. Federal Security Agency, Public Health Service, 1943.
3. Anonymous, Instructors Guide—Sanitary Food Service, U.S. Dept. of Health, Education, and Welfare, Public Health Service, NAVMED p. 133 (1953).
4. Anonymous, Standard Methods for the Examination of Water and Wastewater, 11th Ed., Amer. Public Health. Ass., New York, 1960.
5. Anonymous, National Shellfish Sanitation Program Manual of Operations. Part 1. Sanitation of Shellfish Growing Areas, U.S. Dept. of Health, Education, and Welfare, Public Health Service, 1965.
6. Anonymous, National Shellfish Sanitation Program Manual of Operations. Part II. Sanitation of the Harvesting and Processing of Shellfish, U.S. Dept. of Health, Education, and Welfare, Public Health Service, 1965.
7. Anonymous, National Shellfish Sanitation Program Manual of Operations. Part III. Public Health Service Appraisal of State Shellfish Sanitation Programs, U.S. Dept. of Health, Education, and Welfare, Public Health Service, 1965.

8. Anonymous, Sanitary Standards for Food Processing Establishments, U.S. Dept. of Health, Education and Welfare Public Health Service, Div. of Environment Engineering and Food Protection, Milk and Food Branch, Food Sanitation Section, 1966.
9. Anonymous, *Morbidity and Mortality Report,* U.S. Dept. of Health, Education and Welfare Public Health Service, **17** (1968), 137.
10. Association of Food Industry Sanitations, Inc. and the National Canners Association, "Sanitation for the Food Preservation Industries," McGraw-Hill, New York, 1952.
11. F. D. Crisley and M. J. Foter, The use of antimicrobial soaps and detergents for hand washing in food service establishments, *J. Milk Food Technol.* **28** (1965), 278.
12. J. J. Licciardello, J. T. R. Nickerson and S. A. Goldblith, Destruction of Salmonellae in hard boiled eggs, *Amer. J. Public Health* **55** (1965), 1622.
13. M. E. Parker and J. H. Litchfield, *Food Plant Sanitation,* Reinhold, New York, 1962.
14. I. Shermano and M. Nickerson, Cutaneous Accumulation and Retention of Hexachlorophene—C^{14}, *Fed. Proc.* 13 (1954), 404.
15. H. J. Stafseth, M. M. Cooper, and A. M. Walbank, Survival of *Salmonella pullorum* on the skin of human beings and in eggs during storage and various methods of cooking, *J. Milk Food Technol.* **15** (1952), 70.

Index

The Authors

JOHN T. R. NICKERSON has been intensively involved with modern food technology and the food processing industry for over thirty years. After studying at the Massachusetts Institute of Technology, where between 1932 and 1938 he earned his B.S., S.M., and Ph.D. degrees, he was a research chemist (1937-40) and later (1940-45) Chief Chemist with the Birdseye Frozen Foods Corp. Between 1945 and 1948, he was Chief Chemist with the Hygrade Food Products Corp. In 1948 he returned to his alma mater, M.I.T., where he has successively held the positions of Research Associate in the Department of Food Technology, and Assistant, Associate, and full Professor of Food Technology. Dr. Nickerson is a member of the American Society for Microbiology, the American Chemical Society, the Institute of Food Technologists, Sigma Xi, and Phi Tau Sigma. He is also a fellow of the American Public Health Association.

ANTHONY J. SINSKEY has recently been appointed Associate Professor of Applied Microbiology in the Department of Nutrition and Food Science at the Massachusetts Institute of Technology. Between 1962 and 1966, he was Assistant Professor in the same department. In 1967 and 1968, he was a post doctoral fellow of the Harvard School of Public Health, following his attainment of the Sc.D. degree at M.I.T. during the years 1962 to 1966. Dr. Sinskey is a member of the American Association for the Advancement of Science, the American Society for Microbiology, the Institute of Food Technologists, the Northeast Section of the American Society for Microbiology, Sigma Xi, the Society for General Microbiology, and the Society for Industrial Microbiology.